Constructing a Bridge

Inside Technology
edited by Wiebe E. Bijker, W. Bernard Carlson, and Trevor Pinch

Wiebe E. Bijker, *Of Bicycles, Bakelites, and Bulbs: Toward a Theory of Sociotechnical Change*

Wiebe E. Bijker and John Law, editors, *Shaping Technology/Building Society: Studies in Sociotechnical Change*

Stuart S. Blume, *Insight and Industry: On the Dynamics of Technological Change in Medicine*

Louis L. Bucciarelli, *Designing Engineers*

Geoffrey C. Bowker, *Science on the Run: Information Management and Industrial Geophysics at Schlumberger, 1920–1940*

H. M. Collins, *Artificial Experts: Social Knowledge and Intelligent Machines*

Paul N. Edwards, *The Closed World: Computers and the Politics of Discourse in Cold War America*

Eda Kranakis, *Constructing a Bridge: An Exploration of Engineering Culture, Design, and Research in Nineteenth-Century France and America*

Pamela E. Mack, *Viewing the Earth: The Social Construction of the Landsat Satellite System*

Donald MacKenzie, *Inventing Accuracy: A Historical Sociology of Nuclear Missile Guidance*

Donald MacKenzie, *Knowing Machines: Essays on Technical Change*

Constructing a Bridge
An Exploration of Engineering Culture,
Design, and Research in Nineteenth-Century
France and America

Eda Kranakis

The MIT Press
Cambridge, Massachusetts
London, England

Set in Baskerville by Wellington Graphics.
Printed and bound in the United States of America.

Library of Congress Cataloging-in-Publication Data

Kranakis, Eda
  Constructing a bridge : an exploration of engineering culture, design, and
  research in nineteenth-century France and America / Eda Kranakis.
    p.   cm. — (Inside technology)
  Includes bibliographical references and index.
  ISBN 0-262-11217-5
  1. Suspension bridges—France—Design and construction—History—19th
century. 2. Suspension bridges—United States—Design and construction—
History—19th century. 3. Technology—social aspects—France—History—
19th century. 4. Technology—social aspects—United States—History—19th
century. I. Title. II. Series.
TG71.K73   1996
624′.5′094409034   dc2096–13570
                                                              CIP

To the memory of my father.

Contents

II
Social Determinants of Engineering Practice: A Comparative View of France and America in the Nineteenth Century

Acknowledgements

This book has been long in the making, and along the way I have acquired many debts to family members, teachers, colleagues, institutions, librarians, and archivists. I owe the greatest debt to Edwin Layton, my teacher and mentor, who has aided the realization of this book in countless ways, and whose commitment to the highest intellectual and scholarly standards continues to pose a creative challenge. Thanks go also to James Hippen, who introduced me to the world of history of technology and set me to studying the history of suspension bridges.

One of the greatest joys of the scholarly endeavor has been the opportunity to combine friendship with intellectual exchange. I would especially like to thank Olga Amsterdamska, Gertrud Blauwhof, Stuart Blume, Carolyn Cooper, Don Davis, Nil Disco, Joan Fujimura, Robert Gordon, Jan Grabowski, Tom Hughes, Chunglin Kwa, Paul Lachance, Loet Leydesdorff, Svante Lindqvist, Keith Luria, Antoine Picon, Terry Shinn, and Nicole St.-Onge for stimulating discussions, comments, and advice that materially aided this book. Donald Cardwell and family hosted me in England and took me to libraries and archives, including the museum and archives at Coalbrookdale, for which I am still grateful.

Many other colleagues and acquaintances aided this intellectual endeavor as well. I owe many thanks to the editors of the Inside Technology series, Wiebe Bijker, Bernie Carlson, and Trevor Pinch, for the time and effort they put into reading and commenting on my first draft. Larry Cohen answered my endless questions, and in many other ways aided and encouraged this project. Paul Bethge edited the manuscript with exceptional skill and precision, for which I am very grateful. I would also like to thank David Billington, Dan Dolan, H. de Jong, Hiroshi Tanaka, Robert Vogel, and John H. White for helpful discussions on suspension bridges and steam injectors.

Institutional and financial support has made it possible to carry out the research and writing for this book. A National Science Foundation dissertation research grant launched the project. A summer research grant from Yale University allowed me to carry out further research in Europe, and a Mellon Fellowship at Yale University gave me a semester off for research. A special thanks to Wiebe Bijker, who suggested me as a candidate for a fellowship at Delft University, and to Delft University and its Department of History of Technology for taking up that suggestion and hosting me for six months. I am grateful as well to the Royal Dutch Academy of Sciences, to the University of Amsterdam's Department of Science Dynamics, and to the University of Ottawa for financial and institutional support.

Mme. Thisy was very helpful at the Ecole des Ponts et Chaussées, and she and her family have offered their friendship and hospitality over many years. Jonathan Goldstein responded to my query about American knowledge of Chinese bridges, James Edmonson shared research notes concerning suspension bridges with me, and Ronald L. Michael shared information from his copy of the records of the United States Direct Tax of 1798 for Fayette County before I was able to get a copy. A special thanks to Bernie Carlson and his research assistant, John Bozeman, for help in getting copies of some of the illustrations from Navier's book on suspension bridges. Michael Walsh helped to prepare the bibliography and to rework the table on early American chain bridges, and Lynda deForest assisted in proofreading and indexing. I am grateful for the excellent and thorough work of Susan Yavari, who provided some last-minute research assistance.

Librarians and archivists in Europe and America have been very helpful. The interlibrary loan departments at the University of Minnesota and the University of Ottawa both made special efforts to get books for me on short notice. Many thanks as well to the archivists, librarians, and support staff at the American Philosophical Society, the Archives de l'Académie des Sciences (Paris), the Archives Nationales, the Bibliothèque Forney, the Bibliothèque Historique de la Ville de Paris, the Centre de Recherches sur la Civilisation Industrielle (Le Creusot), the Conservatoire National des Arts et Métiers library, the Delft University Library, the Fayette County Courthouse, the Ecole Centrale library, the Ecole des Mines library, the Ecole des Ponts et Chaussées library, the Franklin Institute, the Hagley Museum and Library, the Historical Society of Pennsylvania, the Historical Society of Western Pennsylvania, the Institut de France library, the Musée de l'Air, the National Library of Canada, the Library of Congress Manuscript Division, the National Museum of American History, the New York Historical Society, the Northhampton County Historical Society, the Presbyterian Historical Society, the Service de la Propriété Industrielle, Purdue University's Goss Library, the Uniontown Public Library, the University of Virginia Library, the University of Minnesota Department of Special Collections (especially Nicole Pettit), and Yale University's Rare Book Library.

Last but not least, I would like to thank my husband, Evangelos, for his continual help and support; our daughter, Mata, for her love and *joie de vivre;* and our nanny and housekeeper, Tellie Joaquin, who kept our house running during the last stage of this project. I will always be grateful to my mother and father, who have helped and inspired me throughout the years. My father launched me into computers early on, which took the drudgery out of writing. And both my parents, each in their own way, have provided examples of lifelong commitment to a profession.

Portions of this book have been published separately. Part of chapter 5 appeared as "Navier's theory of suspension bridges" in *Acta Historica Scientiarum Naturalium et Medicinalium* 39 (1987). Chapter 6 appeared as "The affair of the Invalides Bridge" in *Jaarboek voor de Geschiedenis van Bedrijf en Techniek* 4 (1987). Portions of chapters 7–9 appeared in "Social determinants of engineering practice: A comparative view of France and America in the nineteenth century," in *Social Studies of Science* 19 (1989).

Constructing a Bridge

Introduction

In his classic study of American democracy, Alexis de Tocqueville drew a contrast between the ways in which "democratic" and "aristocratic" societies influenced technology and the production and use of knowledge. He argued that American democracy fostered production of large quantities of low-quality goods whereas aristocracies (such as France) fostered limited production of unique, high-quality goods. With respect to the development of scientific and technological knowledge, Tocqueville argued that American democracy gave rise to empirical, experiential, knowledge whereas aristocratic societies fostered the development of more abstract, theoretical knowledge:

Those who cultivate the sciences among a democratic people are always afraid of losing their way in visionary speculation. They mistrust systems; they adhere closely to facts and study facts with their own senses. . . . Hardly anyone in the United States devotes himself to the essentially theoretical and abstract portion of human knowledge. . . . In aristocratic societies the class that gives the tone to opinion and has the guidance of affairs, being permanently and hereditarily placed above the multitude, naturally conceives a lofty idea of itself and man. . . . These opinions exert their influence on those who cultivate the sciences as well as on the rest of the community. They facilitate the natural impulse of the mind to the highest regions of thought. . . . Men of science at such periods are consequently carried away toward theory; and it even happens that they frequently conceive an inconsiderate contempt for practice. . . . Permanent inequality of conditions leads men to confine themselves to the arrogant and sterile research for abstract truths, while the social condition and the institutions of democracy prepare them to seek the immediate and useful practical results of the sciences.[1]

Many of Tocqueville's ideas and themes have been adapted or echoed in more recent scholarship in the history of technology, notably in studies by John Kouwenhoven, Hugo Meier, Eugene Ferguson, and John Kasson.[2] The merit of these studies is that they confront the possibility that relationships among social structure, ideology, and patterns of technological development exist at a national or a societal level. Yet, because they explore the effects of democracy or republicanism on American technology *in general* (e.g., whether it has especially fostered mechanization and mass production), these studies do not provide a great deal of contextual detail about the specific forces that shape the development of particular artifacts or technological traditions. Nor is there much historical literature that attempts to link the themes and ideas in these studies to more detailed empirical and comparative studies.

In the present study, I try to make such a link by means of a comparative analysis of how specific social and institutional structures in nineteenth-century France and America shaped the production and use of technological knowledge and artifacts.[3] On the one hand, I examine a particular technology, the suspension bridge, in order to see how and why its development was handled differently by specific individuals in these two countries. On the other hand, I assume that technology and knowledge are shaped first and foremost by the *immediate* environment within which technologists are educated and work, and I seek to understand the structures and mechanisms within this environment that shape national technological traditions. In line with this aim, I deliberately avoid framing explanations in terms of such vague and malleable concepts as "democracy," "aristocracy," and "republicanism." These concepts are not irrelevant, yet their influence is necessarily mediated by such lower-level social and institutional structures as technical schools, patent systems, community structures, and work organizations. Moreover, there is no simple rule that specifies exactly what kind of patent system or work organization or technical school either "democracy" or "aristocracy" produces. It is therefore important to look at how lower-level structures actually embody societal values. In summary, then, the present study aims to show how *specific* mechanisms within French and American technologists' immediate environments shaped the development of technological traditions and of specific artifacts within those traditions.

When doing comparative studies at the national level, it is difficult to achieve both breadth and depth. Detailed comparisons of individual technologists and artifacts may not seem to warrant generalizations about national traditions, whereas broader comparisons may not provide adequate insight into the careers of individual technologists and the artifacts they create. For this reason, the present study adopts a "micro/macro" approach. The book consists of two complementary studies, one more narrowly focused than the other. The micro study looks in depth at the processes of technological research and design as they were handled by two individuals in the early nineteenth century, one American and one French; the macro study surveys the French and American technological communities over the entire span of the nineteenth century.

More specifically, the micro study focuses on the technology of the suspension bridge (including design, research, and construction) as it was handled by the American inventor James Finley and the French engineer Claude-Louis-Marie-Henri Navier. Finley was the "inventor" of the modern suspension bridge with a level roadway suitable for vehicle traffic. He patented such a design in 1808, and suspension bridges based on his plan were built in the United States until 1826. The innovation diffused to Britain and then to France, where it attracted Navier's attention. In 1823 Navier published a treatise on suspension bridges that included a design for a suspension bridge over the Seine, in Paris. Construction began in 1824, but Navier's bridge was never completed.

On one level, the study of Finley and Navier is a study of technological diffusion. Yet its primary aim is not to analyze the diffusion process but rather to show how the very different environments in which these two individuals worked led them to develop this particular technology in entirely different ways, both in terms of their design goals and in terms of the kinds of knowledge used in the design process. I examine the backgrounds, careers, and working environments of Finley and Navier and explain how these shaped their respective approaches to the study and design of suspension bridges. My analysis shows that, although both wanted to achieve a more fundamental, "scientific" understanding of this technology, their methods for gaining this understanding were quite different. Finley carried out a series of experiments with loaded cables in an effort to understand inductively the relationship between the weight borne by the cable (i.e., its load) and the cable's tension (i.e., the amount of pull on the cable, directed along its curve). Navier, on the other hand, studied these and other relationships mathematically and deductively, using, among other things, calculus, statics, and the theory of elasticity.

Finley's and Navier's design goals also differed profoundly, in part because they responded to differing national needs and priorities. Finley, who lived in a small town in western Pennsylvania, wanted to adapt the suspension bridge to the needs and conditions of America's poor and sparsely populated interior, and his design represented a kind of do-it-yourself plan that could be readily constructed by carpenters and blacksmiths without special expertise. In contrast, Navier, an engineer in the Corps des Ponts et Chaussées, a prestigious state engineering corps, wanted to adapt the suspension bridge to the traditions and goals of French monumental architecture. His design was for a unique, elegant, monumental structure, the construction of which required considerable money, manpower, and technical expertise.

In effect, this micro study isolates two episodes in the development and diffusion of suspension bridges, analyzing each as an end in itself rather than merely as a point along a "technological trajectory." In doing so, it helps to fill a gap in the Tocqueville genre of literature by focusing on the question of how distinct national environments and research traditions shape the development of specific technologies and the work of individual technologists.[4]

The macro study focuses on knowledge and research traditions among French and American technologists during the nineteenth century. It analyzes certain global differences in the ways French and American engineers defined and studied technological problems—that is, certain differences in their traditions and in their ideologies of technological research. It shows that French technologists devoted more time and energy to mathematical research on technological questions than American technologists, whereas in the case of experimental research and testing just the opposite was true. On one level, the findings of this study mirror the observations of Tocqueville quoted above. Yet Tocqueville not only provided no evidence or justification for his position, but he also said nothing specific about the research practices of French and American technologists.[5] The issue has been largely ignored by scholars since.[6]

Documenting the existence of distinct French and American traditions of technological knowledge and research, I characterize some of the differences between them and examine how these traditions were preserved and disseminated. I do not portray these traditions as inevitable by-products of "democracy" or "aristocracy." Rather, I examine how they were specifically linked to the different structures of France's and America's technological communities and to certain characteristics of those technological communities, including their relations to the state, their reward systems, and their systems of technical education (which shaped research methodologies and the production of knowledge). For example, I show that the French technological community was more stratified and hierarchical than the American, and that each stratum was associated with a distinct form of technical education, focusing predominantly on either mathematical, experimental, or workshop training. The emphasis on mathematical knowledge was specifically characteristic of the top stratum of French technologists and of France's elite technical schools. America's technical schools were not as stratified, and their curricula united these three forms of knowledge.

Finally, the macro study shows—for the case of a specific technology, the steam injector—how the French and American traditions of technological knowledge and research led to different patterns of technological development. It shows that the American research tradition was more oriented toward producing design-oriented knowledge through empirical research and testing. Partly as a result of this orientation, steam injectors evolved more rapidly and extensively in the United States than in France.

While it is important to understand the differing approaches of these two studies, it is also essential to recognize how they are interconnected. In addition to their shared historiographic concerns, the studies are linked in that each provides an interpretive context for the other. The micro study examines French and American technological practices in a relatively early, formative period. By focusing in depth on two individuals and a single technology, and by attempting to reconstruct the distinct worlds in which these individuals worked, this study helps to reveal more sharply some of the social and economic forces that fostered the differential evolution of French and American research and design traditions. The micro study also has the advantage of bringing the issue of national (or regional) traditions down to the level of individuals: it provides a greater understanding of how individuals internalize these traditions. The macro study, for its part, helps to place the research practices of Navier and Finley in a larger context by showing how their work relates to the evolution of national traditions. It makes clear that there were patterns of technological knowledge, research, and design that transcended specific technologies and individual practitioners.

On another level, the studies in this book are linked through a mutual concern with the issue of the social shaping of technology, a topic that has become increasingly central for scholars interested in the relations between technology and society. Two features of the studies help to enlarge our understanding of this issue: their comparative approach, and a shared conceptual framework that takes technological practice as its starting point.

Ulrich Wengenroth cites a lack of comparative studies as one of the major shortcomings of the social constructivist approach in technology studies.[7] Yet, as Wengenroth points out, the comparative approach has a number of important advantages for studying the interlinking of technology and society. First, the comparative approach "presses for theorizing and explicit hypotheses." It forces us to attempt to look beyond the particular, to see more general patterns in the interlinking of technology and society, and to relate the particular to the general. Second, Wengenroth notes that "comparative analysis provides one of the few means to undertake some kind of 'indirect experiment.'" Of course, unlike laboratory experiments in which the investigator is able to set and control parameters, the relevant variables in a comparative historical study cannot be fully isolated from one another or altered at will: they must be taken more or less as they come. Yet even within these limits it is possible to learn a great deal about the interdependence of technology and society. Wengenroth argues, indeed, that the "comparative approach alone can weigh the various social and material components in the construction of a technology."

Wengenroth sees comparative analysis as a "panacea against technological determinism," yet its value may go beyond this. Ideas of technological and social determinism are now rejected by many scholars of science and technology. Few would now accept that technology, in and of itself, determines social patterns, or that social patterns completely determine technological development. Nevertheless, the historical record makes clear that not all changes are possible in all contexts. With respect to technological development, the integration of new technologies into the social fabric depends on what is already there. Change is not precluded, but not all kinds of change are possible or likely in all circumstances: the technologies and technological visions we choose *do* depend on existing socio-technical structures. Thus, one of the major challenges we now face, which follows from a rejection of technological and social determinism, is to find approaches that will allow us to handle structure without losing sight of the network-building and society-rearranging dimensions of socio-technical development.[8] The advantage of comparative analysis is that it is can help to meet this challenge.

The comparative approach provides a means or finding out "what society is made of" at one time and place relative to another.[9] It provides a method for identifying important structures or resources at the disposal of particular societies and seeing how change occurs in relation to these structures. By "structures" I mean not merely social structures but also socio-technical systems and networks (comprising people, institutions, laws, technologies, etc.).[10] The systems and networks found in distinct societies may differ both in the elements they contain and in their organizational relationships. Comparative analysis provides a means of mapping out these systems in specific instances and then seeing how they influence the process of change—in other words, how they shape the growth of new networks and the extension of existing ones. Comparative analysis makes it possible to explain why certain networks and systems become larger, more powerful, or more enduring than others; why a particular network contains some elements and not others; why those elements are linked in certain ways and not others; and how particular network structures shape technological development. For example, comparative analysis makes it possible to map out the very different kinds of networks for bridge building that Finley and Navier made use of and to understand how these partially preexisting networks shaped the specific bridge designs they conceived.

The second feature shared by the studies in this book is an emphasis on contextual analysis of technological practice. This focus involves looking closely at what technologists do in their work and trying to understand why they do what they do and why they organize and direct their work in particular ways. It means taking into account a complex range of activities: goal formulation, design, research aimed at the production of knowledge as well as artifacts, publishing, consulting, applying for patents, negotiating with politicians and officials, managing people, and so on. More generally, it means emphasizing *process history* rather than event history: attempting not only to catalogue or restructure events according to some theoretical or historiographic framework but also to explain the historical processes and the goals that shaped technological events and creations.

An emphasis on practice has several advantages for those who investigate the social shaping of technology. First, because it is concerned as much with what technologists do as with the artifacts they create, this approach calls attention to the importance of research and knowledge production in the realm of technology and raises the question of how these activities are themselves socially shaped. Until recently, technological research and knowledge were treated as though they were simply determined by nature or technical necessity. Yet, as the studies presented in this book show, technological research and knowledge are indeed socially malleable, sometimes with significant consequences.

The second advantage of this approach is that it focuses more attention on the design process. It leads to a contextual analysis in which design is viewed not merely as a cognitive activity but as a process that is simultaneously technical, cognitive, and social—a process by which knowledge, hardware, and social goals are merged to achieve coherent and viable artifacts. Designs should not be viewed simply as technical solutions; they are also distillations of scientific and technological knowledge, socio-economic blueprints, and embodiments of ideologies and values. This concept of design provides a basis for identifying the wide range of factors that together determine the character of an artifact. Social, economic, technical, cognitive, and other factors operate at various levels and in various ways; yet in the end they must come together as a set of criteria guiding the design process. Designers must integrate these criteria or make tradeoffs among them. In order to do so, they must often transform them: social and economic criteria must be transformed into technical parameters and vice versa, and technical specifications of one sort must be transformed into technical specifications of another sort. Almost any category of transformation can be found to occur.[11] In the case of Finley, we will see how the social problem of a lack of trained engineers was transformed into a very simple technical specification for a suspension bridge with a sag/span ratio of 1/7.

Focusing on design as a socio-cognitive process can help us to understand better how technologists (individually and collectively) make such transformations and tradeoffs, and how they integrate considerations at various levels to produce viable, holistic designs. As a rule, any technology or technological system can be designed in more than one way. Careful study of design processes therefore offers a means of explaining the choices made in specific cases. In addition, design processes are themselves socially organized in ways that affect patterns of choice and possibilities for choice. For example, in some cases design work continues even after construction of the artifact has begun, and the two phases are allowed to interact; in other cases a design is "frozen" before construction. Studying design processes comparatively therefore offers a means of understanding how organizational factors shape the contexts in which design choices may be made, and the range of choices that may be made.

The third advantage of a more practice-oriented framework is the historiographic complement of the other two. Specifically, this approach provides a basis for profitably integrating three major lines of research that have so far tended to remain separate: studies about technological knowledge,[12] studies about the design process,[13] and studies about the social shaping of technology (SST).[14] On the one hand, SST studies have largely ignored the subject of technological knowledge and have not focused extensively on the design process. On the other hand, studies of technological knowledge and engineering design have often been limited to a cognitive perspective.[15] To put the matter in slightly different terms: SST studies have not given sufficient attention to research and design processes (although these play crucial roles in technological development), and studies of technological research and design have not given sufficient attention to the social and organizational dimensions of these processes. Focusing more attention on a contextual analysis of technological practice offers a way to overcome these imbalances and to develop a more integrated perspective that draws fruitfully on all three research traditions.[16]

What I have in mind specifically is an approach that (1) seeks to analyze and understand the social and cognitive nature of technological practices and how these practices shape the character and development of technology, (2) looks carefully at both the social and the cognitive dimensions of design processes in technology, and (3) focuses on knowledge as well as artifacts, and on the relationship between knowledge and artifact design. Such an approach provides a means of exploring more fully the complex relationships among goal formulation, research, design practices, the shapes of artifacts, and the relation of all of these to the social and historical environment. Such a program invites a more critical and more detailed assessment of the process of creating technology—of technology in the making, as Bruno Latour would say—and of the patterns of technological practice that emerge and dominate at various times and places. Since these issues bear on the question of the steerability of technology, they are pertinent not only to scholars within technology studies but also to those interested in technology policy and assessment.

It is this focus on practice and on process that primarily distinguishes the studies presented in the two parts of this book from other comparative studies in the history and sociology of technology. And since one of my aims is to illustrate the value of this approach, it is important that I explain briefly how it has shaped these two studies.

Part I, which concerns Finley, Navier, and the suspension bridge, deals specifically with the process of technological design. This process is analyzed in terms of three interacting elements: the socio-technical goals of the designer, the knowledge and research that are brought to bear on the design problem, and the social context and organization of design and construction. The designer must integrate these elements so as to achieve a product that is both socially and technically viable within a specific environment.

Let me give an example of one of these elements: Finley and Navier had to make their plans viable for very different kinds of networks for selecting and constructing bridges. In the case of Finley, the patent system was an important element of this network. It provided a major channel through which prospective builders could learn about Finley's design and how to construct it. Finley's design was accordingly intended to function as a kind of generic blueprint that builders could adapt to a particular site. Builders who chose to follow the plan were to pay Finley patent royalties of $1.00 per linear foot of span. With this system, of course, Finley had little or no control over the actual building process, because communities and builders (with no special expertise in bridge design and construction) could decide to use his plan without using him as a consultant. Finley took this situation into account in working out the details of his design. For example, he provided instructions on how to prepare the design for a specific site, and he deliberately avoided the need for any but the most trivial arithmetical calculations.

In the case of Navier, the processes of design and construction were highly formalized, following rules and guidelines set down by the Corps des Ponts et Chaussées (the state engineering corps for roads and bridges). Bridge designs had to be for specific sites, and had to be approved by a committee of high-ranking engineers within the Corps. Navier knew that these committees placed high value on plans that, in their view, were solidly based on mathematical theory. Navier accordingly linked his plan as far as possible to his suspension bridge theory. Whereas Finley deliberately tried to minimize the need for mathematical calculations, Navier sought, if not to maximize the need, then at least to make mathematical calculations a prominent and indispensable part of the plan. Navier also knew that the Corps would demand a high level of expertise on the part of the builders, and accordingly he made the availability of such expertise an essential element of the plan. Indeed, when the Corps contracted the project out, it bound the contractor to follow the specifications set down by Navier and the committee and to have the work directly supervised by Corps engineers. The point is that these two very different kinds of networks helped to shape the material characteristics of Finley's and Navier's designs, and indeed gave them different social meanings.

Part II, which concerns the distinct research traditions of French and American technologists, deals with the social shaping of technological research practices. In particular, it shows that technologists often follow heuristics that reflect broad social perspectives and concerns as well as specific technological and economic considerations. These heuristics are rooted in ideologies—in the shared ideas and values that serve to orient, evaluate, and justify the research practices and programs of a technical community.[17] My study attempts to understand how such "ideologies of practice" are maintained and reproduced and how they channel research and design efforts. Finally, I show that these ideologies exist at a national level—that is, they reach the high level of aggregation that is presupposed in the works of Tocqueville, Kouwenhoven, Ferguson, Kasson, and others.

The ideologies of practice I specifically examine include ideas about the relationships between science and technology and among mathematical theory, experiment, and practice; ideas about the "proper" way to go about designing technological artifacts, and more generally about the "proper" way to go about formulating and resolving technological problems; and ideas about what technologists should learn and how they should learn it—in short, ideas about what constitutes the "ideal" patterns of technological practice. The French and American technical communities differed in their views on these matters. In France the prevailing view was that mathematical theory should precede and guide both experimental research and design. For French technologists, "theory" meant, above all, quantitative, mathematical theory. In the United States the prevailing view was that experimental research and empirical practice should guide design efforts, and that theory should emerge from experimental and empirical work. Theory for nineteenth-century American technologists could, moreover, just as well be qualitative and explanatory as mathematical and "predictive."

Ideologies can have discernible influence on research practices, on traditions of artifact design, and on the creation of bodies of knowledge and know-how. With respect to the first, I show that the differing ideologies of French and American technologists led them to adopt different kinds of research programs. French programs tended to emphasize the development and evaluation of mathematical theories, whereas American programs aimed more at the development and evaluation of artifact designs. French and American researchers also claimed to follow different methods of invention and design. The epitome of the French approach was to work out the details of a design through mathematical analysis, and from there to build a full-scale prototype. The quintessential American approach was to work out a design by building, testing, and experimenting with a series of models.

John Staudenmaier's historiographic study *Technology's Storytellers* has pointed to the need for a "new paradigm for the history of technology" rooted in a contextualist approach.[18] My hope is that the present study, through its comparative approach and its attention to the social shaping of technological practice, can contribute to the elaboration of such a paradigm.

I

A Tale of Two Bridges: Finley, Navier, and Suspension Bridge Design

Whenever it is a question of tracing the evolution of a technology from its origins to a mature form, there is a tendency to focus attention only on those elements of each intervening design that either influenced what came after or were influenced by what came before. In other words, the designs are not studied in their totality, as ends in themselves; rather, aspects of them are selected to be analyzed in relation to a generalized pattern of evolution (a "technological trajectory"). Histories of this type do fill some practical needs, but they give the misleading impression that technological development proceeds mainly according to an internal dynamic and that it is only secondarily or peripherally affected by social and economic forces. What is lost is the realization that every design and every technological activity is shaped first and foremost by its immediate and intended socio-economic environment.

In recent years, a growing body of research has begun to counteract such Whiggish perspectives by attempting to "deconstruct" technological designs and developments in order to reveal their social contents more clearly. The advantage of this new orientation is that it integrates the history of technology more firmly with general social and economic history. Deconstructing a design necessarily involves *reconstructing* the social and economic environment in which and for which the design was produced. The research must focus as much on the environment as on the technology.

The necessary link between the deconstruction of a design and the reconstruction of the social, economic, and cognitive world in which and for which it was produced is, in terms of methodology, both a weakness and a strength. It is a weakness in that this kind of contextual approach precludes a broad view over space or time. It is a strength in that the more confined view can bring with it a deeper understanding. Who has not felt, at some time, that they gained a better understanding of a period or process of historical change by analyzing a single event in depth than through any number of more general but inevitably more superficial and opaque surveys?

An excellent example of the depth of insight that can be achieved through the careful analysis of a particular episode can be seen in Svante Lindqvist's book *Technology on Trial: The Introduction of Steam Power Technology into Sweden, 1715–1836* (Uppsala Studies in History of Science, no. 1 (Almqvist & Wiksell, 1984)), which concerns the failure of the introduction of one machine into Sweden in the eighteenth century. By analyzing this single "event," Lindqvist opens up an entire world for us. We begin to understand the range of technological issues that were of particular concern in Sweden in this era and why these and not other issues were important, we see how social structure and ideology shaped Sweden's technological endeavors, and we gain insight into the role of the state in technological development. In short, we begin to experience aspects of the social and technological culture of early-eighteenth-century Sweden holistically, focused through a particular chain of events.

Like Lindqvist's book, my study of Finley, Navier, and their work on suspension bridges aims at a holistic understanding of technological change. The goal is to use this form of analysis both as a means to gain more general insight into the differing social contexts of technological change in France and America in the early nineteenth century and as a means to explore the issue of determinism and choice in technological development. More immediately, this form of analysis is intended to reveal the social meaning of the bridges of Finley and Navier with respect to their own immediate environments.

Chapter 1 examines the socio-economic environment of Fayette County, Pennsylvania, around 1800, and Finley's career and experiences within this environment. Chapter 2 looks at Finley's research on suspension bridges and the specific design system he worked out, and explores how these reflected Finley's environment. Chapter 3 looks at how this design system fared in practice on the basis of the histories of the bridges known to have been built following his plan and on the basis of the success of Finley's system relative to competing design systems. Chapters 4–6 provide a similar analysis for Navier. Chapter 4 examines Navier's education and career within a broader social and institutional context. Chapter 5 examines Navier's role in the diffusion of suspension bridge technology from Britain to France, shows how he transformed this technology by theorizing it, and examines how Navier's theorizing was shaped by his social and institutional environment. Chapter 6 considers how the specific bridge design Navier developed was shaped simultaneously by his research program, by his institutional environment, and by the social and technical systems that he relied on for the bridge's construction.

Finley, Technology, and Frontier Society

James Finley was a farmer and a county judge in a small, rural, largely Scottish-Presbyterian community in western Pennsylvania's Fayette County, a region that had only begun to be developed. He was not trained as an engineer. Nevertheless, Finley's invention of the level-roadway suspension bridge was significant enough to inspire Europe's leading civil engineers. In order to explain the conditions that set Finley on his path toward innovative bridge design, it is necessary to look more carefully at the economy and culture of Fayette County at the turn of the nineteenth century and to consider Finley's position within his community.

Finley was born in Maryland in 1762.[1] He was of the stock of Scottish Presbyterians who had emigrated in the early seventeenth century to Ulster (Northern Ireland) and then, in the eighteenth century, to America. A large proportion of these immigrants landed at Newcastle, Delaware, and founded settlements in Delaware and Maryland.[2] In Maryland, Finley belonged to a parish known as East Nottingham, whose pastor was also a James Finley.[3]

Finley the bridge designer was reported by a friend and colleague, Ephraim Douglass, to have had an "English education"—one that emphasized English literature and the sciences rather than the classics.[4] The sophistication of Finley's writing suggests that he did have formal education, but there is no evidence that he attended a major university, and there is some evidence to suggest that he did not attend any university at all.[5] The most reasonable hypothesis, in view of Finley's Scottish Presbyterian background, is that he was educated in a Presbyterian academy, popularly called a "log college."[6] A likely candidate (because of its proximity) is the Nottingham Academy,[7] which was located in Cecil County, either within or immediately adjacent to East Nottingham parish. However, there were two other academies located in adjacent counties, and since these were boarding schools it is possible that Finley attended one of them. The first of these, Fagg's Manor, was in Chester County, close to the Pennsylvania-Maryland border[8]; the second, Pecqua, was in Lancaster County.[9]

If Finley did attend one of these log colleges, his education would have given him, at the very least, a knowledge of the "three Rs" and an introduction to science and philosophy. The curriculum at Nottingham Academy before 1760 included Latin and Greek classics, logic, arithmetic, geometry, geography, "part of ontology," and natural philosophy.[10] Finley may well have had a more extensive science training than this early curriculum suggests, however. In the next few decades, the Presbyterian academies began to place more emphasis on science and technology and less on the classics. Indeed, it is acknowledged that these academies were among the most innovative schools in the United States in the late eighteenth century and the early nineteenth century. For example, the curriculum of the Canonsburg Academy, founded in western Pennsylvania in 1794, included "English, Euclid's Elements of Geometry; Trigonometry, Plane and Spherical, with the later application to Astronomy; Navigation, Surveying, Mensuration, Gauging, Dialing Conic Sections, Algebra, and Bookkeeping."[11]

Sometime before 1784, Finley moved from Maryland to Fayette County, Pennsylvania, where he settled on a farm north of Uniontown.[12] He was a member of a group of 34 families who, between 1765 and 1784, moved from East Nottingham parish to southwestern Pennsylvania, all eventually settling within a 30-mile radius.[13]

Fayette County was at that time still virtually a frontier region. Its population in 1790 was 13,325, and Uniontown, the county seat and the largest town in the county, was a mere village. Finley's friend and colleague Ephraim Douglass complained in the 1780s:

. . . this Uniontown is the most obscure spot on the face of the globe. . . . The town and its appurtenances consist of our president and a lovely little family, a court house and school house in one, a mill and consequently a miller, four taverns, three smith shops, five retail shops, two tan yards, one of them only occupied, one saddler's shop, two hatters' shops, one mason, one cake woman (we had two, but one of them having committed a petit larceny, is upon banishment), two widows and some reputed maids, to which may be added a distillery. . . . I can say little of the country in general but that it is very poor in everything but its soil, which is excellent. . . . But money we have not, nor any practicable way of making it.[14]

By 1820 the population of Uniontown had grown to 1,058. Pittsburgh, the largest town in western Pennsylvania, located about 50 miles from

Uniontown, had grown from 367 inhabitants in 1790 to more than 7,000 in 1820.[15]

The infrastructure and economy of western Pennsylvania, including Fayette County, were primitive before the turn of the nineteenth century. Most people lived in log cabins or log houses, and farms were largely self-sufficient.[16] A detailed picture of housing arrangements in Fayette County can be gleaned from records of the U.S. Direct Tax of 1798, which included an inventory of each property in the county, with a description and a valuation of the dwellings and outbuildings on each property. These records show that 68 percent of the dwellings in the county were log cabins under $100 in value, and in some townships the figure was above 80 percent. The mean value of a dwelling in Fayette County was $89.[17] The overall mean value for dwellings in the United States at this time was $262, nearly three times the average for Fayette County.[18] Union Township, where Finley lived, was the richest township in Fayette County in terms of dwelling values; the average there was $188 at a time when the national mean value for rural dwellings was $181.[19] Still, 91 percent of the dwellings in Union Township were of log construction.[20]

Currency was scarce in western Pennsylvania, and poor transportation limited trade and the size of markets.[21] Before 1800 it took between two and three weeks to travel the 300 miles from Pittsburgh to Philadelphia.[22] These conditions led farmers to distill their surplus grain into whiskey in order to have a product that could be transported to external markets at reasonable cost without danger of spoiling. Within the region, whiskey became a second currency.[23] Western Pennsylvania was also undeveloped industrially. For example, the first blast furnace west of the Allegheny Mountains was established only in 1790, in Fayette County. Before then, iron had to be transported from the east on pack horses. By 1810 there were eighteen blast furnaces in western Pennsylvania, eleven of them in Fayette County.[24]

The social structure of western Pennsylvania was comparatively fluid and open. Slaveholders made up less than 3 percent of the population, and the total number of slaves in the western counties of Pennsylvania decreased from 880 in 1790 to 70 in 1820.[25] In Union Township, slaves made up only 0.3 percent of the population in 1810.[26] Among the white population, there were social distinctions based on wealth, education, and professional or political standing, but these were far less rigid than the social gradations that existed in Europe; and certainly there was nothing approaching the European tradition of aristocracy. Among the individuals with high status in western Pennsylvania were preachers, doctors, lawyers, politicians, and public officials. The latter two groups came predominantly from among the larger landowners of the region.[27] And manual labor did not carry the social stigma that it did in (for example) France—western Pennsylvanians could not afford such a luxury. Virtually everyone had to participate in the tasks of farming, home building, and making clothing and tools. Indeed, the folkways of this and other frontier regions included strong traditions of cooperative labor such as quilting bees and house, church, and barn raisings.[28]

Western Pennsylvanians were known for their radical political tendencies, which were no doubt heightened by the fact that Pennsylvania's state constitutions of 1776 and 1790 extended suffrage to all free white males. Of course not everyone in the region was radical, but the tenor of political discourse tended to be more democratic and republican than in eastern Pennsylvania. When a state convention was held to consider ratification of the federal constitution in 1789, seven of the nine delegates elected from western Pennsylvania opposed ratification partly on the ground that there was no bill of rights.[29] After ratification, western Pennsylvanians were active in the campaign for democratic amendments to the Constitution. Their radicalism became even more pronounced in the early 1790s, when attempts were made to enforce a federal excise tax on whiskey. The move was widely and violently resisted throughout western Pennsylvania. Opposition meetings (including one in Uniontown) were attended by prominent community members. Tax collectors were tarred and feathered and driven out of town. Eventually the federal government sent in a militia of some 13,000 men to quell the revolt—the Whiskey Rebellion, as it came to be known.[30]

It is clear that Finley had a high social and economic standing, linked to his career as a politician and public servant. He was elected as a justice of the peace in 1784, and as a county commissioner in 1789.[31] He was elected to the Pennsylvania House of Representatives in 1790 and in 1791, and to the Pennsylvania Senate in 1793. In 1791 he was appointed Associate Judge for Fayette County, a position he held until his death in 1828.[32]

The U.S. Direct Tax records of 1798 show that Finley was among the top 6 percent of residents of Union Township in terms of the value of his dwelling (although not in terms of overall property ownership). His 287-acre farm was larger than the average in that area, but not exceptional, and he owned no other property in the county. The farm, which included a log barn, a stable and "cow house," and a wheel-wright shop, was valued at $2,412. This valuation places him in the top 10 percent of all adult males in the nation (approximately half of whom owned no property.) Finley lived in a two-story log home, 27 feet by 27 feet, with 13 windows containing 171 panes of glass. (Taxes on dwellings were assessed partly on the number of panes of glass they contained.) On his property there was also a two-story stone kitchen (23 feet by $16\frac{1}{2}$ feet), a two-story stone milk house, and a log smoke house.[33] The total assessed value of his home, with its additional buildings and its two-acre lot, was $750. Out of a total of 337 dwellings listed for Union Township, only 18 others had equal or greater assessed value.[34]

When Finley died, he bequeathed to his wife Rachel "the choice of my horses, with saddle and bridle, the choice of my cows, and one bed and bedding."[35] His debts were to be paid through the sale of the rest of his personal estate (which was not inventoried) and, if that was not sufficient, then through the sale of real estate. Whatever remained was to be divided up equally between his wife and their three children, "share and share alike." Eleven years after Finley's death, in 1839, his entire property except the house and its two-acre lot was sold for $8,501.06.[36] The property, not including the house and its lot, was found at that time to contain a little more than 283 acres. The assessed value of Finley's land in 1798 ($9 per acre), and its sale value in 1836 (roughly $31 per acre) indicate that his land was within the top 20 percent nationwide in quality.[37]

Finley's Social and Intellectual Environment

Brief biographies of some of Finley's friends and colleagues provide an idea of the social circles in which he moved. Ephraim Douglass, Fayette County's treasurer (appointed by the county commissioners), corresponded with Finley while the latter was serving in the Pennsylvania Assembly, which met in Philadelphia, and an extant letter which he wrote to Finley makes clear that he was a family friend. Douglass, who had lived in Pittsburgh before moving to Uniontown and had worked alternately as a "clerk, scrivener, carpenter, cabinet maker, lumberman, blacksmith, gunsmith, stone mason, and shop keeper,"[38] lived in a two-story log home with several outbuildings. The house was smaller than Finley's, but the assessed value with the outbuildings was $1,000.

Another of Finley's colleagues, Thomas Gaddis, was elected Justice of the Peace in 1784 (along with Finley) and served as county commissioner just before Finley. Born in 1742 and married in 1764, Gaddis moved to Fayette County after purchasing land there in 1769. His landholdings were greater than Finley's; his 424-acre farm ranked as the twelfth largest in Union Township. Yet his dwelling was more modest: in 1798 he lived in a log home with a stone milk house, assessed at $300. He remained in that house at least until 1812.[39]

Another colleague and neighbor, Isaac Meason, was also, with Finley, an Associate Judge for Fayette County. Meason, who started out with 500 acres, became the most prominent ironmaster in the region, and by 1799 he ranked as the wealthiest inhabitant of Fayette County. When he died, in 1818, he bequeathed to his wife one-third of both his real estate and his personal estate, half of his house and its furnishings, six cows, a horse, two servants, and $1,000 per annum. His estate included 20,000 acres of land, forges, furnaces, rolling mills, grist mills, salt works, bridges, toll ferries, and a town; it also included a Georgian stone mansion, the largest and most elegant in the region.[40] But in 1798 Meason lived in a two-story log home, with a separate log kitchen, valued at $160.[41]

Finley's circle of colleagues also included his fellow representatives from Fayette County in the Pennsylvania Assembly, Albert Gallatin and John Smilie. Gallatin, born in Switzerland, was aristocratically bred and educated. He had a large estate in Fayette County (nearly 1,000 acres), and he lived in a brick house, "Friendship Hill," valued at $600. Gallatin was said to have made substantial profits from a glass factory he had established, one of the very first in western Pennsylvania. The Fayette County tax lists for 1798 show that his holdings included not only a glass manufactory but also a grist mill, two saw mills, and a warehouse. Gallatin's farm was said to have brought him an income of about $2,000 per year. Smilie, born in Ireland, emigrated to America in 1760 at the age of 19. He had first settled on 80 acres of land in Lancaster County, but eventually he moved to Fayette County, where he acquired a 343-acre farm. In 1798 he lived in a log home with a stone kitchen, valued at $250.

Politically, both Gallatin and Smilie were republicans. Gallatin's views on major issues were close to those of Jefferson, and indeed he served as Secretary of the Treasury under Jefferson and Madison.[42] Both Smilie and Gallatin opposed ratification of the U.S. Constitution. Smilie believed the Constitution to be "a device of despotism,"[43] and

Gallatin opposed its centralizing tendencies and its bicameral legislature (which he held to be undemocratic). Both men also opposed moves in 1789 to alter the ultra-democratic Pennsylvania constitution of 1776. When a constitutional convention was held in 1790, both were elected as delegates, and their efforts were influential in helping to retain some of the democratic features of the 1776 Pennsylvania constitution.[44]

In 1793 Smilie was elected to the U.S. Senate. (Gallatin had been elected, but he was rejected on a technicality.) Since Smilie was serving at the time as a senator in the Pennsylvania Assembly, an election was held to replace him, and James Finley was chosen to fill the post.[45] Although Finley is reputed to have been a Federalist,[46] he must have been regarded as a moderate; otherwise it is hard to imagine why a consistently anti-Federalist constituency would have voted him in as Smilie's replacement at the height of the movement against the whiskey tax, at a time when the political mood in western Pennsylvania had become more radical.[47]

One point that emerges from these sketches is the extent to which men of varying wealth, social background, and education shared and cooperated in the political structure of the community. The activities of this diverse group also merged within the realm of technology, industry, and "practical science" (science oriented toward commercial and technological concerns). We know that Finley was interested in technology, but similar interests were also evident among his political colleagues. For example, Ephraim Douglass, in a letter written to Finley in Philadelphia in 1793, requested that Finley inquire at the American Philosophical Society about an eclipse that had taken place; Douglass was interested to know if it had been the cause of a severe frost that had damaged many fruit trees.[48] Albert Gallatin, apart from his involvement in local industry in western Pennsylvania, became a prominent advocate of "internal improvements" (i.e., improvements in transportation infrastructure, notably roads, canals, and bridges). As Secretary of the Treasury, Gallatin worked out a plan with President Jefferson for a system of roads and canals extending from Maine to Georgia. In 1808, Gallatin presented to Congress a report on roads and bridges in which he mentioned one of Finley's chain bridges.[49]

Isaac Meason sponsored many innovative technological endeavors. He financed the work of Thomas C. Lewis, who introduced the Cort process and the first rolling mills into American iron practice. He was also involved in Finley's first suspension bridge enterprise (1801).

When Fayette County and adjacent Westmoreland County agreed to pay jointly for the project, the commissioners requested a meeting with Meason "to consult and complete contract relative to James Finley, Esq., undertaking to erect an Iron Bridge over Jacob's Creek." The bridge was to be built "at or near" Meason's residence.[50] Meason's forges and furnaces provided the ironwork for the bridge.[51]

Interest in practical science and technology was also a prominent aspect of Scottish-Presbyterian intellectual culture. The Scottish-Presbyterian educational system provided a foundation for a range of scientific and philosophical activities and interchanges. In western Pennsylvania, clergymen and lay graduates of the Presbyterian academies and colleges forged an intellectual network, cemented by ethnic, religious, and family ties, that defined the structure of much of the cultural life in that region. Its influence was far greater than the proportion of Scottish Presbyterians in the population (which amounted to around 25 percent for all of western Pennsylvania) would suggest.[52] A look at some of the participants of this network thus offers insight into yet another important dimension of James Finley's cultural environment.

Finley belonged to the Laurel Hill parish in North Union Township. Its pastor from 1782 to 1803 was Rev. James Dunlap.[53] Dunlap had studied first under Rev. James Finley and subsequently under John Witherspoon at the College of New Jersey.[54] Witherspoon, who had emigrated from Edinburgh in 1768 to take up the presidency of the college, introduced his students to the ideas of the Scottish Enlightenment. His lectures in moral philosophy provided a thorough grounding in Scottish Common Sense philosophy. Moreover, following the prevailing trend in Edinburgh, Witherspoon placed great emphasis on the study of science and technology. Sections of his history lectures focused on the development of science and industry, and he expanded and upgraded the teaching of science in the curriculum to the extent that the third and fourth years were almost entirely devoted to mathematics and natural philosophy.[55] Dunlap brought these intellectual traditions to Fayette County, where he served not only as the minister in Finley's parish but also as an educator. In 1802, when Canonsburg Academy expanded to become Jefferson College, Dunlap was chosen as its principal.[56]

Other students of Witherspoon, including the clergymen Nathaniel Sample, John McMillan, and Thaddeus Dod and the judge Hugh Henry Brackenridge, played equally important roles in shaping the scientific and intellectual life of their communities. Sample settled in

Lancaster County, in southeast Pennsylvania, where he helped to found the Strasburg Scientific Society.[57]

John McMillan and Thaddeus Dod were part of a group known as the "four horsemen" who were chosen by the Presbyterian Church to oversee the progress of education in western Pennsylvania. McMillan, a leading figure in the establishment of Canonsburg Academy, also helped to organize Washington Academy, close to Fayette County, in 1787. (Benjamin Franklin donated £50 to the school and recommended an instructor.) Dod, like Rev. Finley, was a trustee of Canonsburg Academy, and he became the principal of Washington Academy.[58] According to a contemporary, Dod was an outstanding scholar in the classics and the sciences, "with the happy faculty of infusing into those who were capable of it, an intense love of science and literature."[59] Dod was interested in technology as well; extant manuscripts of his include notes on navigation.[60]

Hugh Brackenridge settled in Pittsburgh and became a judge, politician, and writer. He played a central role in establishing the first newspaper west of the Alleghenies, the *Pittsburgh Gazette*. Brackenridge also supported the interests of his relative John Gilkinson, who in 1798 opened the first bookstore in western Pennsylvania.[61] Gilkinson's favorite authors included Reid, Hutcheson, Locke, Newton, and Bacon. According to Brackenridge's son, much of Gilkinson's time was "devoted to natural philosophy and the higher mathematics, for which he seemed to possess an extraordinary aptitude. He sat up until midnight, 'exhausting the lamp of life, in feeding the lamp of science.'"[62] In addition to his sponsorship of newspapers and bookstores, Brackenridge was involved in founding the Pittsburgh Academy, which ultimately became the University of Pittsburgh. The trustees of the school included John McMillan, Rev. James Finley,[63] and Edward Cook, who was an associate judge in Fayette County along with James Finley the bridge designer.[64]

Rev. John Taylor, one of the most respected professors at Pittsburgh Academy, lectured on astronomy, having "brought with him to the school a fine set of globes and astronomical apparatus."[65] Taylor also made the calculations for the local almanac and taught an evening course on that subject. One of Taylor's students, Morgan Nevill, eventually became a lawyer and a newspaper editor and acquired an extensive library.[66]

In 1811 a series of chemistry lectures was given at the Pittsburgh Academy by Frederick Aigster,[67] who had previously taught at Dickinson College, a Presbyterian institution in Carlisle, Pennsylvania.[68]

Aigster's lectures, which were open to the public, focused on the applications of chemistry to "agriculture, mining, tanning, papermaking, brewing, and cooking."[69] Interest in Aigster's lectures led to the founding in 1813 of the Pittsburgh Chemical and Physiological Society, and Aigster became the society's lecturer on mineralogy.[70]

In 1819 the Pittsburgh Academy was rechartered as Western University and came under the direction of Rev. Robert Bruce, a Presbyterian minister trained in Edinburgh. Bruce also became the university's professor of natural philosophy, chemistry, and mathematics. A student reported that Bruce was "an extensive reader, liberal minded, and a most accurate scholar in the several branches he professed. He was not only learned but extremely critical. He had the philosophy of Bacon and Descartes, Hume, Reid, and Dugald Stewart at command—he had himself been a student of Dugald Stewart."[71]

Bruce helped to organize the Pittsburgh Philosophical and Philological Society in 1827, and he became its first president. The society's goal was to investigate and discuss "subjects of practical utility."[72] Bruce lectured in mechanics and mathematics, John Taylor in astronomy. The society also had a lecturer in architecture and civil engineering. Two years later, the Pittsburgh Mechanics' Institute was formed, and Bruce became its first president.[73]

As the foregoing sketches suggest, western Pennsylvania was in many ways supportive of scientific and technological interests. This was due not only to the influence of ideas and values stemming from the Scottish Enlightenment, but also to more immediate economic needs and possibilities. Many western Pennsylvanians believed it was crucial to build up their region's technological and industrial base, in order to have access to larger markets and to lessen their dependence on the East (e.g., for iron); moreover, outstanding economic opportunities existed in the region for those who were willing and able to make the effort and take the risk.[74]

The Emergence of Finley's Interest in Technology

The nature of the economic and cultural environment in western Pennsylvania goes far toward explaining how it was that James Finley, a farmer and a county judge in a small rural community, became actively interested in invention. Yet an equally significant factor in stimulating this interest was Finley's stay in Philadelphia as a state legislator between 1790 and 1794. At that time, Philadelphia was both the state and the national capital. Moreover, it was the largest city in the United States, and one of the nation's leading scientific centers.[75] Evidence suggests that it was in Philadelphia that Finley became interested in bridges and developed the idea of designing a suspension bridge.

Finley probably already had a general concern about "internal improvements" before he went to Philadelphia. As a farmer, this issue would naturally have interested him. As a county commissioner (in 1789) he had to devote some attention to it, because the building of local roads and bridges in Pennsylvania was handled by these officials. The way the system worked was that a proposal was first submitted to the Court of Quarter Sessions, presided over by the county judges. The court then appointed a board of viewers to report on the site and the plan for the prospective road or bridge. If the viewers and the court deemed the project necessary and worthwhile, the court directed the county commissioners to appropriate funds and have the work undertaken. When Finley served as county commissioner, he was requested to view at least one site: that for a road between Brownsville and the Youghiogheny River.[76]

In Philadelphia, increasing attention was being given to internal improvements by the state legislature and by leading business and scientific figures. In 1791 a society was formed in Philadelphia for the "Improvement of Inland Navigation." The next year, partly as a result of efforts by this group, the state legislature granted its first charter to a turnpike corporation, for the construction of a 62-mile turnpike from Philadelphia to Lancaster—the first leg of the route to Pittsburgh, which Finley himself had to travel. This project, completed in 1794, was the first of many; by 1820, at least 1,500 miles of turnpikes had been constructed throughout Pennsylvania. At least eight turnpike corporations and three bridge corporations had been established over the Pittsburgh-Philadelphia route.[77]

The state legislature, since it was responsible for granting charters for bridge and turnpike companies and for approving state funding for important construction projects, became a clearing house for new ideal and proposals. Thomas Paine forwarded a plan for a cast-iron arch to the legislature,[78] and Charles Willson Peale circulated a pamphlet about a wooden arch bridge he had invented. Peale's pamphlet opened with this plea: "Legislatures, and you men of influence in the counties of each State! Turn your attention to this important object—shorten the distance to market for the sale of the product of your lands. I offer you a cheap and easy mode of building Bridges. . . ."[79]

As a member of the legislature, Finley could not have remained ignorant of these developments. Furthermore, he could hardly have been better situated to learn about ideas and plans for bridge construction. Philadelphia was an important center of innovation in bridge design, for two reasons. First, the Schuylkill River ran directly through the city. Bridges were needed to accommodate an increasing flow of goods and people to and from western regions, but the Schuylkill presented difficulties that made traditional bridge designs either impracticable or too expensive. In 1808 a Philadelphia journal, the *Register of Arts,* described the situation as follows:

The Schuylkill which washes the western front of the city of Philadelphia . . . has long been attended with many serious inconveniences. . . . Its borders, to an extent of one hundred miles, are skirted by precipitous mountains and hills. Its tributary streams, suddenly filled in seasons of rains or melting snows . . . without notice or time for precaution, fill the river with frequent floods which no common works of art within their reach have heretofore been capable of withstanding. Although these attributes are not to a certain degree uncommon, yet, in this river they are peculiarly dangerous. They occur at irregular periods and often at seasons of the year when floods are generally unexpected. These circumstances at all times created doubts of the practicality of any permanent erection. The depth of water opposite the city added to the difficulties and apprehensions.[80]

The second reason why so many innovative bridge designs originated from Philadelphia was the presence of the American Philosophical Society there. The society promoted interest in technological innovation in general, and the challenge of spanning the Schuylkill inspired several members to come up with innovative plans. A plan for a chain bridge, first by suggested John Jones in 1773[81] and then worked out in greater detail the following year by Thomas Gilpin,[82] was inspired by the knowledge that chain bridges had been built in China. It is important to trace the history of this example of stimulus diffusion and to examine the Jones-Gilpin proposal in greater detail, because they almost certainly provided the immediate inspiration for Finley's efforts. At the same time, this analysis will provide a basis for understanding how Finley's design differed from its predecessors.

Chain Bridges in Asia and in America

Chain bridges have been built in Asia at least since the sixth century A.D. They evolved from an earlier type of suspension bridge, made with bamboo cables, that was found throughout the mountainous Tibetan massif by the first century B.C. The wild rivers and steep cliffs of this region called for bridges that could be built without a centering (false-work) and without intermediate piers; suspension bridges were the best (and in some cases the only) alternative. Evidence for the transition from bamboo cables to iron chains in the sixth century is incomplete, but the innovation probably originated in the provinces of Szechuan and Yunnan, east of Tibet, where a mature iron technology had developed and bamboo suspension bridges were common.[83]

Chinese iron suspension bridges were composed of a number of parallel chains held taut and anchored in abutments or in solid rock. The chains were forged by hand from wrought-iron bars 2 or 3 inches in diameter. Floor planking was laid directly upon the chains, so that the roadway followed their curve. Additional chains, anchored separately, served as hand railings. Because these bridges did not have level, rigid roadways, they swayed and undulated and were thus unsuitable for vehicular traffic.[84]

Over the centuries dozens of chain suspension bridges were built in western China, in Tibet, in Bhutan, in northwest India, and in northern Pakistan—regions remote from major trading ports. Jesuit

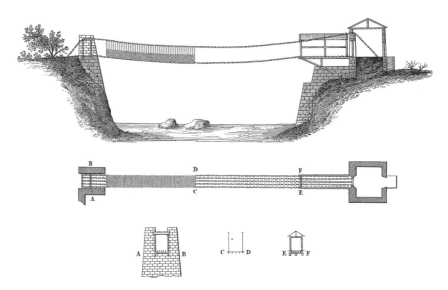

Figure 1.1
A traditional Chinese chain bridge with a curved deck. Source: Navier, *Rapport et mémoire sur les ponts suspendus,* second edition (Paris: Carilian-Goeury, 1830). Courtesy of Special Collections Department, University of Virginia Library.

Figure 1.2
The Chin-Lung Ch'iao chain suspension bridge over the Yangtze River.
Source: Joseph F. Rock, *The Ancient Na-Khi Kingdom of Southwest China*, volume 1 (Harvard University Press, 1947), plate 111.

missionaries brought accounts of them back to Europe in the seventeenth century. For example, in 1667 Athanasius Kircher published a description and an illustration of a chain bridge over the Mekong River in Yunnan. Illustrated accounts of this bridge were also published by J. B. Fischer von Erlach in 1725,[85] by Carl Christian Schramm in 1735,[86] and by others subsequently.[87] By means of such publications, knowledge of Asian chain bridges passed from Europe to the American colonies.[88]

An active China trade, centered in Philadelphia, fueled American interest in all aspects of Chinese civilization. As a result of the China trade, Chinese designs and objects became popular in the United States. It is estimated that at the beginning of the nineteenth century between 10 percent and 20 percent of the furnishings in a Philadelphia home were of Chinese origin. Wealthy landowners often had their houses and gardens decorated "*à la Chinois*." Thomas Jefferson planned a Chinese pavilion for Monticello, the architect Benjamin Latrobe designed a Chinese-style summer house in 1806, and a plan

Figure 1.3
A fanciful European portrayal of a chain bridge over the Mekong River in
Yunnan. Source: Carl Christian Schramm, *Merkwürdigsten Brücken* (Leipzig:
Breitkopf, 1735).

from William Chambers's *Designs of Chinese Buildings* (London, 1757)
was used for a pagoda built in Philadelphia's Fairmount Park.[89]

Jonathan Goldstein, in his book, *Philadelphia and the China Trade,
1682–1846,* has pointed out that members of the American Philosophical Society were among the Americans most interested in Chinese
culture, technology, and husbandry. Indeed, the preface to the first
volume of the society's *Transactions* specifically encouraged the adoption of Chinese techniques and products: "By introducing the produce
of those countries which lie on the east side of the old world, and
particularly those of China, this country may be improved beyond
what heretofore might have been expected. And could we be so fortunate as to introduce the industry of the Chinese, their arts of living
and improvements in husbandry, as well as their native plants, America
might in time become as populous as China."[90]

In view of the attention paid to Chinese achievements by the American Philosophical Society, it is not surprising that John Jones thought
to adapt a Chinese idea to the problem of spanning the Schuylkill,

especially since this problem did not appear to be solvable by means of traditional Western bridge designs. Jones's bridge was to consist of a single 400-foot span. By Western standards this appeared quite daring, but Jones asserted that it was small compared to "the bridges which some of the most Judicious Oriental Travellers tell us are to be found in China."[91] Unfortunately, a model of the bridge which Jones sent to the American Philosophical Society has not survived, and his written description was meant to be interpreted with the aid of the model.[92] He did state, however, that the design included iron chains and iron columns.

It is evident that Jones's proposal was no more than a preliminary sketch. Jones did not survey the intended site, and he provided neither cost estimates nor design specifications for the bridge. Thomas Gilpin extended Jones's ideas in this direction, however.[93] Gilpin surveyed the most likely site for the structure, and he did prepare a cost estimate and design specifications. Acting on a suggestion by Jones, Gilpin proposed to construct large embankments on either side of the river so as to decrease the span of the bridge to 300 feet. Gilpin's plan reveals unmistakable Asian influence. Like the Asian bridges, the roadway was to be made up of a number of parallel chains held fairly taut between the embankments. (Gilpin specified a sag/span ratio of only 1/20.) The roadway planking was to be attached directly to the chains so as to follow their curve.[94]

The designs of Jones and Gilpin were formulated more than 15 years before Finley served in the Pennsylvania legislature. Yet they were not forgotten; in 1808 they were mentioned in the prominent Philadelphia journal *Port Folio*.[95] Finley could have learned about these designs through conversation or by inquiring at the American Philosophical Society.[96] Information about Chinese chain bridges was also readily available in published sources, such as William Chambers's book *Designs of Chinese Buildings,* and at least one bridge designer of the era—Thomas Pope—was convinced of their influence on Finley.[97]

It is reasonable to conclude, therefore, that sometime during the period 1790–1793 Finley learned about the plans of Jones and Gilpin and their Asian antecedents. This hypothesis is consistent with what we know about the chronology of Finley's work on suspension bridges. His last term in Philadelphia ended in 1793, and his first trial suspension bridge was built in 1801.[98] In 1808 he obtained a patent for a comprehensive plan to build chain suspension bridges.[99] The fact remains, of course, that in his publications Finley never admitted any

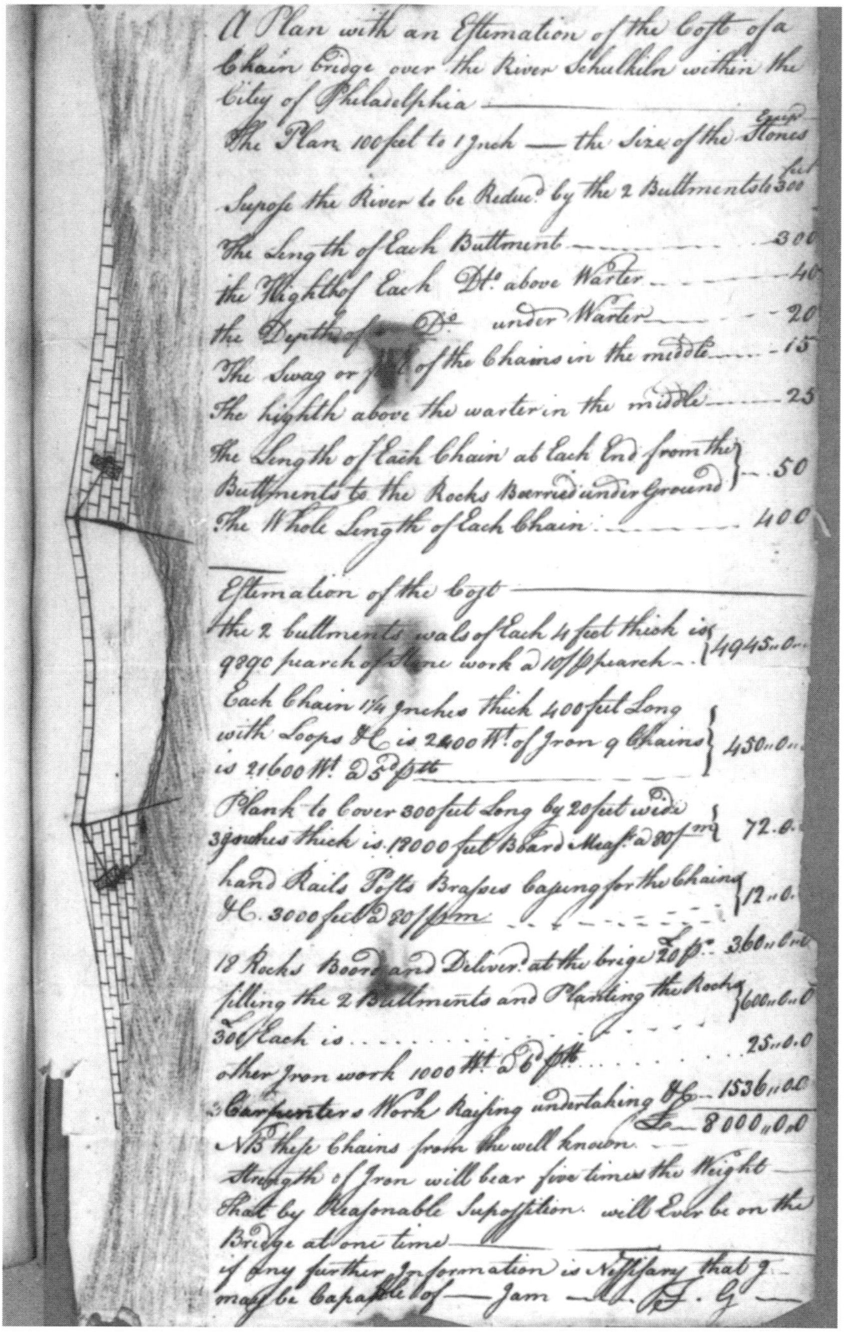
A Plan with an Estimation of the Cost of a
Chain Bridge over the River Schulkiln within the
City of Philadelphia ——
The Plan 100 feet to 1 Inch — the Size of the Stones Equal
Supose the River to be Reduc'd by the 2 Buttments to 300 feet
The Length of Each Buttment ---- 300
the Highth of Each Dto. above Warter ---- 40
the Depth of Dto. under Warter ---- 20
The Swag or fall of the Chains in the middle ---- 15
The highth above the warter in the middle ---- 25
the Length of Each Chain at Each End from the Buttments to the Rocks Barried under Ground } ---- 50
The Whole Length of Each Chain ---- 400

Estimation of the Cost
the 2 buttments wals of Each 4 feet thick is 9890 pearch of Stone work @ 10/ p pearch } 4945..0..
Each Chain 1 1/4 Inches thick 400 feet Long with Loops &C is 2400 wt of Iron 9 Chains is 21600 wt @ 5d p lb } 450..0..
Plank to Cover 300 feet Long by 20 feet wide 3 Inches thick is 18000 feet Board Measr @ 80/ p m } 72.0..
hand Rails Posts Brasses caping for the Chains &C. 3000 feet @ 80/ p m } 12..0..
18 Rocks Boor'd and Deliver'd at the bridge 20/ p — 360..0..0
filling the 2 Buttments and Planting the Rocks } 600..0..0
30/ Each is ---- 25..0..0
other Iron work 1000 wt @ 6d p lb ----
Carpenters Work Raising undertaking &C — 1536..0..0
£ — 8000..0..0
NB these Chains from the well known Strength of Iron will bear five times the Weight that by Reasonable Supossition will Ever be on the Bridge at one time ——
if any further Information is Nessisary that I may be Capable of — I am — T. G —

Figure 1.4
Thomas Gilpin's 1774 plan for a chain bridge over the Schuylkill River at Philadelphia. Source: Thomas Gilpin Letterbook, Maury Family Papers, University of Virginia Library. Courtesy of University of Virginia Library.

debt to Jones and Gilpin or any knowledge of Asian chain bridges. Yet it was not in his interest to do so, because his publications were all intended to promote his chain-bridge patent. Moreover, Finley's design, although founded on a knowledge of these precedents, differed from them in two major ways.

First, Finley's design had a level, rigid roadway suspended from chains, rather than a roadway laid down upon chains so as to follow their curve (figures 1.5 and 1.6). In retrospect this may seem an obvious or minor innovation, but its importance cannot be overestimated. A suspension bridge with a roadway that follows the sag of the chains can never effectively serve as anything but a footbridge, whereas a suspension bridge with a level roadway can be made to support heavy vehicular traffic. Moreover, the latter is a qualitatively different type of structure, one that has potential for great length. (The maximum length between supports is much greater for a suspension bridge than for any other type of bridge and is greater for a suspension bridge with a level roadway than for one with a curved roadway.) In short, Finley's innovation represented a basic new structural type.

The other way in which Finley's design differed from its predecessors, in particular from Gilpin's, was that it was a standardized, universally applicable design system rather than a proposal intended for a specific site. Moreover, Gilpin's proposal (described in longhand on a single sheet of paper) gave no indication of how specific elements of the bridge, such as the roadway, were to be constructed. Finley's plan (described in a ten-page article in *Port Folio*) gave precise details for every element. It explained how to estimate the amount of iron needed for the supporting chains, how to fabricate the chain links, how to determine the proper lengths for the main chains and for the hangers, how to connect the hangers to the main chains and the roadway, how to design the roadway, the towers, and the anchorages, and how to adapt the design for multiple-span bridges.[100] It evidently took Finley years to work out this design system, and for him it must have represented a contribution at least as important as his original insight for a level roadway.

In retrospect, and even in comparison with many contemporaneous European bridges, Finley's design system might seem quaint and crude. The anchorages were merely pits filled with stones. The towers were simple wooden frames with no pretense to beauty or grandeur. Builders were instructed to use a "board fence" to prepare a model of an intended bridge. It could even be argued that, apart from his idea

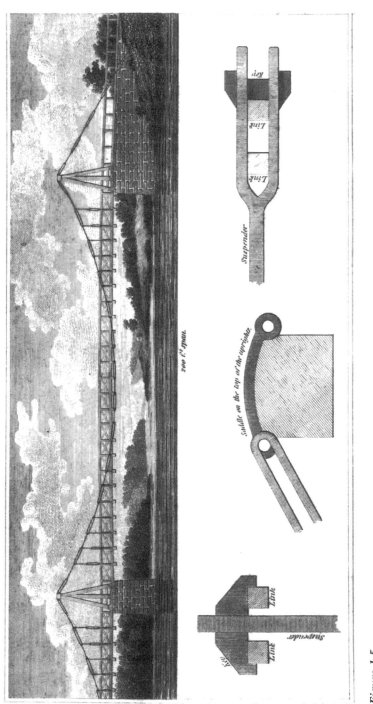

Figure 1.5
James Finley's patent chain bridge. Source: *Port Folio*, n.s. 3 (June 1810), p. 441.

Figure 1.6
Paul Svinin's painting of a Finley chain suspension bridge, ca. 1811. Source:
Avrahm Yarmolinsky, *Picturesque United States of America, 1811, 1812, 1813;
Being a Memoir on Paul Svinin* (New York: W. E. Rudge, 1930). Courtesy of
Department of Special Collections, University of Minnesota. This bridge,
identified in Yarmolinsky's book as the Newburyport suspension bridge, was
almost certainly the Falls of Schuylkill bridge—see figures 3.1, 3.2, and 3.11.

for a level roadway, Finley contributed nothing to the cumulative stock
of suspension bridge technology. The suspension bridges built only 15
or 20 years later had entirely different cables, anchorages, towers, and
roadways.

Yet such an interpretation presupposes a "trajectory" view of the
history of technology, whereas the aim of this study is to understand
Finley's design on its own terms, in relation to its own environment,
and in relation to Finley's goals. Viewed in this way, Finley's design
was plain and simple not because he was unable to produce better but
because he chose to make it that way. That it did involve conscious
choice on his part is suggested by the way Finley contrasted the
character of his design with that of a stone arch bridge: "May I venture
to glance at the grand, majestic arch of solid stone, with any idea of
the contrast between it and our simple contrivance? Happy for me,
utility economy and despatch, are the ruling passions of the day, and

will always take preference of expense, idle elegance and show, until the minds of men become contaminated with vanity or some worse passion."[101] As this comment reveals, Finley's design efforts were guided in part by an ideology that favored simple functionality over "idle elegance and show." On a more concrete level, Finley's ideological position served to justify a set of design goals rooted in the needs and conditions of the rural, frontier society in which he lived. These design goals and the means by which Finley implemented them will be analyzed in detail in the following chapter.

2

Designing the Chain Bridge

Finley's Design Goals

In his efforts to develop a viable plan for a suspension bridge, James Finley was guided by four general design goals. He sought a bridge design that would be economical to build and maintain, a design that would be uncomplicated enough to permit construction and maintenance by blacksmiths and carpenters without specialized training, a standardized design that would be universally applicable, and a scientifically valid design. What the last goal meant can be understood by considering Finley's critical speculations about wooden truss bridges:

What then is a wooden frame bridge, consisting of two or three hundred tons of timber to every 200 feet span? View its upper and under timbers of the framing, and its hundreds of ties and bracings in every direction—what is the task of each, and how much more can they bear—what burden can the bridge support—or is the huge mass of which it is composed, burden sufficient for itself, and the whole structure so complex that nothing but loose conjecture can say anything about it?[1]

As this comment suggests, a scientifically valid design meant, for Finley, a design that had some rational basis, a design that permitted a reliable estimation of the "burden" that could be safely supported—in short, a design based on something more solid than "loose conjecture." What Finley specifically had in mind was a design based on the "natural laws" of equilibrium of such a structure.

Finley's four design goals were not conceived arbitrarily; they did not reflect merely whim or subjective preference on his part. Rather, they were a considered response to the social, institutional, and economic possibilities and constraints within which Finley (and ultimately the design itself) had to function. Of course, in an absolute sense, Finley could have chosen whatever design goals he wished; yet in order to have a reasonable chance of technical and economic success a bridge design had to be well adapted to existing conditions, and Finley was astute enough to recognize this fact.

The most immediate considerations for Finley were economic. Finley's effort to work out a suspension bridge design represented an entrepreneurial undertaking within the context of the patent system. In the United States, a patent gave an inventor a 17-year monopoly on the right to exploit an innovation. Finley's aim was to patent a complete design system and then earn profits from it by selling the right to build bridges from his plan to anyone who asked, for the price of a dollar per linear foot of span. This idea was not unique to Finley, and the patent system rapidly became the primary forum of competition among American bridge designers after its establishment in 1790. All the leading bridge architects contemporary to Finley, including Timothy Palmer,[2] Theodore Burr, Lewis Wernwag,[3] and Charles Willson Peale,[4] patented their designs. Likewise, a group or a community that intended to construct a new bridge reviewed current patents in the process of selecting a design.[5]

This particular institutional-economic context, involving competition through the patent system, had several implications for Finley's design. First, to earn much profit, Finley's design would have to be sold, or reproduced, as widely as possible, and this made it essential for the design to be economical relative to its competitors, standardized, and universally applicable. It also had to be technically viable: if bridges built according to his plan consistently fell down, it would not be long before no new buyers could be found. Since Finley's concept of a level-roadway suspension bridge was so radically new, and since iron had only recently begun to be used for building bridges, Finley could not rely on experience or on craft know-how. If the design could be justified scientifically, however, this would provide a basis for its technical viability.

A more general factor that Finley had to consider was the undeveloped state of America's economy and material infrastructure. Expansion into new frontier areas created a need for more and better roads, sturdier bridges, and a better-developed canal system. At the same time, most communities, particularly frontier communities, had very limited economic resources with which to undertake such projects.

Fayette County was on a major route of travel to the west, and a disproportionate number of heavily laden vehicles consequently passed through the region, which heightened the need for good roads and bridges. Yet the population density was low, and Finley was conscious of the region's limited financial resources. In 1791 Fayette County's treasurer complained to Finley of being a "treasurer without money, and of consequence without credit."[6] Finley's posts as county commissioner and associate judge also made him privy to information about the county's financial situation. The main functions of the county commissioners were to prepare a yearly budget (including appropriations for public works), to review the county tax lists (which gave the value of all taxable property), and to set the tax rate. County judges, for their part, were responsible for directing a yearly audit of the tax books. As a member of the state legislature, Finley was in a position to learn about the financial situations in other counties too.[7]

In view of Finley's immediate knowledge about county and state finances, it is not hard to imagine that he would have felt it important to design bridges that would be economical to build and maintain. Of course, not all road and bridge projects depended on public funding. After 1790 more and more of these projects were sponsored by groups of businessmen and citizens for profit. But here again there was little willingness to spend more than was absolutely necessary. As I have already noted, a further reason for pursuing an economical design was that it promised to be an important advantage in the competition with other inventors who were also patenting and attempting to sell their bridge designs.

Just as important as the quest for economy was the need for technical simplicity. Finley realized that, to achieve widespread application, his design should not require the availability of builders with advanced knowledge of mathematics or with specialized technical qualifications (beyond general carpentry or blacksmithing skills). Unlike France, the United States at the time had very few formally trained engineers. It has been estimated that there were no more than 30 civil engineers in the entire country in 1815.[8] Before the founding of West Point (in 1802) there was no institution in America that provided comprehensive training for civil engineers, and even West Point did not offer a solid course in the subject until 1816.[9] What few engineers there were generally became involved with large-scale public works (such as canals) rather than with comparatively small enterprises like bridge building.

Under the circumstances (and particularly in sparsely settled rural and frontier areas, with their limited resources), bridge construction remained largely in the hands of local craftsmen. Although some craftsmen who specialized in bridge building were extraordinarily creative and capable,[10] the average carpenter or blacksmith could not be expected to undertake very large or complex projects. For example, it could not be assumed that a local craftsman would have anything more than a rudimentary knowledge of arithmetic.[11] As a consequence, for a design to be widely applicable it had to be relatively straightforward and uncomplicated to build. Finley took particular care to create a standard plan that could be adapted to a particular site with no need for any but the most trivial arithmetic or geometric calculations. His plan was supposed to "enable any person to make a rough estimate for any particular case."[12]

The net result of these conditions, and of Finley's response to them, was a design that can best be characterized as a standardized do-it-yourself plan. It was a distinctly utilitarian design, with no attempt to achieve aesthetic elegance or monumentality. In many respects it was an archetypical frontier technology. Yet it embodied significant experimental research, and some of the characteristics of the design plan that made it easy to follow were discovered by Finley through this research.

Finley's Research Program and Its Roots in Common Sense Philosophy

The aim of Finley's research was to learn more about the laws governing the behavior of suspension bridges. His methodology—which was to discover these laws through systematic experimentation—drew on the Scottish Enlightenment ideas that were being widely discussed at that time throughout America. Particularly influential were the Common Sense philosophers Frances Hutcheson, Thomas Reid, and Dugald Stewart, who created an intellectual synthesis that emphasized the use of observation, experiment, and inductive reasoning to expand man's understanding of himself and of the universe. They believed that applying inductive scientific methods would make it possible to uncover fundamental laws in any field, from agriculture to moral philosophy.[13]

At the College of New Jersey, John Witherspoon echoed these ideas: " . . . perhaps a time may come when men, treating moral philosophy as Newton and his successors have done natural, may arrive at greater precision. It is always safer in our reasoning to trace facts upwards than to reason downward upon metaphysical principles."[14] Witherspoon's successor at the college, his son-in-law Samuel Stanhope Smith, went even further. Drawing heavily from Thomas Reid, Smith developed a five-point scientific method, which he presented in his lectures on moral philosophy. The essentials of his method can be summarized as follows: laws, in any field, were to be formulated by induction from an ample empirical data base (derived through observation or experiment); the laws so formulated were to be assumed universal until proven otherwise; in all cases the testimony of the senses was to be taken as valid.[15]

Comments by Finley also reflected a preference for knowledge rooted in observation and experiment. "The well informed will not so lightly treat any information obtained or supposed to be obtained by actual experiment," he wrote in his article in *Port Folio* (1810).[16] In an expanded version of the article, published in pamphlet form the following year, Finley went further: "Theorists may talk as large as they please, but the lessons of experience are more wholesome."[17] In a later section of the pamphlet he referred to the need to satisfy "phylosophy and common sense."[18] These remarks served as signposts for readers; they signaled Finley's acceptance of the Common Sense intellectual tradition.

More important, however, Finley adopted the empirico-inductive methodology advocated by the Common Sense philosophers and their American followers. In particular, he carried out inductive experiments with cables and pulleys in order to discover laws for proportioning suspension bridges so that they would withstand the forces acting on them. Most important, the main cables—which were, in Finley's words, "the whole skeleton" of the bridge and its "whole strength"[19]—had to be able to resist tremendous tension.

An essential conceptual basis for Finley's experiments was the recognition that the tension in a cable supported at both ends (as in a suspension bridge) is not necessarily equal to the distributed load it supports (i.e., the weight of the roadway and the traffic upon it). The experiments he carried out helped him to determine quantitative relationships between these two parameters—load and tension—for a cable with a given sag/span ratio. (Figure 2.1 illustrates what is meant by sag and span.) The fundamental parameters of Finley's experiments were thus sag, span, tension, and load. The main procedure he followed was that of systematic parametric variation.

One of the experimental setups Finley designed is illustrated schematically in figure 2.2. One end of a cable was secured at A and the other end was extended over a pulley at B. Finley attached a weight to the free end of the cable, at C. He then loaded the portion of the

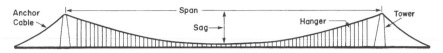

Figure 2.1
A schematic showing the principal elements of a suspension bridge.

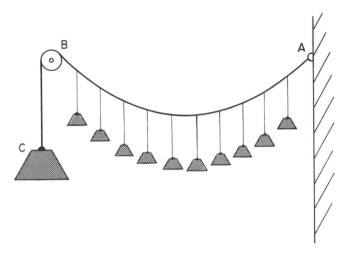

Figure 2.2
Finley's method of studying the behavior of cables bearing distributed loads, in particular the relationships among sag, span, load, and tension.

cable between A and B uniformly along the horizontal (as ideally would be the case in a suspension bridge).[20] With this setup, the span (i.e., the horizontal distance between A and B) remained constant. This left three variables: the tension in the cable, the load between A and B, and the sag of the cable. To vary the tension in the cable it was necessary to vary the weight at C, for with this experimental setup the weight at C determined the tension. (The cable's tension at C had to be equal to the weight at C, and because pulleys equalize tension the tension had to be the same on both sides of the pulley.) The sag of the cable changed when either the weight at C or the load between A and B was altered.

Finley systematically varied the weight at C and the load between A and B in order to determine the general relationships among the three variables. First, with the tension (i.e., the weight at C) held constant, Finley varied the load between A and B. He found that the sag of the cable increased as the load increased. Then, with the load between A and B held constant, Finley varied the weight at C. He found that the sag of the cable decreased as its tension increased. In other words, Finley found a direct relation between the cable's load and its sag, and an inverse relation between its tension and its sag.[21]

Although these generalizations were significant, they were still quali-tative and therefore not sufficient as guides in the design of suspension bridges. To gain quantitative information about the relations among sag, span, tension, and load, Finley devised a set of experiments that were ingenious and yet remarkably simple. He began once again with the experimental setup illustrated in figure 2.2. First he attached a weight to the cable at point C (assume 10 pounds). He then distributed an equal amount uniformly along the horizontal between A and B (e.g., ten one-pound weights, evenly spaced). Finley knew that the tension in the cable between C and B was equal to the weight sup-ported at C (that is, 10 pounds). And he knew that the tension in the cable was 10 pounds just on either side of the pulley, again because pulleys equalize tension. On the basis of symmetry, Finley concluded that the tension at A was equal to the tension at B and hence to the tension at C.[22]

But if the tension in the cable was 10 pounds at A and at B, was it also 10 pounds everywhere in between? Since the cable formed a curve and therefore continually changed direction, might not the tension vary accordingly? Finley devised a clever experimental setup to inves-tigate this question, illustrated schematically in figure 2.3. A cable, loaded uniformly along the horizontal, passed over two pulleys at F and G so as to represent half of a suspension bridge's cable (the portion from the center to one end of the span). The ends of the cable were weighted at E and H. Finley discovered that, to maintain equilibrium, the load at E had to be greater than the load at H. This demonstrated that the tension of such a cable was least at the center and greatest at the supports. (The difference between the weights at E and H when equilibrium was established gave the variance in tension. Finley ob-served that the tension was "about an eleventh less at the middle of the bridge than at the ends."[23]) Within the context of the original problem, Finley now understood that the 10 pounds of tension at the supports A and B in figure 2.2 represented the cable's maximum tension.

To reiterate: Finley had loaded a cable uniformly along the horizon-tal so that the total load was equal to the maximum tension in the cable. Under these circumstances he found the sag/span ratio to be

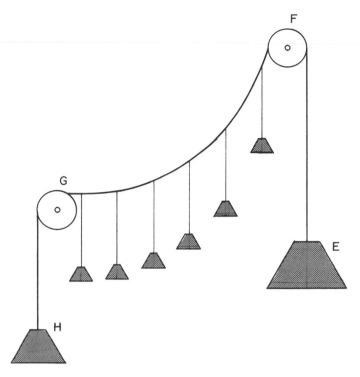

Figure 2.3
Finley's method for determining the variance in cable tension between middle
of span and supports.

approximately 1/6.5 or 1/7. Finley realized that this ratio provided an
easy means of estimating cable tension: if the sag/span ratio of a
suspension bridge were set by the designer at about 1/6.5 or 1/7, then
the maximum overall tension that the cables would have to bear
would simply be equal to the total load they supported (i.e., the weight
of the bridge itself and the loads upon it, assumed to be uniformly
distributed).[24]

Finley went on to examine the conditions prevailing for sag/span
ratios of 1/9, 1/14, and 1/30. He found that, for a given load (held
constant), the tension in a cable increased as the cable's sag/span ratio
decreased. For example, in a cable with a sag/span ratio of 1/9 the
tension was 1.33 times the load. However, in a cable with a sag/span
ratio of 1/14 the tension was 2.0 times the load.[25]

Finley suggested that builders always adopt a sag/span ratio between
1/6 and 1/7,[26] which would allow the amount of iron needed for the

chains to be calculated directly from the load that the chains had to bear. In fact, adopting this ratio freed builders from having to distinguish between load and tension. Finley established a simple procedure for making the necessary calculations. First, the weight of the roadway was to be estimated. This required knowledge of the span and the width of the prospective bridge and data on the unit weight of the timber to be used. Finley provided an average value for the latter. Second, in order to reach an estimate for the total load the bridge cables had to support, the weight of the roadway was to be multiplied by 5 or 6. This factor was intended to allow for all the loads to which the bridge would be subjected (e.g., traffic and snow). It was also a safety factor—an allowance for weaknesses or flaws in the cables, or other imperfections. Finally, the total load calculated in the second step was to be used to calculate the size and the number of the chains needed for the bridge. This calculation required a value for the strength of iron; Finley used 60,000 pounds per square inch, which he got from a table in an article on the strength of materials written by John Robison for the third edition of the *Encyclopaedia Britannica* (1797).[27] Finally, with a knowledge of the weight per linear foot of the chains (for which Finley again provided estimates), the builder could calculate the total weight of the iron needed to make the chains.

In establishing quantitative relationships between load and tension, and in using these relationships to proportion the cables of his bridge, Finley had gone beyond the ideas of John Jones and Thomas Gilpin. Neither Jones nor Gilpin appears to have made a conceptual distinction between load and tension. Jones asserted that "there is no doubt but chains might be made Sufficiently Strong, to suspend a Bridge of this or even a greater Length, supposing it had no Curve."[28] Finley's conceptual framework, however, led to the conclusion that diminishing the curve of a chain would diminish its load-carrying capacity, since the tension produced by a given load would become greater. In short, Finley's conceptual framework indicated that chains with "no curve" would be weak, whereas Jones seemed to suggest that such chains would be strong.

Gilpin's proposal shows that he did not distinguish between load and tension, and that he did not link these parameters to a cable's sag/span ratio. This conclusion follows from the fact that when Gilpin's bridge plan is analyzed on the basis of Finley's conceptual framework the data do not correlate with Gilpin's own conclusions about the strength of the bridge. Gilpin's bridge was to be composed of nine chains, and the links of each chain were to be formed from single bars having a cross-sectional area of 1.56 square inches. (The bars were hooked at each end.) If a value of 60,000 pounds per square inch is assumed as the maximum working stress for iron (which was the accepted value in that era), then each chain could support a tension of 46.8 tons (U.S. measure) and all the chains together could support a tension of 421 tons.

Now suppose that Gilpin did distinguish between cable load and tension in the manner of Finley; then Gilpin would have determined that the 1/20 sag/span ratio of his bridge would produce a tension roughly 2.5 times the load borne by the chains. Hence, although the chains could support a total tension of 421 tons, they could support a total load of only 168 tons.

According to Gilpin, the weight of the chains amounted to 21,600 pounds, or roughly 11 tons. Gilpin also indicated that the chains were to be covered with a roadway of wood planking 3 inches thick, although the kind of wood to be used was not specified. Since the bridge was to be 300 feet long and 20 feet wide, this amounted to a total of 1,500 cubic feet of wood. Oak, often used for planking, weighs about 60 pounds per cubic foot; hence, the total weight of the planking for Gilpin's bridge amounted to 90,000 pounds, or 45 tons; if a lighter wood like pine were used, this total would be somewhat less. Subtracting the weight of the chains and the planking leaves a load capacity of 112 tons.

Gilpin indicated that the chains "from the well known Strength of Iron, will bear five times the Weight that by Reasonable Supposition will Ever be on the Bridge at one time."[29] This statement implies that no more than one-fifth of 112 tons would ever be on the bridge, which amounts to 22 tons. It is hard to imagine, however, that Gilpin believed 22 tons to be the largest load to which his bridge could ever be subjected. Snow loads often weigh many tons,[30] and a crowd of people on the roadway could easily exceed 22 tons. On the basis of an estimate given later by Navier, the weight of a crowd could amount to 40 pounds per square foot[31]; this would mean a total load on Gilpin's bridge of 120 tons, more than the entire remaining load capacity of the bridge. Alternately, the hypothesis that no more than 22 tons could ever be on the bridge leads to the conclusion that the weight of traffic loads on the bridge would never exceed 66 pounds per square yard. But this would amount only to the equivalent of one child per square yard of the bridge surface, which is hardly the greatest weight that could be imagined.

Determining Lengths of Hangers and Links

Since Finley's bridge was to have a level roadway suspended from chains with a substantial sag, the builders would have to be able to determine the proper lengths for the hangers. (The hangers are the vertical iron bars by which the deck is suspended from the main cables. Their lengths increase from mid-span to each end of the cable. See figure 2.1.) In addition, each link of the main chains was to span the entire distance between two hangers. Yet as the chains curved upward from the center of the span to the towers, the angle between links and hangers necessarily varied. Consequently, the links also had to vary in length.

As a solution to these problems, Finley worked out an empirical method based on the use of a two-dimensional scale model of the bridge to be built. It is very likely that this method was suggested to him by the article on bridges which he consulted in Abraham Rees's *Cyclopaedia*.[32] In that article, an analogous method for proportioning arch bridges was outlined in the context of a discussion of the "theory of equilibration," a theory that had arisen in the eighteenth century out of attempts by scientists to determine the optimal shape for an arch. Robert Hooke, who offered the first solution to the problem, indicated that the optimal arch must have the same curve as a freely hanging chain suspended from two points at equal height—i.e., a catenary curve. Hooke reasoned that a chain suspended in this way would be in complete equilibrium. By analogy, an arch formed by the exact inversion of this curve would also be in equilibrium.[33]

The article in Rees's *Cyclopaedia* presented the theory of equilibration not as a mathematical theory from which the proper shape of an arch could be deduced but as a conceptual framework that justified the use of an inductive, empirical method for proportioning arch bridges—namely, the method of using suspended chains, loaded in various ways, as two-dimensional models from which to determine the shape and proportions of the desired arch. As the Rees article noted, "the analogy between the standing and the hanging arch has been traced out, not so much for the purpose of corroborating the true theory of equilibration, as for the sake of deducing from it a very popular and general mode of construction; strictly accurate in its principle, and yet so simple in its application, that the most illiterate artist may safely practice it."[34]

As the passage above suggests, those who developed the theories of equilibration did not invent the chain method of proportioning. An artisan, one Isaac Gadsdon, gave the first hint of it in a work he published in 1739, *Geometrical Rules Made Easy for the Use of Mechanicks Concerned in Building*.[35] The chain method was thus the product of a craft tradition, but through its subsequent link with theories of equilibration it became widely known to scientists and inventors, including Thomas Jefferson,[36] Thomas Paine, and, not least, James Finley.

One application of this method, which was meant to take into account the weight of the stonework that would have to be built up between the inside curve of the arch and the roadway, was particularly relevant to the problem of proportioning a suspension bridge. In this application, a chain was first suspended from two points at equal height. A horizontal line was then drawn below the suspended chain to represent the roadway. (Everything had to be in scale, and the whole model was inverted.) Next, pieces of chain were suspended from the principal chain at equal intervals as measured along the horizontal. The pieces increased in length from the center to the ends so that each one just reached the line marking the roadway of the inverted arch. The increasing lengths (and hence weights) of chain represented the increasing weights of the stonework that would be built above the arch from the center to the ends.[37]

Finley recognized that the chain method could provide a simple yet valid means of proportioning suspension bridges, for what was a suspension bridge but a "hanging arch?" Finley therefore adopted the method in the set of design procedures that builders were to follow. He instructed:

To find the proportions of . . . a bridge of one hundred and fifty feet span, set off on a board fence or partition, one hundred and fifty inches for the length of the bridge, and draw a horizontal line between these two points representing the underside of the roadway. . . .[38]

Next, a cable was suspended so as to represent the main chains of the bridge. Finley indicated that the end points of the cable should be placed $23\frac{1}{2}$ inches above the line marking the roadway. (This established a sag/span ratio between 1/6 and 1/7.) Weighted cables were then suspended from the main cable at equal intervals as measured along the horizontal, so as to represent the hangers of a suspension bridge. The weights all had to be equal (in contrast to the arch proportioning method where the weights increased from the center to the ends). Finley advised that, when thus loaded, the lowest point of the main cable should reach just below the roadway line. By extrapolating from the model, the builder could determine the proper length for each hanger as well as the overall length of the main chains between supports, and the length of each link.[39]

Finley adapted this method to proportion not only single-span bridges but also multiple spans. He devised the arrangement illustrated schematically in figure 2.4. A cable was extended over three pulleys (B, C-C', D),[40] with equal weights placed at the ends (A and E). The cable sections between the pulleys were loaded uniformly along the horizontal. These sections represented the spans of the bridge, while the pulleys represented the intermediate and end supports.

Finley specifically analyzed the situation in which one span (B-C) was longer than the other (C'-D). This meant that the load between B and C was greater than that between C' and D, but Finley explained that the cables would slide over the pulleys, altering the sag of each cable section, until the tensions at B, C, C', and D all became equal. In this way, the correct sag/span ratio for each span could be determined, as well as the proper length for each hanger. The model showed that the sag/span ratio had to be greater for the longer section (B–C) in order to compensate for the greater load it supported.[41]

Finley's experiments on the proportioning of chain bridges aroused his interest in the nature of the curve formed by the cables. He understood that this curve was not an arc of a circle, and he knew that it was not a catenary, because (as the Rees article indicated) a catenary resulted when a load was distributed uniformly along the length of a cable rather than along its span. Finley's approach enabled him to reproduce the curve empirically but not to characterize it analytically,

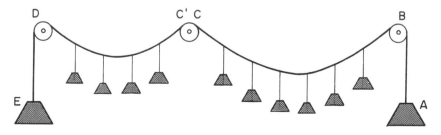

Figure 2.4
Finley's method for proportioning multiple-span bridges.

and he did not acquire the mathematical techniques necessary to undertake such an analysis.[42] In principle these techniques, which included calculus, were available,[43] and Finley could have used them not only to characterize the curve of a suspension bridge cable but also to deduce quantitative relationships among the key variables of his investigations: sag, span, load, and tension. However, in view of his background, career, and environment, it would have been unusual if Finley had taken this approach. The educational and intellectual structures in Finley's environment did not encourage this kind of research, and Finley, with a farm to manage and public responsibilities, was probably not inclined to devote years to studying advanced mathematics. In contrast, an empirico-inductive approach could often generate reliable information in a short period of time and without the need for long preparatory study.

The Suspension Bridge and Its Rivals

With the knowledge acquired through his research, Finley sought to make a rational assessment of the advantages of the suspension bridge over its major rivals, the truss and the arch. As we have already seen, Finley took exception to trusses on the grounds that there was little rational basis for their design. He asserted that for a suspension bridge there was, in contrast, "sufficient data for the strength and burden of all its parts."[44] Finley criticized arch bridges on the basis of their structural behavior relative to suspension bridges: "Every part of any arch bridge is ready, indeed is struggling to escape from its proper position, urged by the general pressure or by any unequal bearings at the joints."[45] In contrast, he explained, the greater the distributed load supported by the chains of a suspension bridge, "the more firmly do [the chains] adhere to their proper station"[46]—that is, the more stable their equilibrium. Suspension bridges were, moreover, cheaper and easier to construct than arches, because they required no expensive centering.

Cast-iron arches met with particular disapproval from Finley because he felt they represented an inefficient use of materials in comparison with suspension bridges. Finley noted, for example, that the Coalbrookdale bridge, with a span of 100 feet, contained $378\frac{1}{2}$ tons of iron. He calculated that a suspension bridge of equal span would require only a small fraction of that amount of iron. Finley concluded: "May I not with some degree of exultation ask, who ever thought of the skeleton of a bridge so light and so strong, so permanent, and so easily erected and repaired or renewed in such parts as may require it?"[47]

Finley's final argument in favor of suspension bridges concerned their potential for long spans:

In the business of bridge building, one of the first and most important inquiries is, which of all the modes that have been proposed is capable of the greatest extension. I wish it be understood that in this point the chain bridge claims a clear and decided preference.[48]

This insight arose from Finley's consideration, using available data on the strength of iron, of the distance an iron wire or chain could be extended before it would break under its own weight. He found this distance to be more than 3 miles. He also calculated that an iron bar 1 inch square and 2 miles long, fastened at both ends, could support a uniformly distributed load of 20,000 pounds. Extended only a mile, the same wire bar could support a uniformly distributed load of 40,000 pounds. "I say when contemplating on these things," Finley wrote, "it is impossible to resist the idea that something further may be done in the art of bridge building than has yet been accomplished."[49]

Finley's speculations led him to calculate the amount of iron needed to construct suspension bridges with single spans of 1,000 and 1,500 feet—visionary calculations for that era! In the case of the 1,000-foot span (assumed to be 15 feet wide) he determined that the chains would have to support a distributed load of 1,200 tons. With a 1/7 sag/span ratio, the maximum overall tension in the chains would then be 1,200 tons also. Finley estimated the strength of iron at 60,000 pounds per square inch and calculated that 20 chains of 1-inch-square iron bar would be needed for the bridge—about 163,500 pounds of iron.[50]

Finley acknowledged that a primary argument against long-span suspension bridges was the concomitant need to construct very high towers. Apparently some argued on this basis that suspension spans could never exceed 400 feet. But Finley remained adamant:

It will readily be granted that there is something novel and forbidding in this part of the structure; but it will afford some relief to reflect on the effects of experience and habit of thinking. He who built the first canoe, discovered more nobleness of mind than he who now builds a first rate ship of the line; and on the score of genius, is entitled to more credit. What appears at first thought altogether impracticable, frequently becomes plain and easy. Shall we not then hope that our puny canoe, with a little cultivation of genius, will soon spring up into a formidable ship?[51]

Finley argued that to build such high towers would actually be no more difficult than the construction of the piers for Timothy Palmer's Schuylkill bridge, which were sunk "40 feet deep in mud and water." He suggested that high towers might be built "after the manner of ship-masts."[52]

To those who still hesitated at the prospect of suspension bridges of great span, Finley offered the further assurance that these bridges actually became more sturdy and secure the greater their extent. For the longer and hence heavier they became, the less would be their response to the moving loads upon them (because these loads would become an ever smaller proportion of the total load supported by the chains). "The conclusion cannot be resisted, that fifty tons will make no more impression on the bridge of 500 feet, than ten tons will make on the bridge of 100 feet." The building of long-span suspension bridges was indeed practicable, Finley insisted, "and as soon as an opportunity offers, the thing shall be done, and in such a manner as to satisfy both phylosophy and common sense."[53]

The Technological Elements of Finley's Design

Whatever advantages Finley's bridge design acquired through its basis in experimental research could be realized only by means of hardware and technique. With this perhaps obvious truth as a guideline, and always with an eye to simplicity, economy, and ease of construction, Finley worked out how each portion of the chain bridge was to be fabricated: the chains, the hangers, the roadway, the towers, and the anchorages. It is important to review these technical details because the choices they represent reflect not only Finley's social and economic goals but also his broader ideas about the suspension bridge as a structural form. This overview provides a necessary foundation for comparisons with the choices made by Navier.

Finley paid particular attention to the way in which the main chains of the bridge should be fabricated and treated. The chains were to be composed of ordinary links, oblong in shape (figure 2.5). A link was to be formed by two iron bars welded together so that each bar became one side of the link (with the welds at the curved ends of the link). Finley noted that the strength of a link was theoretically equal to the sum of the strengths of its two sides, despite the fact that its curved ends were single where they supported the ends of adjacent links. Yet Finley also realized that, in practice, tension was not distributed uni-

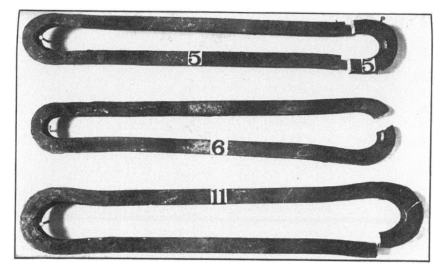

Figure 2.5
Photograph of original links from the Essex-Merrimack suspension bridge, constructed under license from Finley in 1810. The links are shown after a loading test. Source: *Cornell Civil Engineer* 19 (1911), p. 225.

formly throughout a link; the greatest stress occurred at the ends. Finley therefore emphasized the need to make the curved ends of each link two or three times thicker than the sides.[54]

Thickening the ends of each link entailed a significant increase in the quantity of iron needed to make the chains. To compensate for this increase, Finley sought to reduce the total number of links in the chains—that is, to make the chains from a small number of long links rather than a large number of short links. That is why he wanted the length of each link to be made equal to the distance between two hangers.[55]

Finley discussed how to finish the inside edges of the links and how to ensure a close fit between links. He also outlined a fabrication method that smiths would find convenient. To preserve the chains, Finley suggested that they be either painted or, preferably, coated with canvas and pitch.[56]

Finley's attention to the details of link fabrication suggest that he consulted knowledgeable practitioners in his community, such as the ironmaster Isaac Meason or one of Meason's employees. (Recall that Meason was involved in Finley's trial suspension bridge, constructed in 1801.) Certainly it would have been in Finley's interest to seek out such advice—particularly in regard to the use of iron, since the potentials and limitations of iron for bridge construction were not yet well understood. Iron bridges in the West dated back only to 1781, when the Coalbrookdale cast-iron arch was completed in Britain.

Thomas Paine had been in a position similar to Finley's several years earlier. Paine sought to extend an idea for an iron arch bridge into a viable construction plan, and to do so he solicited the aid of eminent iron manufacturers in Britain. They not only provided encouragement for Paine's work, but also confirmed the overall practicability of his idea and gave advice on matters of design and construction, such as the best method to join the sections. It is not unlikely that Meason (or craftsmen who worked with Meason) played an analogous role for Finley.[57]

Finley's chain design required that each link support one hanger. Yet each link was turned 90° with respect to the one preceding. Thus, Finley had to devise two methods to attach the hangers to the links (figure 2.6). In the first method, the top of a hanger terminated in a "U" whose arms embraced a vertically positioned link. The ends of the hanger were then bolted to a key supported by the vertical link. In the second method, the sides of a horizontally positioned link embraced the end of a hanger. Again, the hanger was bolted to a key supported by the link. However, the keys were different in each case. One was designed to rest only on the top arm of a vertically positioned link; the other was designed to rest on both arms of a horizontally positioned link.[58] The hangers were composed of iron links or bars, and they terminated at their lower ends in stirrups that supported transverse wooden floor beams (5 × 10 inches).[59]

The transverse beams constituted the bottom level of a three-tiered roadway. The second tier, laid upon the transverse beams, consisted of four longitudinal stringers that ran the entire length of the bridge. These longitudinal stringers were designed to add stiffness to the deck. Each stringer was composed of several members; to connect them, the members were made to overlap, and they were bolted together at either end of the overlapping section. The overlaps extended between adjacent transverse floor beams. Finley explained that the stringers needed to be only about 3 inches wide, but that they had to be at least 12 inches deep in order to "stiffen the bridge as much as possible." The third and top level of the roadway consisted of the floor planking, which was nailed to the stringers.[60] (We will see later that Navier made a different choice, with different economic implications, and was little concerned with providing stiffness in his roadway.)

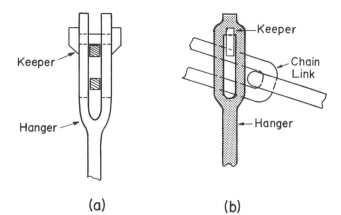

Hanger Suspended From Vertical Link

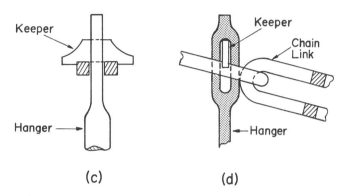

Hanger Suspended From Horizontal Link

Figure 2.6
A system devised by Finley to suspend hangers from vertically and horizontally positioned cable links.

Finley asserted that his system of continuous stringers was twice as strong as a system whose individual members spanned no more than the distance between adjacent floor beams. A failure in the latter system would occur with a single break in the center of a member. Finley explained that in his system failure could occur only if there were breaks in the center and at the ends.[61]

Heavy latticed railings were to be located on either side of the roadway. In addition to their protective function, the railings were intended to help stiffen the span. Other bridge builders of Finley's era likewise used trussed railings to provide stiffness. Charles Peale, in a description of the railings on his wooden arch bridge published in

1797, explained that "an important advantage in these numerous railings, besides insuring safety to the passengers, is in their adding great strength to the arch." Further on, Peale emphasized again that "the railings are . . . a very important part of the Work . . . , for a Bridge may be made so light, or in such inconsiderable thickness of materials, as to be left very elastic or springy. But the railings being substantial . . . will render the bridge *stiff and strong.*"[62]

The towers of Finley's bridge were A-shaped wooden frames. On top of the towers, wooden blocks with convex upper surfaces supported fixed, convex iron saddles (figure 1.5, bottom center). The main chains and the anchor chains were fastened to the saddles by means of pin connections so that the saddles functioned like curved links. On a multiple-span bridge, however, a saddle might support the main chains of two spans. The chains designed by Finley were thus not truly continuous from anchorage to anchorage, since the saddles were fixed. This was necessary because continuous chains would have had too much play, so that traffic loads would produce considerable deflections. As Finley noted, a load on one span would "tend to sink it and raise the rest."[63]

Fixed iron saddles, however, could allow no movement to equalize the tension in the chains on either side of a tower. Tension imbalances could arise from expansion and contraction of the chains due to temperature variations or from moving loads upon the bridge. Finley therefore devised two measures to counteract these effects. First, he gave advice on how to bolt down the wooden towers to prevent them from being overturned. Second, he emphasized that the chains descending from either side of a tower should have the same angle of inclination.[64]

In order for the main chains and the anchor chains to have the same angle of inclination, the anchor chains had to be quite long: for a single-span bridge, the overall length of each chain measured from anchorage to anchorage would be nearly twice the length of the roadway. Finley saw that this potential waste of iron could be avoided if the anchor chains supported portions of the roadway. Then the overall length of a chain from anchorage to anchorage would be only a little greater than the length of the roadway. Finley stressed that these anchor spans (as they are now called) should be true half-spans with lengths about half that of the central span they flanked. He added that their proportions could be determined experimentally with an appropriate system of loaded cables and pulleys. The drawback of anchor spans, he noted, was that they might entail the construction of additional piers. Moreover, they tended to increase the amount of deflection produced by moving loads.[65]

For the purpose of securing the cables into the ground, Finley stipulated that each anchor chain should terminate in four branch chains, and that each branch chain should be securely bolted to a large, flat rock placed at the bottom of a pit. Other rocks were then to be laid around and above this flat rock, so as to create a kind of platform. Upon this platform, enough stones were to be piled to resist a tension at least 30 percent greater than the tension of the branch chain. To calculate the tension of the branch chains required only basic principles of geometry. Finley noted, for example, that if an anchor chain terminated in two branch chains making angles of 120° with the anchor chain and with each other, then the tension of each branch chain would be equal to the tension of the anchor chain. But Finley advised any builder with doubts about calculating these tensions that the problem could be resolved experimentally using a system of cables, weights, and pulleys. He claimed to have carried out such experiments, although he provided no details.[66]

Evidently some builders who used Finley's design wanted to abandon his anchoring system and adopt shortcut methods. One suggestion was merely to bolt the anchor chains to very large rocks. Finley responded: "This is wandering from our favorite principles. . . . To my mind, all these methods have something in them too precarious and unsafe to be depended on. Give me sufficient hold of a platform of some kind, and let me know the weight of the materials that rest on it, and I shall know on what I depend."[67] Finley was referring here to the rock platform prepared in each anchorage pit and to the calculated weight of stones placed upon it. The point was that the strength of Finley's anchorages could be determined and controlled to a significant degree, whereas the strength of the alternative systems could not.

The care with which Finley dealt with technological elements such as links, hanger clips, joists, railings, and anchoring systems show that he devoted considerable time to them, in the process acquiring a great deal of know-how. His attention to these elements and their interactions provides further evidence that his design was a reasoned response to social and economic goals, involving conscious technical tradeoffs. In the following chapter we will examine how Finley's choices affected the technical and commercial success of his system.

3

Finley's Design System in Practice

The precise number of chain bridges constructed from James Finley's plan is not known. In 1820, when the French engineer Joseph Cordier toured America to learn of advances in transportation technology, he reported that 40 chain bridges had been built.[1] I have been able to collect details on 18 constructed before 1820 (table 3.1). In addition, information exists for three chain bridges built between 1824 and 1827 and for two bridges that were proposed but never built. Although the quantity and quality of information on individual bridges vary, this information provides a basis for assessing how Finley's social, economic, and technical choices affected the viability of his design system in early-nineteenth-century America.

It is important to recognize the complexity of turning a design on paper into a built object. In this chapter I will examine the "career" of Finley's design system as it was translated into a series of built objects. My goals are to explain the pattern of this career and to show that its degree of success or failure over time cannot be explained merely by recourse to an abstract idea of the design's technical adequacy. This is not to suggest that the notion of technical adequacy is irrelevant. Indeed, it will be taken into account in the following analysis. Yet it will be shown that the career pattern of Finley's design system can be explained more fully through an understanding of its position within a market niche than through an abstract assessment of its technical adequacy.

Finley of course did not use the term "market niche," yet he did have an analogous concept in mind. He consciously directed his design toward a certain kind of builder, and toward a certain capacity to pay (expressed in overall cost per linear foot of building the bridge). Together these two elements structured a set of relationships between the technical dimensions of the design and their social functions and

Table 3.1
Early American chain bridges. (For sources, see appendix to chapter 3.)

Year built	Location	Size data	Builder	Cost	Comments
1801	Fayette County, Pa. (Jacob's Creek)	one span L 70′ W 13′	Finley	$600	County bridge. 1833: replaced by timber span for $267.
1808	Georgetown, Md. (Potomac R.)	one span L 130′ W 15′	Templeman	$5,000	1810: damaged by flood. 1832: damaged by ice floe. 1840: damaged by flood. 1853: damaged by flood, replaced by truss.
1808–1809	Philadelphia, Pa. (Schuykill R.)	two spans L 153′ × 2 W 15′	Templeman and local builders	$20,000	Toll bridge. 1811: fell (cattle and weak iron clip). 1816: fell (snow and rotted timber).
1809	Wilmington, Del. (Brandywine R.)	one span L 145′ W 30′			
1809	Brownsville, Pa. (Dunlap's Creek)	one span L 120′ W 15′			County bridge. 1820: fell (snow and wagon). Replaced by wooden span.
1809	Brownsville, Pa.	one span L 112′ W 15′			

Year	Location	Dimensions	Builder	Cost	Notes
1810	Frankfort, Ky. (Kentucky R.)	two spans L 167' × 2			Toll bridge (Frankfort Bridge Co.).
1810	Doylestown, Pa. (Neshaminy R.)	two spans L 100' × 2	John Parker	$5,250	County bridge.
1810	Reading, Pa. (Schuylkill R.)	two spans L 150' × 2	Ulrich Kissinger		Toll bridge. 1830: replaced by covered wooden span.
1810	Delaware County, Pa.	Small			
1810	"Pauling's Ford" (Schuylkill R.)	one span (?) L 200'			Toll bridge. 1820: damaged by ice.
1810	Cumberland, Md. (Will's Creek)	one span L 115' W 15'	Valentine Shockey		County bridge. 1831: posts gave way. 1837: abutment gave way. Replaced by Town (?) truss.
1810	Newburyport, Mass. (Merrimack R.)	one span L 244' W34'	Templeman and Samuel Carr	$25,000	Toll bridge. 1827: snow and wagon damage. c. 1900: strengthened. 1909: removed (traffic burden).
1811	Kentucky	one span L 104'	Local contractor		
1811	Kentucky	one span L 200'	Local contractor		1811: fell (insufficient anchoring).

Table 3.1 (continued)

Year built	Location	Size data	Builder	Cost	Comments
1811	Bedford, Pa. (Juniata R.)	two spans (?) L 260' W 15'			1841: damaged by flood. Replaced by wooden bridge.
1811	Easton, Pa. (Lehigh R.)	two spans L 200' × 2	Jacob Blumer (?)		Toll bridge.
1813–1814	Northampton, Pa. (Lehigh R.)	two full spans + two half-spans L 175' × 3 W 32'	Jacob Blumer	$20,000	
1824	Allentown, Pa. (Lehigh R.)		Jacob Blumer		Toll bridge, repl. w/covered wooden truss. 1841: swept away by flood. 1850: Unstable (removed).
1826	Palmertown, Pa. (Lehigh R. Gap)	one full spans + two half-spans L 160' × 2	Jacob Blumer		Toll bridge. 1841: flood damage to abutment. 1857: ice damage. 1826: broken dam. 1926: fire. 1933: replaced (traffic burden).

Date	Location	Designer	Dimensions	Cost	Notes
1826–1827	Newburyport, Mass. (Merrimack R.)	Thomas Haven	three full spans + two half-spans + one draw span L 1000'		
	Pittsburgh, Pa. (Monongahela R.)		two full spans + two half-spans L 1200' W 30'	$32,326 (est.)	Proposed only. Covered wooden span (Wernwag) built on site in 1818.
	Columbia, Pa. (?) (Susquehanna R.)		25 spans L 5600' W 35'	$200,000 (est.)	Proposed only. Probable site was spanned by Burr arched truss.

meanings, such that each social or economic decision had technical repercussions and vice versa. Tracing the career of Finley's design system as a series of built objects can therefore help to reveal more fully the structure of social, economic, and technical relationships in Finley's design system and to show how this set of relationships shaped the competitive position of Finley's design system over time. This analysis will also show why the very notions of success and failure can be rather problematic in the realm of technology.

The Essex-Merrimack and Palmerton Bridges

Two bridges built to Finley's plan remained in service for a century or more: the Essex-Merrimack Bridge, built over the Merrimack River at Newburyport, Massachusetts, and the Palmerton Bridge, built over the Lehigh River near Palmerton, Pennsylvania. Their histories show that Finley's plan could be made to work in the environment of early-nineteenth-century America.

The Essex-Merrimack Bridge (figures 3.1 and 3.2) stood for 99 years, from 1810 to 1909. Built under the general supervision of John

Figure 3.1
The Essex-Merrimack chain bridge. Source: Llewellyn Edwards manuscripts, National Museum of American History. Courtesy of Smithsonian Institution.

Figure 3.2
The Essex-Merrimack chain bridge. Source: Llewellyn Edwards manuscripts, National Museum of American History. Courtesy of Smithsonian Institution.

Templeman, with local carpenter Samuel Carr in direct charge of construction, it had a single span of 244 feet and was 30 feet wide. The flooring, the latticed railings, the hangers, and the wooden frame towers all followed Finley's design. The bridge was supported by ten chains of 1-inch-square iron bar.[2] Their number and cross-sectional area, calculated according to the method outlined by Finley, provided a safety factor of 6.

Templeman introduced and patented several modifications to Finley's design for the chains, however. Rather than the very long links (5–10 feet) recommended by Finley, Templeman used links about 2 feet long. He also abandoned the curved saddles devised by Finley. Instead, he trebled each chain where it passed over the tower and then spiked the chains in place. At the anchorages, Templeman used six branch chains rather than the four indicated by Finley. Templeman also covered the wooden towers to protect them from decay, an idea adapted from the practice of covering wooden bridges.[3]

The Essex-Merrimack Bridge suffered one accident during its years of service. In 1827, only 17 years after it was built, the main chains on one side of the bridge ruptured under the combined forces of a frost, a heavy snow load, and a heavily laden wagon drawn by six oxen and two horses. The wagon plunged 40 feet into the river. All the oxen were swept downstream and drowned, but the two drivers and one of the horses managed to swim to shore. The bridge was rebuilt with twelve chains, which proved sufficient. In addition, the deck was divided into two independent roadways, each supported by chains on either side, so that if the chains of one roadway broke the other roadway would remain unaffected.[4]

An observer of the Essex-Merrimack Bridge in 1810 noted that a horse and carriage crossing at a trot produced almost no perceptible motion in the roadway.[5] However, by the end of the nineteenth century the Essex-Merrimack Bridge was subjected to increasingly heavy traffic loads, including electric streetcars. Under these conditions the roadway deflected between 2 and 3 feet. An engineer observed at this time that a streetcar "crossing the bridge as viewed from a boat upon the river" gave "a beautiful illustration of wave motion."[6] Subsequently, the side of the roadway that carried the electric streetcars was strengthened by the Roebling Company. In 1909 the bridge was entirely replaced by a wire-cable suspension span; in response to popular demand the new structure was made to resemble the old.[7]

The Palmerton Bridge (figures 3.3–3.5), built in 1826 by the clockmaker Jacob Blumer,[8] remained in service 107 years. A thorough examination made at the time of its removal by W. H. Boyer and Irving Jelly (the latter an engineering draftsman employed at a nearby zinc company) reveals that Blumer largely followed Finley's plan.[9]

Figure 3.3
An etching of the chain bridge over the Lehigh River at Palmerton, Pennsylvania, constructed in 1826. Source: W. H. Boyer and I. A. Jelly, "An early American suspension span," *Civil Engineering* 7 (1937), May, p. 338.

Figure 3.4
The Palmerton chain bridge, photographed in April 1926. Source: Llewellyn Edwards manuscripts, National Museum of American History. Courtesy of Smithsonian Institution.

Figure 3.5
The Palmerton chain bridge, photographed after the bridge partly burned in
1826. Notice that trussed side railings have been added. Source: Llewellyn
Edwards manuscripts, National Museum of American History. Courtesy of
Smithsonian Institution.

The Palmerton Bridge was composed of a 160-foot central span and
two 80-foot anchor spans. The sag/span ratio of the cables was 1/7. The
bridge had two chains made of bar iron with a cross-sectional area of
2.25 square inches, providing a total strength of 270 tons, assuming
(in accordance with Finley) a maximum permissible stress of 60,000
pounds per square inch. If we estimate the weight of the roadway
according to the general data provided by Finley, we get 34 tons, which
would imply that Blumer adopted a safety factor of around 8—
significantly higher than the factor of 5 or 6 suggested by Finley.

The chains of the Palmerton Bridge were fabricated according to
the specifications in Finley's patent. There was one link per hanger,
and the hangers were connected to the main chains precisely as illustrated in Finley's patent drawing. The roadway was composed of three
layers, as Finley had suggested; however, as photographed in 1926
(figure 3.4), it did not have trussed side railings. Instead, the railings
were composed of wooden joists supported by the hangers. The

wooden towers and tower saddles did follow Finley's design. The only other detectable deviation from Finley's plan was the substitution of links bent into V shapes (figure 3.6) for the branch chains leading to the anchorage pits.[10]

During its 107 years, the Palmerton Bridge was damaged several times by flood, and once by fire, but otherwise it withstood the forces to which it was subjected. Boyer and Jelly stated that during the First World War "army truck trains frequently used this crossing, and one

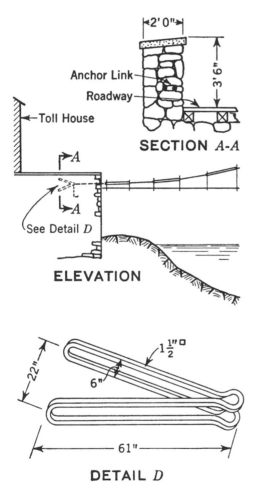

Figure 3.6
V-shaped anchorage links used by Jacob Blumer for the Palmerton bridge. Source: W. H. Boyer and I. A. Jelly, "An early American suspension span," *Civil Engineering* 7 (1937), May, p. 340.

particular truck with a net lading of 12 tons crossed without any indication that the bridge was overloaded."[11]

The Eclipse of Finley's Design System

Apart from the histories of the Palmerton and Essex-Merrimack bridges, Finley's design system appears to have met with an initial burst of enthusiasm, a rapid decline, and a final eclipse after the 1820s. Out of the 21 chain bridges for which I have data, 18 were built before 1815 and the remaining three were built between 1824 and 1827.[12] (The later building spurt may reflect the fact that Finley's patent had by then expired, so builders no longer needed to pay royalties.) Existing chain bridges that needed repair or rebuilding were often replaced instead by wooden arch or wooden truss bridges.

The eclipse of Finley's design system was nearly total. Not only did the design system itself fall into disuse, but Finley's patent was linked to only one other patent in the United States (issued in 1810 to the contractor of the Essex-Merrimack Bridge, John Templeman). Later American suspension bridges did not derive immediately from Finley's design system; they derived from European adaptations of the basic structural form.[13]

Inadequate marketing by Finley may have been one reason for the eclipse of his system. It is clear that at first Finley paid attention to marketing: in March 1808 he signed a five-year contract making the aforementioned John Templeman the "sole Agent of the Patentee for the sale of Patent Rights throughout the United States."[14] Templeman was to receive 50 percent of the royalties for any sale he made.[15] Templeman distributed a circular (figure 3.7) in which he explained Finley's system to prospective customers, and presented examples to demonstrate its low cost relative to other kinds of bridges. He also indicated his status as Finley's agent, and offered his services as a consultant or contractor. In addition, as has already been pointed out, Finley published a description of his system in the *Port Folio* and later in a pamphlet. Finley also published an open letter concerning his bridge system in the *United States Gazette,* a Philadelphia newspaper.[16] Yet all this marketing activity took place before 1813; I have no evidence of such activity after that date.

It is possible that the marketing efforts were hampered by conflict between Finley and Templeman. According to Finley, just after the contract was signed Templeman tried to patent the application of

Georgetown, Dist. Col. March 14th 1809.

SIR,

BELIEVING as I do, that it will afford you pleasure to communicate useful knowledge generally, as well as to your particular friends—And well knowing that one letter in the hands of a man of consideration and influence, will have more useful effect than 1000 news paper advertisements—has induced me to take the liberty to address you, on the subject of the CHAIN BRIDGE—Supposing that you have seen the Bridge at the Little Falls of Potomac, three miles above the city of Washington, or that of the falls of Schuylkill, 5 miles above Philadelphia, I shall not trouble you with a general description of this new and invaluable invention ; but merely contrast the cost of other sorts of bridges with them. For instance, the span of chain bridge now in use at the Little Falls cost only 5000 dollars, and replaces a wooden arch which cost 23,000 dollars, and did not stand seven years from the time of its erection before it rotted and fell into the river. The chain bridge at the falls of Schuylkill cost only 20,000 dollars including two abutments and two piers of hewn stone, while the Market-street Bridge over the same water, and of the same length cost $ 300,000.— Both Bridges have two abutments and two piers each, but the piers of the Chain Bridge are built on solid rock, while those of the lower Bridge are built with the assistance of coffer dams, in deep water and mud—But allowing that those two piers cost $ 80,000, then add $ 20,000, the cost of my bridge at the falls, and deduct the amount from $ 300,000, will leave $ 200,000 in favour of the chain principle—the chains and uprights of which will last a couple of hundred years—the first being painted, and the latter covered with a roof.— There will be of course only the flooring and railing of the Bridge, subject to decay—the renewal of which will amount to mere nothing..

The Bridge at the Little Falls has but one span, and is 130 feet long on a straight line, the one at the falls of Schuylkill has three spans, two of which are 153 feet long, and could have been extended much farther.

The price of the Patent Right is One Dollar per Foot between the Faces of the Abutments on a line.

There are now five of these Bridges in use, not one of which has failed in the smallest degree.

I have a concern in the original Patent Right, and am sole Agent of the Patentee for the sale of Patent Rights throughout the United States.——Any one wishing to construct this species of Bridge, will do well to transmit to me a description of the scite of the contemplated Bridge—say, the width of the water, the depth, and what sort of Bottom it has, whether mud, gravel or sand—and particularly, whether the water to be bridged is subject to high and rapid freshes—the price of foundation or other stone, and the price of lumber, from which I could make a calculation of the probable expence, & send it by post—and then I will, if required, either go on, & superintend the construction, or build the Bridge or Bridges on contract.

I am, Sir, with sentiments of high respect, your humble Servant.

John Templeman.

Figure 3.7
A circular prepared by Finley's agent, John Templeman, to interest potential clients in Finley's system. Source: Circular addressed to Nicholas van Dyke, LMSS Group 4, Box 10, Hagley Museum and Library, Wilmington, Delaware. Courtesy of Hagley Museum and Library.

Finley's system to multiple-span bridges, before Finley could add this claim to his own patent. Ultimately Finley's patent did cover multiple-span bridges, but Finley's displeasure at Templeman's action is evident in his brief account of the affair.[17] Moreover, as was noted above, Templeman went on to get a patent for modifications to Finley's chain design. This patent put Templeman in a position to construct or approve bridges using Finley's general plan plus his own patented chain design, and thus to collect full royalties on his own patent, 50 percent of the royalties on Finley's patent, and possibly also payments for planning and administering the construction of the bridges. It seems, therefore, that Templeman may have been in a position to earn more from the use of Finley's patent than Finley himself. Under these circumstances, it would be understandable if Finley chose not to renew Templeman's contract, or to stop relying on agents altogether. If Finley refused to allow Templeman to use the patent rights to the suspension bridge, Templeman would not be able to exploit his own patent either, since it was an add-on to Finley's broader patent. Of course such a hypothesis must remain merely speculative, but I have found no evidence of a continuing relationship between Finley and Templeman, or of subsequent activity by Templeman after his initial contract with Finley.

It is also possible that the War of 1812 interfered with Finley's business, and that after the war he did not attempt to renew his marketing efforts. A proposal by Finley for a bridge over the Monongahela River in Pittsburgh was approved before the War of 1812; however, the bridge was not built at the time (purportedly because of the war), and the bridge that was built at that site after the war, from 1816 to 1818, was a wooden rather than a chain bridge.[18]

The case of the Monongahela River Bridge points to another factor in the declining fortune of Finley's design system: increasing competition from designers of wooden bridges. As we have seen, Finley developed his design system to respond to what he perceived as a demand for cheap, widely applicable, easy-to-build bridges. Designers of wooden bridges also hoped to exploit this market niche, and were increasingly successful at doing so.

Wood had certain advantages over iron as a building material. It was cheaper and more plentiful than iron. In addition, builders generally had a better understanding of its structural capabilities, both because they used it so extensively and because they benefited from knowledge accumulated through centuries of experience. This knowledge included refined techniques for judging strength and performance. For example, carpenters recognized that certain types of wood suffered changes in texture when overstressed: "[Those types of wood] take a set, and they do not break so suddenly, but give warning by *complaining*, as the carpenters call it; that is, by giving visible signs of a derangement of texture."[19] By paying attention to such warning signs, builders had learned much about how to proportion wooden structures safely.

In the case of iron, a condition similar to the phenomenon of "complaining" in wood would be the point beyond which permanent deformation begins to occur (now referred to as the *elastic limit* or the *yield point*). To understand the concept of elastic limit, think of a bar of iron as being like a spring. When a weight is hung from it, the bar stretches; when the weight is removed, the bar returns to its original length. At some point, however, if enough weight is added, the iron bar remains permanently elongated, even though it does not actually break. The phenomenon of permanent deformation has come to be regarded as an indication of overstressing. Bridges are therefore designed so that their structural elements are not stressed beyond their elastic limit: "In designing a structure such as, for example, a bridge, it is essential to know the elastic limit of the material, for if a girder were subjected to a stress above its elastic limit it would *permanently* change in length, and this might dangerously increase stresses in other girders connected to it, and lead perhaps to collapse of the whole structure."[20] In Finley's era, however, American designers working with iron did not employ a concept of elastic limit or yield point, and perhaps did not even fully recognize such a phenomenon for iron. Finley calculated the strength of iron purely on the basis of its ultimate breaking strength—i.e., the point at which a bar of iron actually breaks or ruptures under tension. Yet in the case of wrought iron the point of rupture may occur at a value substantially above the elastic limit.[21]

We will return to this issue later, when examining the technical factors that affected the success of Finley's design system. The point to be made at present is simply that builders in Finley's environment generally felt more confident in using wood as a structural material than in using iron, and that qualitative judgments of the strength of wood derived from traditional know-how were likely in 1810 to have been more sensitive and refined than quantitative judgments of the strength of iron derived from available experimental data. In particular, by paying attention to phenomena such as "complaining," carpenters could avoid overstressing wood.[22]

Significant advances in the design and construction of wooden bridges after 1790 provided a further basis for their competitiveness in the market. There were two stages in this development.

In the first stage, builders created designs that combined the principles of the arch and the truss.[23] The bridges of Timothy Palmer, Theodore Burr, and Lewis Wernwag (all leading bridge designers contemporary to Finley) belong to this category. Although these structures could not be rigorously analyzed in the sense of determining the forces acting on individual structural elements, their designs were nevertheless founded on insights that had validity—for example, the understanding that trusswork would add stiffness to a wooden arch. Many of these bridges proved to be quite sturdy, and some achieved international renown. Many also proved to be long lasting, especially after builders began covering them (after 1800). Some covered bridges lasted well over 100 years.[24]

Combining the arch and the truss made possible a marked increase in the length of wooden spans. Whereas earlier wooden bridges had maximum clear spans of about 60 feet, some of the new arch-truss combinations were extended beyond 330 feet.[25] Data compiled by the Pennsylvania Legislature on 20 wooden bridges erected between 1801 and 1820 reveal that distances between piers varied between 100 feet and 334 feet, with the average at 195 feet.[26] Data on 16 chain bridges, however, indicate that distances between supports varied between 100 feet and 244 feet, with the average at 155 feet. Thus, arch-truss bridges were made to span greater distances than the early chain bridges, and this no doubt benefited Finley's rivals. Lewis Wernwag, in particular, achieved international acclaim for his "Colossus" bridge in Philadelphia, which had a clear span of 340 feet. This bridge was completed in 1812, during the period of greatest diffusion of Finley's design system, and Wernwag's growing stature as a bridge designer no doubt cut into Finley's business. At least one of Wernwag's bridges was built over a river that was also spanned by a Finley chain bridge.[27]

Arch-truss bridges were not competitive with Finley's chain bridges in price, however. Besides using very large quantities of wood, they required the construction of massive stone abutments to resist the tremendous thrust of the arches. In addition, they often took several years to build, which substantially increased labor costs. For example, the "Permanent Bridge" built by Timothy Palmer over the Schuylkill in Philadelphia had three spans with a total length of around 500 feet. The bridge was constructed over a period of five years, from 1801 to 1805. Some 1,500,000 board feet of timber were required, and nearly 550,000 cubic feet of stone. The cost of the bridge was $300,000, which amounts to about $607 per foot of span.[28]

Cost data exist for seven other arch-truss bridges built by Palmer, Wernwag, or Burr between 1804 and 1817. They ranged from one span up to six spans, with overall lengths of between 244 feet and 1,064

feet. Their construction costs per foot of span ranged from a minimum
of $92 to a maximum of $300, with the average at $173.[29] Cost data
exist for four Finley chain bridges built between 1807 and 1814. They
ranged from one span to three spans, with overall lengths of between
130 feet and 475 feet. The costs of construction per linear foot of span
ranged between $42 and $78, with the average at $56, which amounts
to only one-third of the average cost for an arch-truss bridge.[30] Even
the most expensive chain bridge was only 60 percent of the unit cost
of the cheapest arch-truss bridge. Moreover, in no instance for which
data are available did the construction of a chain bridge take more
than 2 years.

The second stage in the development of American wooden bridges
involved the transition from arch-truss combinations to pure truss
structures. (The difference is that arch-truss bridges always exert outward thrust whereas truss bridges do not; the latter, therefore, require
only end supports rather than massive abutments.) Lewis Wernwag
appears to have been among the first to develop such a design, for he
built such a bridge in 1810. He named it "Economy," and in an
advertising circular (figure 3.8) he claimed that "for simplicity,
strength, and cheapness [it was] exceeded by none."[31] The 1810 bridge
was built over the Neshaminy River, and Wernwag built another over
Frankfort Creek, both in Pennsylvania. Probably he built others as
well, because he is known to have constructed at least 29 bridges
during his career.[32] I have not found cost data for these structures, but
the simplicity of the design suggests that construction costs per linear
foot must have been comparatively low, certainly much lower than for
Wernwag's arch-truss bridges.

The truss form was widely popularized not by Wernwag, however,
but rather by Ithiel Town. In 1820 Town patented a truss design that
became very popular because of its exceptional cheapness, simplicity,
and ease of construction. It could be put together in a matter of weeks
by any builder or mechanic, using only standard sizes of lumber. It
required no special hardware (such as stirrups or keys), nor did it
require the fabrication of complex mortise-and-tenon joints. The
members were joined simply by means of pin connections using tree
nails.[33]

The fact that the Town truss needed only intermediate and end
supports rather than massive piers and abutments also made construction both simpler and cheaper. One Town truss built in South Carolina, a 104-foot span, cost only $2,500, which amounted to $24 per

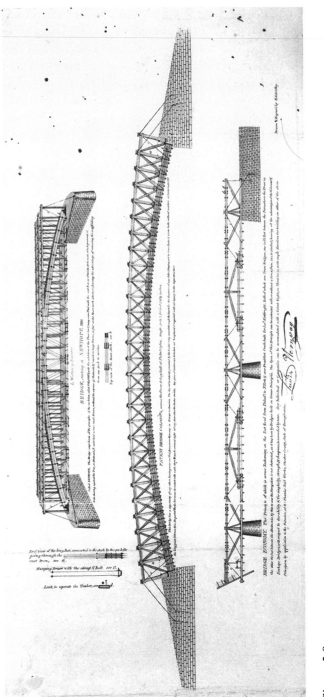

Figure 3.8
An advertising sheet prepared by Lewis Wernwag and illustrating three of his designs for wooden bridges. Courtesy of American Philosophical Society.

linear foot. This represented less than 60 percent of the cost of the cheapest chain bridge for which we have data, and a mere 24 percent of the cost of the cheapest arch-truss bridge. The distance between piers appears to have been generally little more than 110 feet, which means that more piers had to be built to span a given distance than with an arch-truss bridge or a chain bridge, but even so the overall cost of a Town truss remained less.[34]

Over the years, hundreds and likely thousands of Town trusses were built throughout the United States, both as highway bridges and as railway bridges. As late as 1900, there were 100 extant Town trusses in the Boston and Maine Railroad's network alone.[35]

The line of development initiated by Town was taken even further by Stephen Long in a truss design he patented in 1829. Long's truss was much simpler and sparer than Town's, yet it was more rigid. It was also substantially cheaper. The first span he built, known as the Jackson Bridge, had a length of 109 feet, and it cost only $1,670, or a little over $15 per linear foot. In contrast, the cheapest chain bridge for which we have data cost nearly 3 times as much per linear foot. Long's bridge was moreover built in only 6 weeks, by six men (a smith, a carpenter, and four laborers).[36]

Another factor that contributed to the popularity of truss bridges was the fact that builders generally had some feel for and knowledge of the principles of truss construction, which were embodied in the designs of houses, roofs, barns, churches, and many other structures. As Finley himself recognized, the situation with respect to chain suspension bridges was just the opposite. In reference to a suspension bridge that collapsed in 1809 as a result of faulty construction, Finley acknowledged that "great allowance must be made for the undertakers . . . the principles being but little understood at that time; the workmen had never seen anything of the kind, and had scarcely the shadow of information."[37]

What conclusion can be drawn from this overview of the development of wooden bridges? The data suggest that, with the emergence of cheap, effective wooden-truss designs, Finley's system was no longer seen to offer clear advantages in cost or in ease of construction. In contrast, the initial decline of Finley's design system from 1815 to 1820 was probably due not so much to cost competition as to competition with regard to span lengths and to the advantages of wood over iron.

The fact that Finley's chain bridge was competing directly against truss and arch-truss designs raises a further issue. Bridges occupy niches defined by the properties inherent in their basic forms. Today suspension bridges are unrivaled, both technically and economically, as long single spans, whereas trusses function best as relatively short single or multiple spans. This difference can be seen in the fact that, as of 1973, the distance between supports of the longest suspension span was 4,300 feet, while the longest single truss span in North America was only 800 feet.[38] What this means in practical terms is that suspension bridges and truss bridges no longer compete within the same niche.

Finley's writings show that he understood the unique potential of the suspension bridge for long spans, since he hypothesized on the possibility of constructing clear spans of 1,500 feet. Yet his writings show that he did not believe such long spans could be built in his era. But why not a span of 300–500 feet? A design for spans of this length could potentially have given Finley a competitive edge. And the fact that Wernwag succeeded in building a wooden arch-truss bridge with a clear span of 340 feet—the longest wooden span in the world at the time—indicates that a comparable suspension span was possible in principle. Finley's writings implied that he believed his plan to be feasible for a span of 300 feet, but data concerning the 23 bridges known to have been proposed or built according to his plan show that he did not actually attempt a span length beyond 260 feet. It is not hard to understand why: Finley's decision to adapt his design to cheap, simple construction by ordinary craftsmen effectively limited the possibility of building longer spans. Significantly longer spans would have demanded a more complex planning and building process and a somewhat different design. For example, Finley's wooden A-frame towers would not have worked well for a 500-foot span. Long spans require high towers, and A-frame towers high enough for a 500-foot span would have to be too wide at the base.

Finley's decision to adapt the suspension bridge form to the predominantly rural market for cheap, simple bridges meant that he had to make technical tradeoffs, some of which affected the structural behavior of his bridges and perhaps also their competitiveness. For example, Finley's decision to rely on short spans meant that he had to rely extensively on multiple-span bridges, since many rivers were wider than 260 feet. Out of 18 bridges for which we have sufficient data, 10 were constructed with side spans or multiple spans. All of the post-1812 bridges for which data exist were of multiple-span construction. The last bridge built on Finley's patent (1826–1827) had three full spans, two side spans, and one draw span.[39] Even more remarkable was Finley's proposal for a 5,600-foot bridge over the Susquehanna River that was to consist of 25 spans of about 225 feet each (figure 3.9).[40]

Multiple-span suspension bridges were less stable than single-span ones, however. As was explained in chapter 2, Finley observed that the greater the number of spans, the greater the deflection produced by a load of given size and the greater the forces tending to overturn the towers. Finley could not eliminate this problem; as we saw, however, he did take technical measures to minimize it. His method of using curved saddles at the tower tops, which ensured that the chains would not be able to slide back and forth between spans, helped to limit the magnitude of deflections. And by advising builders of multiple-span bridges to bolt the towers in place, he hoped to counterbalance the increased forces tending to make the roadway deflect and the towers overturn: " . . . to resist this tendency the framing [of the towers] must be bound down to the stone work."[41]

Finley's decision to adopt a 1/7 sag/span ratio also sacrificed stability for simplicity of construction. The reason is simple: the greater the sag of a cable, the greater the deflection produced by a given load. A sag/span ratio of 1/7 is large—few if any other suspension bridge designers since Finley have chosen such a large ratio. A ratio of 1/10 or less would have provided more stability by increasing the tension of the chains in relation to their load and concomitantly decreasing the magnitude of deflections. A further implication of Finley's 1/7 ratio was that the towers for a given span had to be taller than for a smaller sag/span ratio. For example, with Finley's ratio of 1/7, a span of 300 feet needed towers about 43 feet high, whereas a ratio of 1/12 would have required towers only 25 feet high. Since Finley saw tower design as the major challenge in constructing longer spans, it is even more understandable why, having stipulated a ratio of 1/7, he did not propose spans beyond 260 feet.[42]

The "do-it-yourself" market toward which Finley oriented his design meant another significant tradeoff: in order to maximize "sales" of his bridge design, he effectively surrendered administrative and technical control of the construction process. Finley allowed local builders to use his design and pay royalties. Such a system opened the possibility that the builders might cut corners or otherwise deviate from the general plan. Finley's contract with John Templeman helped to guard against such a possibility, and to allow for some technical and administrative

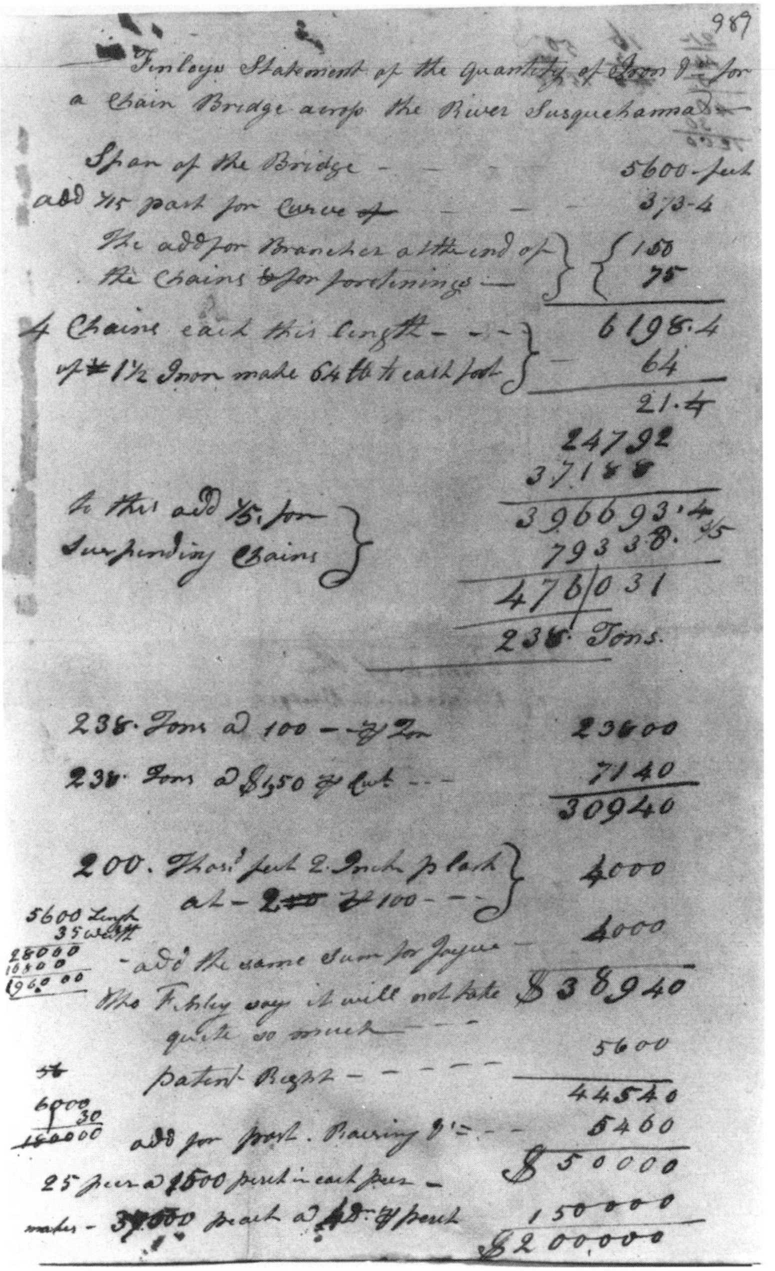
989

Finley's Statement of the Quantity of Iron &c for a Chain Bridge across the River Susquehanna

Span of the Bridge — 5600 feet
add 1/15 part for Curve of — 373-4
The add for Branches at the end of the Chains & for fastenings — 150 / 75
4 Chains each this length — 6198.4
of #1½ Iron make 64 lb to each foot — 64
21.4
24792
37188
396693.4
To this add 1/5 for suspending Chains — 79338.4/5
476031
238 Tons.

238 Tons at 100 — per Ton — 23800
238 Tons at $1.50 per Cwt — 7140
30940
200 Thous.d feet 2 Inch plank at 2.00 per 100 — 4000
add the same sum for Joyner — 4000
Mr Finley says it will not take quite so much — $38940
Patent Right — 5600
44540
add for Pier. Raising &c — 5460
$50000
25 piers at 1500 perch in each pier — makes 37500 perch at $4 per perch — 150000
$200000

5600 Length
35 width
28000
16800
196000

6000
30
180000

Figure 3.9
Finley's cost estimate for a 25-span suspension bridge across the Susquehanna River with a total length of 5,600 feet. Source: Stauffer Collection, Historical Society of Pennsylvania, volume 14, p. 989. Courtesy of Historical Society of Pennsylvania.

control over the use of his bridge design. The contract made Templeman not only a promoter but also something of a technical and administrative overseer of Finley's design. In his advertising circular, Templeman attempted to convince potential customers to rely on his technical expertise: "Anyone wishing to construct this species of Bridge, will do well to transmit to me a description of the site of the contemplated bridge—say, the width of the water, the depth, and what sort of Bottom it has; whether mud, gravel or sand—and particularly, whether the water to be bridged is subject to high and rapid freshes— the price of lumber, from which I could make a calculation of the probable expense, & send it by post—and then I will, if required, either go on, & superintend the construction, or build the bridge or bridges on contract."[43]

Templeman did in fact supervise the construction of a number of chain bridges (most notably the Essex-Merrimack Bridge and a 130-foot span over the Potomac in Washington), but in other cases local builders carried out the planning and construction of a bridge on their own. The clockmaker Jacob Blumer, who built the Palmerton Bridge and at least two other chain bridges in the Lehigh Valley, is one example.[44] Another example is a 115-foot chain bridge at Cumberland, Maryland, built by the local blacksmith Valentine Shockey with the aid of several local workmen.[45]

CHAIN BRIDGE OVER THE POTOMAC NEAR GEORGETOWN, D. C

Figure 3.10
A chain bridge over the Potomac near Georgetown, built in 1808 by John Templeman. Source: Robert Sears, *Wonders of the World in Nature, Art, and Mind* (New York: E. Walker, 1851), p. 460.

Some builders, like Blumer, closely followed the procedures outlined by Finley, but neither Finley nor Templeman could enforce adherence to the plan. As a result, those who wanted to save money or who misunderstood the purpose of various elements in Finley's design sometimes made detrimental changes. For example, in place of the heavy latticed railings that Finley had insisted on to provide stiffness and minimize deflections, Shockey substituted a "slight hand railing of iron" that was "hinged so as not to be bent by the undulations of the Bridge."[46] One of the wooden towers of this bridge later gave way. Although it was said that this was probably due to a weak or rotted timber, the disaster was undoubtedly hastened by increased stresses on the towers that resulted from movements of the cables and the roadway.

Another case in point concerns the construction of two chain bridges in Kentucky in 1811. The builders used a smaller safety factor than suggested by Finley, and took several shortcut measures in constructing the bridges, despite warnings that there might be dangerous consequences. One of the bridges collapsed shortly before the roadway was completed. Finley discussed the incident in the letter he published in Philadelphia's *United States Gazette*. His comments illustrate that administrative innovations were a necessary concomitant of technical development:

Two bridges have been undertaken in the state of Kentucky (besides the Frankfort bridge), one of which is almost two hundred feet in length, the chains are made of inch and a quarter iron, single bar, end to end, or what is called half chain—and the workmen found out another method of anchoring the ends quite novel to the plan hitherto adopted, that is, by letting a bolt into the rock, having an eye projecting a small distance above the surface of said rock to receive the end of the chain. But when the timbers of said bridge were nearly all on, the anchoring gave way and left the whole in ruins. Fortunately for the workmen, they had gone off the work a few minutes before it fell. The other bridge 104 feet in length, inch iron, half chain, has also some surprising departures from our plan, and is uncommonly extravagant in its symmetry, and I suppose must share the same fate. The men were conversant with the Frankfort bridge company, and had every information on the subject; but unfortunately they possessed too much skill of their own to receive any instruction from the experience of others. Although scarce half the iron is employed in these two bridges that ought to have been, yet I have no doubt that had the work been done carefully upon the common plan but that it would have outlived its builders.

It must be expected that these mismanagements will have a tendency to impede the introduction of the chain principle; but a full knowledge of all the circumstances will be, I presume, a sufficient antidote with every candid and well informed mind. . . . It is studied and obstinate deviation, and that alone, that has proved deficient in Kentucky.[47]

Today, public works administrations help to ensure that bridges are built according to approved technical specifications. Yet in the United States in Finley's era there existed for the most part only rudimentary public works administrations (or analogous hierarchies within the private sector). In Fayette County, the elected county commissioners who were in charge of public works were citizens with no special technical expertise. In short, the organizational structures into which Finley's design system was integrated gave considerable scope to local builders to adapt the design as they saw fit, and their mistakes had the potential to damage the reputation of the original design.

The Safety-Factor Problem

Every bridge must be "overbuilt" to some degree to compensate for unknown variables that might affect the strength or performance of the structure (e.g., flaws in materials, unusual climatic conditions). By definition, a safety factor is a factor of ignorance, a "guesstimate." It is always a challenge to minimize overbuilding—that is, to minimize the safety factor, in order to save money—without jeopardizing the integrity of the structure.[48] Finley had an additional challenge in that he wanted to keep the design process mathematically and technically simple, so he presumably did not want a multiplicity of complicated safety factor calculations. For example, although Finley might have chosen separate safety factors for roadway loads and cable stresses, he chose a single, aggregate safety factor, determined by multiplying the weight of the roadway by 5 or 6. This safety factor had to compensate for all additional loads on the roadway (traffic and snow), for the weight of the cables and hangers, for inadequacies in the calculated strength of the cables, for temperature changes that affected the tension of the cables, and for other defects of materials, workmanship, and so forth. In short, it had to cover a lot of ground.

Was the safety factor adequate? Although data on the causes of collapse of Finley chain bridges are not as detailed or comprehensive as one would like, these data, together with considerations about the nature of the safety factor, about Finley's methods of calculating the strength of the cables, and data about the iron used in his bridges show that the safety factor was too low.

Three bridges built on Finley's plan collapsed less than 20 years after completion when chains in their main cables ruptured. Finley and Templeman had expected their bridges to remain in service much longer; indeed, durability was one of the advantages they specifically claimed for the chain bridge. Finley suggested in his *Port Folio* article that a chain bridge might stand for 500 years; Templeman's circular suggested a service life of 200 years. And when Finley constructed his first chain bridge, he formally warranted it for 50 years (except the flooring).[49]

The three chain bridges that collapsed because of ruptured chains were the Falls of Schuylkill Bridge, which fell after less than 5 years in service,[50] the Brownsville Bridge, which collapsed after 11 years in service,[51] and the Essex-Merrimack Bridge, which collapsed after 17 years in service. In each case, the collapse occurred during the winter, under the influence of cold weather and a heavy snow load. In the last two cases, there were also heavily laden vehicles on the bridge. Observers of the Essex-Merrimack collapse reported that "five of the ten chains, which supported the bridge, were snapped in different places" and that "at the instant of the crash, the light evolved by the friction of the chains resembled the vivid streaming of a meteor." The cause of the collapse was also discussed at the time: "Various causes are assigned for the accident, and none with more probability than the united effect of the incumbent pressure of the immense body of snow lying upon the bridge, and the frost which had contracted the particles of iron. These produced a tenseness in the chains which was incapable of resisting the additional pressure of the loaded team, and the whole gave way."[52]

The circumstances of the Essex-Merrimack collapse provide strong evidence that the chains of these two bridges did not have the requisite strength, and thus that Finley's safety factor was too low. This hypothesis gains further weight when we recall that the Essex-Merrimack Bridge was subsequently rebuilt with 12 chains instead of 10, and that the structure gave no further trouble until it was dismantled in 1909. The one other chain bridge that remained in service for many decades, the Palmerton Bridge, was initially built with a safety factor of 8 rather than 6. Its chains never ruptured under snow and traffic loads.[53]

It might be asked, of course, whether the true causes of those collapses can ever be known, and whether extenuating circumstances—rusting of the chains, defects in workmanship, or failure to build the bridge according to plan and with the required safety fac-

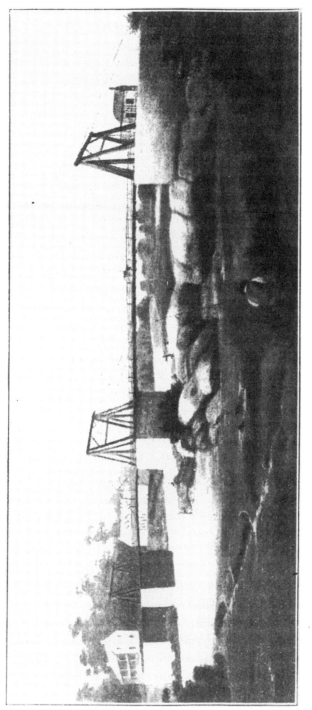

Figure 3.11
The Falls of Schuylkill suspension bridge, from a painting by Thomas Birch. Source: Llewellyn Edwards manuscripts, National Museum of American History. Courtesy of Smithsonian Institution.

tor—might not have been responsible. At least two of the three bridges were built largely according to plan. The Essex-Merrimack Bridge was built under the direct supervision of John Templeman. According to the diary of a Brownsville resident, the Brownsville Bridge (in Fayette County, close to Uniontown) was built under Finley's supervision.[54] The case of the Falls of Schuylkill Bridge is less certain. We know that it replaced another chain bridge that had not been built entirely according to plan and that had collapsed after only a couple of years in service. In response to this earlier collapse Finley had published an open letter in the *United States Gazette* explaining the details of the collapse, and the nature of his comments suggests that he had personally inspected the bridge after its collapse. The collapse of the first bridge was caused by using an open hook to couple two links of one of the chains, something Finley had specifically cautioned against in his *United States Gazette* article of January 1811. Finley would have had every reason to do his best to ensure that the second chain bridge there was properly rebuilt, and the investors would have had an incentive to ensure that the mistakes of the first bridge were not repeated in the second. In view of the first collapse, therefore, it would be a little surprising if the second attempt did not follow the plan.

With respect to the other two possibilities—corrosion and defective workmanship—it can only be said that safety factors are supposed to accommodate such contingencies, and Finley expected that his safety factor would do so:

Although I have taken considerable pains to ascertain and to demonstrate the strength of the chains, it must not be forgot that they may be overloaded. Some books when stating the strength of metals, advise not to load more than half, for fear of the injury by a continued tension. There are two other considerations that must be attended to—the corrosion at least of some of the parts, and we must not expect the execution of the work as complete as the mind could conceive it. And here I would earnestly recommend all those concerned, to adopt it for a maxim, that the chains in all cases should be able to support five or six times the weight of the bridge.[55]

One of the ways in which builders try to ensure the adequacy of a safety factor is by eliminating from the "unknowns" of the safety factor all problems and eventualities that can be predicted and calculated in advance. Finley included snow and traffic loads as unknowns in his safety factor, but in principle they can be estimated in advance. A snow load is easy to calculate because it is a static, distributed load—snow falls more or less evenly over the entire roadway, and just sits there (except for drifting, of course). One has only to determine the weight of snow that might be expected on each square foot of roadway, and a value for snow loading on a given bridge can be determined by simple arithmetic. A snow load could amount to 10 pounds per square foot,[56] which for the Essex-Merrimack Bridge would have led to an additional load of 36 tons. Since the roadway of the Essex-Merrimack Bridge weighed only 100 tons, it is clear that snow loads were substantial. An approximation of traffic loads can be calculated in the same manner as snow loads, by simply assuming the bridge to be uniformly covered with traffic and then estimating what the weight of this traffic per square foot would be.

With hindsight it seems reasonable to conclude that Finley should not have treated traffic and snow loads as unknowns to be accommodated by his safety factor. But it must not be forgotten that his suspension bridge was the first kind of bridge for which such calculations could be carried out in a straightforward manner. With Finley's chain bridge, a knowledge of such loads could be directly translated into an appropriate, quantitative value for the extra strength needed by the cables to support those extra loads. But in the case of arch or arch-truss bridges, existing theory did not permit such calculations.[57] In part as a result of this situation, there was no existing tradition for making such calculations among bridge designers in America.[58] Moreover, they would have added an extra degree of complexity to the design process, which Finley wanted to avoid.

If snow and traffic loads were the only variables that affected the adequacy of Finley's safety factor, then a value between 5 and 6 would have been sufficient. But there was another variable: Finley's method of calculating the strength of the chains. Finley's adopted value for the strength of iron, 60,000 pounds per square inch, was a minimum average value based on empirical data listed in the encyclopedia he consulted. He claimed to have verified this value empirically, although he provided no details of the tests. But how reliable was this figure? With the help of an analysis made in 1909 of the links from Finley's Essex-Merrimack Bridge, we can answer this question. The bridge was dismantled at that time, and an engineering professor at Cornell University, A. P. Mills, carried out tensile tests on bars cut from the original links.

Although Mills's data were gathered 100 years after the bridge was built, they can still tell us quite a lot about the strength of the iron and the chains at the time the bridge was constructed. Since these were thick chains made of wrought iron rather than steel (steel is more prone to rust than wrought iron), they were quite resistant to corrosion. When the engineers Boyer and Jelly examined the original chains of the Palmerton Bridge after they had been in service in the open air for more than 100 years, they found that, although the links had never been painted, there were "no signs of rust." Indeed, the original hammer marks could still be seen.[59] In the case of the Essex-Merrimack Bridge, Mills likewise found that most of the links exhibited very little corrosion. Corrosion would be the major cause for a decrease in strength of the iron over the course of 100 years,[60] but such a decrease would not have affected Mills's assessment of the strength per unit area of the iron, since he based his determination on a direct measurement of the cross-sectional area of the links at the time of the test.[61] Further evidence for the reliability of Mills's findings is that they agree (within 10 percent) with other data on the strength of wrought iron, such as those of Thomas Telford.[62]

Mills's tests yielded average values of approximately 48,000 pounds per square inch for the ultimate strength (i.e., stress at rupture) of the iron, and 32,000 pounds per square inch for the yield point. Mills's data thus showed that the average value of the elastic limit was 67.3 percent of the average value of the breaking stress. This degree of divergence between the yield point and the ultimate strength is further evidence that Finley's value for the strength of iron—which was based on the stress value at the point of rupture, rather than on the elastic limit—was too high.[63]

The deviation of the strength of the links from Finley's calculated value was due not only to the inherent strength of the iron but also to the efficiency of the links. Finley had assumed a link efficiency of 100 percent—i.e., that a link made up of two iron bars welded together at their ends would have fully twice the strength of a single bar. But Mills's tests on chain links from the Essex-Merrimack Bridge showed that the average efficiency of Finley's links was only about 75 percent.[64] It should be noted, however, that Finley gave specific instructions on how to fabricate the ends of the links to ensure they would have the full strength intended. He insisted that the links should be much thicker at the welded ends, to ensure that those parts "be well fortified": "Let the quantity of iron in this place [the weld] be two or three times as much as in any other part of the link."[65] Mills's examination of the links from the Essex-Merrimack Bridge showed, however, that Finley's guideline was not precisely followed. The links were made thicker at the ends, but not to the degree Finley had suggested. And Mills's tests showed that the welded ends were indeed the weak points of the chain: although the links were each about 2 feet long, every link he tested broke within 3 inches of the weld.

If we accept Mills's value of the average elastic limit of the wrought iron from the Essex-Merrimack Bridge (32,000 pounds per square inch), and if we accept his value of 75 percent for the efficiency of the links, then we get an average value for the unit strength of iron chains of 24,000 pounds per square inch. Yet this value would still have been too high in practice, because not all the links attained this average strength, nor did they all attain the average link efficiency. Indeed, two other significant conclusions that emerge from Mills's tests are that the iron of the links varied significantly in strength and that the links varied significantly in efficiency. The elastic limit of the weakest specimen was 65 percent of that of the strongest. The efficiency of the links varied between 47.3 percent and 94.6 percent. If the lowest numbers for elastic limit and link efficiency are used to calculate the unit strength of the chains, then the calculated value becomes much lower than the 60,000 pounds per square inch used by Finley: it becomes 12,300 pounds per square inch, which is only 21 percent of Finley's estimate.[66] If we assume that the effective strength of the Essex-Merrimack Bridge was somewhere between this lower limit and the average, then it appears that the reserve strength of the chains was not 500 tons, as Finley assumed, but somewhere between 20 and 140 tons. This strength would have had to accommodate snow and distributed traffic loads, the weight of the cables and hangers (at least 12 tons, according to Finley's data), increased stresses to the cables due to temperature changes (which can be quite significant), concentrated loads, movement of the roadway, and so on. Since a snow load alone could amount to over 35 tons, it is clear that the actual reserve strength of the Essex-Merrimack Bridge could not have been more than barely adequate.

Mills's data on the variance in the strength of the iron and on the efficiency of the links of the Essex-Merrimack Bridge also raise the issue of quality control. The only way to ensure that a wrought-iron chain will be as strong as intended is to test all its links and eliminate the weak or defective ones. (The alternative is to take the lowest possible link strength as the standard for calculation for all the links, but this option is wasteful of money and materials.) In the history of suspension bridges, materials testing soon became a standardized part of bridge construction. For example, the British engineer Thomas Telford tested each link for the chains of the Menai suspension bridge, constructed in the early 1820s.[67]

Testing is another way to eliminate an unknown that would otherwise have to be included in a safety-factor estimate. Finley did not pursue this option, however: he did not suggest or advocate the testing of links. It is possible that the idea never occurred to him. Yet, in view of the nature of the other tests and experiments he performed and his careful attention to the details of link fabrication, the idea of link testing would have been a reasonable next step. Finley also acknowledged that different samples of iron could exhibit significant variation in strength, and this awareness might also suggest the idea of link testing as a means to eliminate links of low strength.

Even if Finley did think of the possibility of testing links, however, there are several things that help to explain why he would not have chosen to advocate such a procedure as part of his design system. First, the design of his chain did not easily permit the testing of individual links, because the final weld of each link could be made only after it was hooked up to the following link. Second, link testing would have made the building process more complex, expensive, and time consuming, which in turn would have diminished Finley's claims regarding the economy and simplicity of his system (the advantages he particularly emphasized in his *Port Folio* article.) Third, Finley was already faced with the challenge of convincing citizens and craftsmen to build bridges on a radically new principal. How might they have reacted to the stipulation that all the links of the chains be tested in advance? Such advice might have seemed to suggest a potential weak spot in the design system, and might have inspired more doubt than confidence. Finally, there was no tradition of routine quantitative testing of materials used for bridge construction in America in Finley's era, so he had no social precedent or base to build on. For example, the encyclopedia article which Finley consulted, written by John Robison, made clear that research on the strength of wood was more advanced than corresponding research on the strength of iron. And builders everywhere were aware that the strength of wood varied from species to species, from tree to tree, and even within individual trees.[68] Despite this knowledge, American builders did not carry out routine tests of wooden structural members as part of the construction process. In the case of iron, research and testing began in the 1830s—not as a routine part of construction, but in response to growing safety hazards.

In particular, the first extensive American tests on the strength of iron were undertaken only in 1831, by the Franklin Institute, specifically in connection with the problem of boiler explosions.[69] Yet these tests by no means exhausted the subject: in the 1840s, when iron truss bridges came into use for railroads, builders still had to cope with a lack of adequate knowledge about iron, and many of their bridges collapsed.[70] But whatever the reasons for Finley's failure to advocate link testing, the inevitable result was that the possibility of variations in the strength of iron had to be accommodated through a larger safety factor.

Although this discussion of Finley's safety factor has cast doubt on its adequacy, it should not be forgotten that there was a very simple, practical way to solve the problem: more chains. Thus, even if Finley's safety factor was not sufficient, that does not mean that his bridge system should be regarded as a practical failure or that his contemporaries regarded it as such. At the time there does not appear to have been any general consensus concerning the technical adequacy of his system. Those who wanted to reject it could point to collapses as evidence for its inadequacy; but those who wanted to use it and to make it workable could do so—as is shown by the cases of the Essex-Merrimack and Palmerton bridges. And those who wanted to support Finley's plan could point to collapses and disasters involving many other kinds of bridges as proof that Finley's system was as good as any—many wooden bridges collapsed, and many others burned or were carried away by floods.[71]

In short, I do not believe that the decline of Finley's design system can be explained by the issue of the safety factor. Rather, the primary explanations for the decline lie in the nature of the market niche chosen by Finley (which should be understood as having social, economic, and technical attributes), in the decreasing competitiveness of his bridge design relative to other kinds of bridges that were competing within this niche, and in the technical and social tradeoffs Finley had to make as a result of choosing this market niche (reliance on multiple spans, reliance on a large sag/span ratio, and limited technical control of the design and building process).

More broadly, this analysis of the history of Finley-type bridges shows that the technical characteristics of the design were intimately bound up with social and economic decisions. Finley's technical decisions—including his choices of a sag/span ratio and of a safety factor, and his decisions about span lengths—were shaped by his social and economic priorities, and these decisions in turn had other social, economic, legal, and technical repercussions. In short, Finley's decisions were made in relation to his knowledge and perceptions of the existing structures and networks in his society—the patent office, the bridge-building community, local political structures, the existing body of scientific and technological knowledge, the legal system, and so on—and of how best to mobilize these to permit technological innovation. The specifics of Finley's design represented a creative harnessing of these elements to achieve new ends. As is often the case, however, the very ability of his design system to mobilize these elements effectively in the short run limited its competitiveness in the long run, as the institutions and networks of his society—and designers' responses to them—continued to evolve.

Appendix: Sources for Table 3.1

(See notes for bibliographic details.)

Albright, *Two Centuries of Reading,* p. 104.

Boyer and Jelly, "Early American suspension bridge," pp. 338–340.

Condit, *American Building Art,* p. 167.

Day, *Historical Collections of Pennsylvania,* pp. 511, 541.

Edwards manuscripts, National Museum of American History, Smithsonian Institution.

Elbridge, *Diary,* p. 80.

Ellis, *History of Fayette County,* pp. 434–435.

Engle, "1000 feet of continuity," p. 651.

Finley, *Description of the Chain Bridge.*

Finley, "Patent chain bridge."

Finley's Statement for Chain Bridge across Susquehanna, Historical Society of Pennsylvania.

Howe, "Old chain suspension bridge," p. 5.

Jakkula, "History of suspension bridges in bibliographical form."

Johnson's New Illustrated Cyclopaedia, p. 617.

Lowdermilk, *History of Cumberland,* pp. 304–305.

MacReynolds, *Place Names in Bucks County,* pp. 75–76.

Martin and Shenk, *Pennsylvania History Told by Contemporaries,* pp. 513–514.

Mills, "Tests of wrought iron links," pp. 251–281.

Montgomery, *History of Reading, Pa.,* p. 41.

Pennsylvania Correspondent (Doylestown).

Pennsylvania Assembly, *Reports on Roads and Bridges.*

Powers, "Historic bridges of Philadelphia," pp. 300–305.

"Reinforcing the Newburyport Bridge," pp. 314–315.

Roberts, *History of Lehigh County, Pa.,* volume 1, pp. 380–381; volume 2, p. 118.

Shenk, *Historic Bridges of Pennsylvania,* pp. 5–6.

Sherman, "Reconstruction of chain bridge at Newburyport," pp. 585–587.

Slocum, "Unique pioneer bridge," p. 691.

Templeman, circular letter, Hagley Museum and Library.

Wilson, *Standard History of Pittsburgh*, pp. 112–113.

4

Navier: The Making of an Engineer-Scientist

Navier's background and environment differed profoundly from Finley's. Whereas Finley lived in a rural community at the edge of civilization, Navier lived in a major European capital. Whereas Finley was an amateur inventor, Navier was a professionally trained engineer. And whereas Finley was an independent entrepreneur hoping to earn a profit, Navier was a salaried employee in a hierarchical bureaucracy: the Corps des Ponts et Chaussées, one of the national French engineering corps. In view of these important differences, it is not surprising that Navier's way of problematizing the suspension bridge—his response to the technology—was different from Finley's. The link between Navier's environment and the technology he created was indirect: his environment shaped his ideology, goals, and methods of practice, and these in turn influenced his manner of studying and designing suspension bridges.

In this chapter I will trace the specifics of Navier's career within its broader cultural context, with the eventual aim of relating this context to the general character and the specific details of his work on suspension bridges. My aim is not to suggest that Navier was merely a passive cultural sponge; it is to show that, although he acted upon his environment and innovated, he did so from a particular vantage point, and he attempted to give his work meaning in relation to particular communities and their norms. An analysis of the community of engineers of which Navier was a part—the Corps des Ponts et Chaussées in particular—will show that it was an evolving entity. Navier became a part of this community at a time when it was undergoing a fundamental transition, exemplified perhaps most notably by the restructuring of the system of engineering education that was carried out between 1795 and 1800. Navier's response to these changes was reflected in his work on suspension bridges.

Navier, Gauthey, and the French Engineering Corps

Navier (1785–1836) was drawn to an engineering career through family ties. His father, a prominent lawyer in Dijon, died when Navier was an adolescent.[1] At the age of 14, Navier was sent to Paris to be raised in the home of Emiland Gauthey, his mother's uncle. Gauthey was at that time a renowned and high-ranking engineer in Corps des Ponts et Chaussées. It was he who inspired and helped Navier to become an engineer in that corps. Navier explained: "I entered the Corps des Ponts et Chaussées through the attention and under the auspices of my uncle, having as sole merit the good fortune to have been brought up under his eyes, and the ardent desire to prove that, if it was difficult to put the lessons and examples that he provided into practice, it was at least impossible to forget them entirely."[2] Navier's deep respect for his uncle was evidenced not only by statements of admiration but also by the effort he made to obtain and publish, after Gauthey's death in 1806, an unfinished treatise on bridges and canals which Gauthey had been writing. The manuscript had become the property of other relatives of Gauthey, and Navier made considerable financial sacrifices to acquire it.[3] The year of Gauthey's death marked the year of Navier's entry into the Corps des Ponts et Chaussées, and Navier sought and obtained permission from his superiors in the Corps to complete and publish the manuscript. He worked on this project between 1807 and 1816, and the three-volume published work included, besides the original treatise, extensive notes and additions and a biographical éloge on Gauthey.[4]

Because of Navier's close personal and intellectual ties to Gauthey,[5] it is important to look more carefully at Gauthey's career and at the nature of his influence. This analysis will also provide a better understanding of the French engineering corps and their traditions.

Gauthey's influence on Navier had two dimensions, one related to Gauthey's exemplification of the occupational role of the engineer in French society and the other related to the specific character of his career within this occupational group. With regard to the first dimension, Gauthey represented for Navier a certain position in society, a certain social role. A growing body of historical research on French engineers in the state corps (the Corps des Ponts et Chaussées being one among several) makes it possible to characterize the role of these corps in some detail. An important starting point is Anne Blanchard's studies of the military engineers who made up the Corps des Fortifications (also referred to as the Corps du Génie).[6] This corps acquired a formal structure in the latter part of the seventeenth century, in part through the efforts of Louis XIV's minister Jean-Baptiste Colbert. Its organization was rooted in military culture and traditions rather than in the culture of the liberal professions: it was a hierarchically organized elite, and a majority of its members were drawn from the nobility. Blanchard's detailed study of some 1,500 engineers known to have served in the Corps des Fortifications between 1691 and 1791 reveals that 52 percent were from the nobility. This percentage increased further toward the end of the *ancien régime:* 75 percent of those recruited into the Corps between 1778 and 1791 were from the nobility, and 40 percent of these could trace their nobility back at least to the sixteenth century.

The professions of fathers of Corps members give an idea of the latter's social standing. Some 41 percent (out of a sample of 1,256 engineers) had fathers who were either officers or high-ranking officials and administrators. Another 21 percent had fathers who were themselves engineers, while 8.4 percent had fathers who were lawyers, doctors, or professors. Another significant proportion (17.5 percent) had fathers who were "without a profession," which in general meant that they were large landowners able to live off the income from their property.[7]

The procedures for selecting members of the Corps des Fortifications were informal throughout the first half of the eighteenth century, but this changed with the foundation of the Ecole du Génie de Mézières in the years 1748–1751. From then on, entry into the Corps was limited to graduates of this school. Prospective students had to pass an entrance examination and had to certify their family origins and their family's ability to support the cost of their education. Explicit preference was given to candidates from military or noble families. This new system had the effect of upgrading and standardizing the education of fortification engineers, and of enhancing their status.[8]

The history the Corps des Fortifications and the data from Blanchard's study—which concerns not only the social origins of the fortification engineers but also their level of wealth, the kind of houses they lived in, the furniture and books they owned, and their participation in scientific societies—confirm that this group, although not always associated with great wealth, nevertheless had a high social standing, built on a long and distinguished tradition of service to the state and on an acknowledged status as an intellectual elite. Blanchard explains:

Proud of the immensity of their works which involved the moving of incalculable volumes of earth, further still tons of documents, innumerable hours of intellectual reflection and preparatory studies, [the engineers of the Corps des Fortifications] were conscious of carrying out a great undertaking. Bound by their occupation to strict reasoning, mathematical calculations, and precise draftsmanship, they felt themselves assuredly a superior species. . . . Selected from among many candidates after a difficult trial, exceedingly gifted, they deemed that they were the best suited to serve in the most honorable corps. . . . They felt themselves to be wholly in the service of the King, the incarnation of the State. They demanded from that a respectable life [*vie décente*] and honors. In that sense they were the ancestors of the "grands corps de l'Etat."[9]

Available evidence shows that the status of members of the Corps des Ponts et Chaussées was roughly comparable to that of engineers in the Corps des Fortifications. There were, moreover, significant parallels in the evolution of the two corps in terms of both organization and training. The Corps des Ponts et Chaussées also emerged as a definable entity during the reign of Louis XIV, albeit a little later than the Corps des Fortifications.[10] And, like its counterpart, the Corps des Ponts et Chaussées followed a hierarchical, military model of organization.[11] The training and selection of engineers for the Corps des Ponts et Chaussées was informal until the second half of the eighteenth century. In 1747 (just about the time that the Ecole du Génie de Mézières was founded), a Bureau des Dessinateurs and Dépôt des Cartes et Plans was established in Paris as part of the Corps des Ponts et Chaussées, under the direction of Jean-Rodolphe Perronet, an engineer and architect renowned especially as a bridge builder.[12] One of its explicit functions was to train prospective engineers in mathematics, science, drawing, and other subjects necessary for their work, and the Bureau soon came to be referred to officially as the Ecole des Ponts et Chaussées. In contrast with the Ecole du Génie, however, there were no established professorships and no formal lectures. Rather, following Perronet's preference, education was based on the tutorial system, in which the brightest students helped to teach the rest. The mathematical and theoretical training was fairly rudimentary. The students mainly did design projects and fieldwork, spending about half their time assigned to help various engineers around the country with their work.[13]

There was less emphasis on military and noble pedigrees within the Corps des Ponts et Chaussées than within the Corps des Fortifications, and a greater percentage of its members came from families of merchants and entrepreneurs (23 percent for the Corps des Ponts et Chaussées as opposed to less than 4 percent for the Corps des Fortifications). Nevertheless, it was expected that prospective engineers would come from respectable and well-off families. Indeed, a 1777 pamphlet explicitly noted the unsuitability of craftsmen, artisans, and the like, "among whom education and feeling are rarely to be encountered, qualities absolutely essential in the personnel of the Ponts et Chaussées."[14] Prospective students, moreover, had to provide confirmation of their family's ability to bear the cost of their education, as was the case with the Ecole de Mézières. In view of these requirements it is hardly surprising that in the years 1767–1788 less than 1 percent of the students at the Ecole des Ponts et Chaussées came from families of peasants or workers. A further 2.5 percent came from families of artisans. In contrast, 42 percent of the students had fathers who were officers or members of the royal administration, and another 14.4 percent had fathers who were in the liberal professions; 28.5 percent had fathers who were merchants, capitalists, or wealthy landowners.[15]

Blanchard's generalizations about the elite social and intellectual status of members of the Corps des Fortifications, and about their *esprit de corps,* may be extended to the Corps des Ponts et Chaussées. First, members of the Corps des Ponts et Chaussées formed part of the socio-political elite of their regions. They functioned as high-level administrators with considerable technical, financial, and managerial responsibilities. In the eighteenth century, until the French Revolution, members of the Corps des Ponts et Chaussées were involved in directing the *corvée* labor[16] that was used in some regions of France for road construction and repair. Corps members worked closely with the political elites of their regions; on large-scale projects, they interacted with high-ranking government administrators in Paris too.[17] Second, members of the Corps des Ponts et Chaussées were among the intellectual elites of their regions. Many were members of local scientific societies. Gauthey and other engineers in the state corps in the region of Bourgogne, for example, became members of Dijon's prestigious Académie des Sciences, Arts et Belles Lettres. This academy was the focal point of Enlightenment thought in Bourgogne in the eighteenth century. One of its prize competitions helped to establish Jean-Jacques

Rousseau's reputation, and its membership included Voltaire and the scientists Prieur de la Côte-d'Or and Guyton de Morveau. Gauthey was elected to this academy in 1761, 3 years after entering the Corps, and he presented at least a dozen papers there over the course of his career. Through these papers he gained national recognition—one, concerning the theory of arches, inspired Louis XVI's Minister of Finance, Turgot, to write to Trudaine about Gauthey: "You have on the premises an engineer who has made himself known by his good memoirs on physics."[18]

By the end of the eighteenth century the members of the Corps des Ponts et Chaussées had a strong *esprit de corps*. They saw themselves as servants of the state with an important mission, and as guardians of long and distinguished traditions. They saw themselves as the bearers of a learned, rationalist tradition within engineering—a tradition embodied both in scientific research and in the many reports, memoirs, and detailed project plans they prepared over the course of their careers. They also saw themselves as contributors to a tradition of monumental architecture—a tradition embodied in the elegant stone bridges and public buildings they created.[19]

Elite status, intellectual prowess, service to the state, *esprit de corps*, and the preservation of great traditions were among the most important hallmarks of the Corps des Ponts et Chaussées toward the end of the *ancien régime*. Navier's *éloge* for Gauthey reveals that he was conscious of the Corps' prestige and that he wanted to participate in its traditions. Such a goal was almost expected of the adopted nephew and protégé of one of the highest-ranking engineers within the Corps des Ponts et Chaussées, because family ties counted for a great deal in the state engineering corps. It is possible to trace dynasties of fortification engineers extending over generations; and the Corps des Ponts et Chaussées was informally referred to as "la grande famille" because there were so many family ties among its members.[20]

Nevertheless, Gauthey was not just an abstract representation of a type of career. The organizational framework of the Corps des Ponts et Chaussées permitted both a range and an evolution of career patterns. Therefore, in order to understand Gauthey's particular significance as a role model for Navier—and the degree of convergence or divergence of their careers—it is important to look also at the specific details of Gauthey's background and activities within the Corps.

Gauthey: Architect-Engineer

Gauthey was born in Chalon-sur-Saône in 1732. His father was a respected doctor. Gauthey studied at a local Jesuit school from 1740 to 1748, but his father died in the latter year, and Gauthey was sent to live and study with his uncle, who taught mathematics to the pages of Versailles. Subsequently Gauthey studied with a well-known architect in Paris, and then, from 1757 to 1758, at the Ecole des Ponts et Chaussées under Jean-Rodolphe Perronet. Gauthey helped to pay for his education by tutoring other students in mathematics. (Such tutoring was a part of the educational system of the Ecole at the time.) Gauthey divided his time between giving mathematics lessons in the winter and visiting important engineering projects in the summer.[21]

Upon completing his training under Perronet in 1758, Gauthey was appointed to the post of *3e sous-ingénieur* (third assistant engineer) in Bourgogne, his region of origin. There he worked for many years on a wide range of engineering and architectural projects. In 1782, the year Gauthey turned 50, Bourgogne's chief engineer, Thomas Dumorey, died, and Gauthey was chosen to replace him. He held this position for 9 years. The final stage of Gauthey's career began in 1791 with the reorganization of the Corps des Ponts et Chaussées that was carried out in the wake of the French Revolution. In that year, at the age of 59, Gauthey was called to Paris and appointed one of the *inspecteurs généraux,* the top stratum within the Corps. Thereafter, until his sudden death in 1806, nearly every major civil engineering proposal in the country passed through his hands.[22]

More revealing than Gauthey's formal career path is an examination of his activities and projects over that long career. The recent work of Antoine Picon concerning Gauthey and the Corps des Ponts et Chaussées in the eighteenth century shows that Gauthey was a transitional figure between two eras of French engineering: "Gauthey . . . represents a sort of compromise between the figure of the artist-engineer, closely related to the architect characteristic of the classical age, and that of the engineer-scientist, of the technological engineer, contemporary of industrialization."[23] In the earlier era, engineers in the state corps were mainly *hommes de projet*—practitioners and entrepreneurs (in the broad sense of that term)—who were closely in touch with classical architectural traditions. Their designs drew on empirical rules and Vitruvian architectural canons. In the later era, engineers came to identify increasingly with what Picon has termed "the analytical ideal." This ideal embodied a program to make constructive practice more fully deducible from mathematical theory; it sought to make theory a powerful and creative tool that would free engineers from the constraints of architectural tradition. Gauthey was transitional in the sense that he accepted the idea that mathematical theory should hold a significant place in the engineer's "tool kit," but he did not fully subscribe to the analytical ideal. His own work remained strongly rooted in architectural tradition and practical experience; to the extent that he relied on mathematical theory, he used it more as a heuristic than as a formal guide to practice.[24]

An examination of Gauthey's methods of practice shows that he was an engineer of wide-ranging talents. Much of his work contributed to the tradition of monumental architecture, such as a beautiful small church he built in the town of Givry, close to Chalon-sur-Saône. Responding to an emerging trend, Gauthey sought in its design to integrate Gothic and classical elements in a dramatic new way.[25] Gauthey's bridges, although not especially innovative, were noted for their elegant, majestic designs. One of the most beautiful, constructed between 1781 and 1790, consisted of a series of stone arches spanning the Echavannes at Chalon-sur-Saône. Picon suggests that Gauthey's approach to bridge design was "less radical, but also more subtle in some respects than the approach of Perronet." In particular, he notes that Gauthey's approach was more architectural, and that his "decorative vocabulary" was more extensive and refined than Perronet's.[26]

Gauthey was not only an artist; he was also a highly skilled technical practitioner, having spent 25 years in the field as an assistant engineer. In this capacity he planned, directed, reviewed, and participated in dozens of engineering and architectural projects, including canals, roads, bridges, churches and other religious buildings, public buildings, chateaux, quays, and mills. In some cases his direct involvement in construction projects continued after his promotion to chief engineer. This was true of his church at Givry and also of one of his most significant technological works: the ambitious Charolais Canal (also known as the Canal du Centre), which linked the Loire and the Saône in the region of Chalon-sur-Saône.

In terms of organization, administration, and finance, the Charolais Canal has a long and complicated history that need not be recounted here.[27] What is important in the present context is the technical challenge it posed. The Charolais Canal was a "summit-level" canal, which means that the middle part of the canal was at a higher level than the two ends. Such a canal is possible only if there is enough water to feed it at its summit. Complicated administrative interactions delayed construction of this canal for years, in part because the high-level administrators in Paris who needed to approve the project (including Perronet) doubted that an adequate water supply existed. Yet Gauthey, through careful observation and surveying, found evidence to the contrary. He discovered that a small lake in the region (l'étang de Longpendu) could be used as a reservoir to feed the canal at its summit. This lake alone did not provide enough water; however, Gauthey found that it was so situated that it could be made to hold a large additional volume of water, and that the additional water could be made available by diverting several streams into the lake.[28] To make sure such a plan would work, Gauthey systematically measured the discharges of all the creeks in the area "five to six times per year over the course of three years."[29] Later, when he developed the technical details of the Charolais Canal project, Gauthey also took time to visit and study other major canals around the country so as to keep fully abreast of the latest improvements in construction techniques.

Through his long experience with projects such as the Charolais Canal, the church at Givry, and the bridge at Echavannes, Gauthey acquired a deep respect and concern for the details and development of constructive practice—for the kind of tacit knowledge that comes through experience and observation.[30] The importance Gauthey attached to tacit knowledge can be seen in his response to the dominant theory of arches of his day, as developed by the French academician Philippe De La Hire. This theory relied on a particular model for the mechanism of arch failure, yet Gauthey rejected the hypotheses underlying it as being "entirely arbitrary," and "disproved daily by experience."[31] The "experience" to which Gauthey was referring was the experience gained by years of observing the behavior of arches in the field. When he directed bridge construction, Gauthey observed the behavior of the arches as they were built upon their centerings and then when the centerings were removed. He learned that, as an arch was built up, different parts of the centerings alternately raised or settled, revealing the joints of rupture of the arch. These observations helped Gauthey to visualize the balance of forces in the arch and to develop a better understanding of the mechanism of arch failure.[32]

Gauthey's concern for the details of constructive practice also figured in his contributions to soil mechanics, in particular the theory of retaining walls. He extended the existing theory of retaining walls and its application, as developed notably by Augustin Coulomb and Gaspard Riche de Prony, by taking into consideration the kinds of technical parameters designers had to work with: the type of soil, the height and angle of inclination of the inner and outer faces of the retaining wall, and the type of material of which a wall was to be made.[33]

Gauthey was also a dedicated experimentalist; he has been referred to as "a partisan of the experimental method."[34] According to oral tradition within his village of Bissey-sous-Cruchaud (18 kilometers from Chalons), Gauthey carried out many hydraulic experiments in the creek that ran by his estate.[35] His concerns about arch theory also led him to carry out many tests to determine empirically how arches failed. He loaded old arches, and new ones made expressly for experimentation, until they failed, in order to observe how failure would occur.[36] Gauthey also carried out an extensive set of experiments on the strength of stone in compression, in response to a debate concerning the stability of the dome of Paris's Pantheon, and he developed a new testing machine for this purpose.[37]

What were Gauthey's credentials as a theorist? His formal scientific education did not bring him to the forefront of knowledge in mathematics and theoretical mechanics. He never learned calculus, and Antoine Picon has characterized his education in hydraulics and mechanics as "very much behind" the limits of knowledge of his day.[38] Nevertheless, Gauthey did not remain theoretically ignorant. As was suggested above, he followed, participated in, and even contributed to major theoretical debates within his field, although his contributions did not embody major new theoretical initiatives.

The way in which Gauthey related theory, experiment, and practical knowledge in his work can be seen more fully through his involvement in a dispute over the design of the Pantheon, created by his friend the architect Soufflot.[39] The pillars and arches that were to support the dome appeared inordinately slim to many observers at the time. Soufflot's design was attacked in particular by an architect and engraver, Pierre Patte, who used La Hire's theory of arches to argue that Soufflot's plan was defective.[40] Gauthey came to Soufflot's defense by showing that Patte's conclusions did not actually follow from the theory, and that the theory was in any event flawed and could not be relied upon owing to its arbitrary and inaccurate portrayal of the mechanism of arch failure.[41] Gauthey did entertain certain doubts about Soufflot's design, however. In particular, he was not convinced that the stone of which the pillars were to be made would be able to resist the loads they were supposed to bear. It was in order to investigate this matter further that Gauthey undertook his experiments on the resistance of stone to crushing, and they led him to conclude that the stone was indeed strong enough. Later, when the Pantheon was built, the pillars suffered some deterioration. Gauthey made a careful study of the damage and concluded that the problem lay in the way the columns were constructed. Ultimately, repairs were made successfully on the basis of Gauthey's recommendations, which included the construction of additional buttresses to support the dome.[42]

The example of the Pantheon controversy shows that Gauthey evaluated theory and made practical recommendations in the light of experiment, experience (tacit knowledge), and careful observation. His critique of Patte's theoretical attack on Soufflot rested partly on a reconsideration of the implications of La Hire's theory, but also on a rejection of the hypotheses on which the theory was based, on the grounds that these hypotheses did not accord with everyday experience in the field. More generally, Gauthey's use of theory reveals that he saw it as a tool to be used pragmatically and heuristically to help extend the limits of the building art, or to justify practical extensions of these limits after the fact. He saw value in theoretical analysis and debate, but he did not believe that practice could or should be entirely subordinated to theory.[43]

Navier's Training as an Engineer

It was said that Gauthey "used every opportunity to temper his nephew's theoretical studies with practical knowledge regarding bridge and channel construction."[44] Indeed we may suppose that, initially, Navier did intend to model his own career after that of his uncle. Nevertheless, he did not entirely follow in Gauthey's footsteps. Whereas his uncle's career remained solidly rooted in the world of constructive practice, Navier's career came increasingly to straddle the two worlds of engineering and mathematical science. His involvement in these worlds was rooted in his education as well as in his social, intellectual, and work environments, all of which differed in some important ways from the environment experienced by his uncle during much of his long career.

One important factor in Navier's career was his education. In Gauthey's generation it was possible to become a member of the Corps des Ponts et Chaussées after only a couple of years of training at the

Ecole des Ponts et Chaussées, with no lecture courses at all. By Navier's era, the system for training engineers to serve in the state corps had changed profoundly. Now the prerequisite to a career in the Corps des Ponts et Chaussées was 5 years of training, including two years of lecture courses at the Ecole Polytechnique (which prospective engineers for all the state corps had to attend) and 3 years of lecture courses, projects, and fieldwork at the Ecole des Ponts et Chaussées. A brief overview of the emergence of the Ecole Polytechnique and its role in the system of French engineering education will help to explain the profound impact of this transformation.

The Ecole Polytechnique was founded in 1794 in the wake of the Revolution. Initially the school had a fairly practical orientation, yet it underwent major changes between 1799 and 1812. The educational program was reorganized in 1799: the duration of studies was reduced from three years to two, and the curriculum was given a predominantly theoretical focus. Curricular changes continued to be made in the following decade, resulting in an increasing emphasis on mathematics (including calculus), and theoretical mechanics. Prospective students had to pass a competitive examination in order to gain entry to the school. Equally significant, the school was organized along strict military lines. Students were armed, uniformed, and subject to military discipline. The school's military culture served to reinforce both an *esprit de corps* and the engineering corps' ties to military culture and traditions.[45]

With the reorganization of the Ecole Polytechnique in 1799, a new relationship was established between this school and the older engineering schools, such as the Ecole des Ponts et Chaussées. Before 1799 the curriculum of the Ecole Polytechnique overlapped with those of the older engineering schools to the extent that it began to compete with them. With the reorganization in 1799, however, a new system was established whereby students wishing to enter any of the state engineering corps would first attend the Ecole Polytechnique to receive a common grounding in science, mathematics, architecture, and drawing and would thereafter attend one of the traditional engineering schools in order to receive more specialized training in a particular field. The significance of this development cannot be overestimated: four to five years of professional training instead of two or three meant that the education of members of the state engineering corps in France reached a much higher level of scientific sophistication than was previously the norm.[46]

In the case of the Ecole des Ponts et Chaussées, as we have seen, prospective students prior to the Revolution needed only a rudimentary knowledge of arithmetic and geometry, and their subsequent training was in some respects like an apprenticeship, since there were no formally organized courses. However, with the establishment of the new system linking the Ecole Polytechnique and the *écoles d'application*, the curriculum of the latter was also upgraded. The transformation was made in 1799 under the leadership of Gaspard Riche de Prony. Formal lecture courses were instituted at the Ecole des Ponts et Chaussées in that year, and three chairs were established: one in construction, one in applied mechanics, and one in "stereotomy" (analogous to descriptive geometry). Navier entered the Ecole Polytechnique in 1802, at the age of 17, so he and his fellow students were among the first to experience this qualitatively new system of education at the Ecole Polytechnique and the Ecole des Ponts et Chaussées.[47]

It is possible to get a fairly specific idea of what was taught at the Ecole Polytechnique when Navier attended. The first-year curriculum for 1801, just a year before Navier entered the school, was as follows.[48]

Subject	Lessons	Percentage of time
Calculus	60	16
Mechanics	40	10
Descriptive Geometry & Calculus Applied to Geometry	153	40
Physics	25	8
Theoretical Chemistry	54	10
Drawing	100	16

This table reveals that by 1801 two-thirds of the students' time in the first year was spent on mathematics and theoretical mechanics. The proportion for second-year students was only about 27 percent in 1801, but by 1812 it had reached nearly 50 percent. Navier spent part of his second year away from the school, however. As one of the top ten students in his first-year class at the Ecole Polytechnique, Navier was sent to Boulogne to serve as an Officier du Génie Maritime (officer of naval engineering) in January 1804.[49]

A more detailed picture of Navier's education at the Ecole Polytechnique can be gained by considering the professors who taught there and the textbooks they used. The professors of mathematics and theoretical mechanics included Gaspard Monge, Gaspard Riche de Prony, Sylvestre Lacroix, Jean Pierre Hachette, and Siméon Denis Poisson. All were renowned engineers or scientists, and all were at some point elected to the Académie des Sciences. Monge (1746–1818), the leading figure of this group, played a central role in the establishment of the Ecole Polytechnique. His most noted scientific contributions were in geometry, particularly descriptive geometry and analytic geometry. The former, which might be described as the mathematics of drawing and technical drawing, involved problems in projective geometry, the theory of shadows, perspective, and so on. Monge's text on this subject, *Traité de géométrie descriptive*, was required reading for *polytechniciens*, including Navier. Another required text was Monge's *Feuilles d'analyse appliquées à la géométrie*, which dealt with analytic geometry: the generation, analysis, and theory of various curves and surfaces, the theory of coordinate transformations, etc.[50]

Hachette, a close collaborator of Monge's, taught the descriptive geometry course at the Ecole Polytechnique.[51]

Lacroix (1765–1843), who taught calculus, published his lectures as a textbook, *Traité élémentaire du calcul différentiel et du calcul intégral*, in 1802. This work brought together in a coherent and systematic way the advances in the subject made by Euler, Lagrange, Laplace, Monge, Legendre, and other leading mathematicians of the period. It was the most comprehensive and up-to-date calculus textbook in the world at the time. Years later, in 1816, when the English mathematicians Charles Babbage, George, and John Herschel attempted "to breathe a new spirit into the nation's science," one of the first things they did was translate Lacroix's text.[52]

Poisson (1781–1840), the youngest of the group, was acting professor of analysis and mechanics in 1802. Poisson had graduated from the Ecole Polytechnique only two years earlier, but he had done so well that he was immediately made a *répétiteur* (tutor) at the school.[53]

Prony (1755–1839), a leading engineer in the Corps des Ponts et Chaussées and a protégé of Jean-Rodolphe Perronet, was of the generation between Gauthey and Navier. From 1783 to 1791, Prony assisted Perronet in his duties as head of the Corps; in 1791 he was appointed chief engineer in Paris. Prony also became director of the Ecole des Ponts et Chaussées in 1798, and it was he who planned the reorganization of that school's curriculum. Prony represented a further turn toward the "analytical ideal." His course in mechanics at the Ecole Polytechnique drew extensively on the work of Joseph-Louis

Lagrange. His own research contributions involved some of the earliest systematic applications of calculus and Lagrangian mechanics to problems in engineering. One of his most influential works was *Nouvelle architecture hydraulique* (published in two volumes in 1790 and 1796), which brought together and systematized the era's major advances in hydraulics and hydrodynamics. It was one of the texts used at the Ecole des Ponts et Chaussées, and Navier certainly read it.[54]

As these brief sketches suggest, Navier and his fellow students at the Ecole Polytechnique received an education in mathematics (calculus, differential equations, analytic geometry) and in theoretical mechanics (statics, dynamics, hydrodynamics) that could hardly have been more advanced or more thorough. Indeed, the training there gave bright students the opportunity to move very rapidly to the forefront of research in these areas.

After completing his studies at the Ecole Polytechnique, Navier spent two years at the Ecole des Ponts et Chaussées. When Prony reorganized the curriculum of this school in 1799, he set down three goals: to eliminate any duplication with the Ecole Polytechnique, to develop a curriculum that would build on the knowledge students acquired at the Ecole Polytechnique, and to "occupy [the students] uniquely with problems of application [*objets d'application*] related to the science and the art of the Ponts et Chaussées engineer." The second and third goals testified to Prony's vision of mathematical theory as the foundation upon which engineering practice must be built. As Picon explains in his history of the Ecole des Ponts et Chaussées, "Prony's ambition was to transform building into a rigorous, hypothetico-deductive body of knowledge."[55]

The training of students at the Ecole des Ponts et Chaussées was not as closely integrated with the Ecole Polytechnique as Prony's "analytical ideal" would suggest, however. The education given at the Ecole des Ponts et Chaussées after 1799 in fact continued many traditions and kept many priorities that were prominent in the time of Perronet. Formal lectures occupied less than half the students' time during their three years at the school.[56] The remainder of the time was devoted to essay and design projects and competitions, and to on-the-job training gained during summer "campaigns" in which the students assisted Corps engineers in their work. The on-the-job training helped to give students entrée into the world of constructive practice, from its administrative, financial, and managerial procedures to its tacit technical knowledge. The design and essay projects gave them additional practice and also served to transmit the tradition of monumental architecture—often to the point of directing students more toward the past than the future. In essay projects, students were asked to discuss such issues as the beauty of engineering monuments and the kinds of decoration that most befit them. An orientation toward the past is particularly evident in the design and drawing competitions. The competitions focused on reviewing existing solutions to design problems rather than on developing new ones, and those involving bridges were concerned mainly with stone arches rather than bridges of metal or wood.[57]

Nevertheless, despite the gap between ideal and reality in the educational program of the Ecole des Ponts et Chaussées, the ideal of deducing technological solutions from mathematical theory had a substantial impact on the ideology and practices of Corps engineers. This ideal had practical repercussions: the degree to which an engineer was seen to contribute to its realization could affect his path of promotion, his status, and his reputation within the Corps. The ideal was, moreover, promoted through the very structure of the educational system: all prospective engineers in the state corps were required to learn calculus and theoretical mechanics, and the educational system formally presented these subjects as the foundation for the more specialized training of the *écoles d'application,* as the traditional engineering schools came to be called. The analytical ideal was also promoted in the *écoles d'application* themselves, sometimes in lecture courses and sometimes in essay projects that required students to defend the theoretical approach.[58]

Navier's Transitional Years

Navier was among those who took the analytical ideal to heart and acted on it. In the process of doing so, he became so involved in the French mathematical community that he can be said to have had a "hybrid career" which moved between the worlds of science and technology.[59] This is where his career diverged most profoundly from Gauthey's. The theoretical debates in which Gauthey participated always centered around engineering questions. Gauthey never became closely affiliated with the mathematical community and never took up the mathematicians' research interests, whereas Navier did. Further, although Gauthey used theory, his use of it was always guided by tests, experiments, and tacit knowledge. In contrast, mathematical theory became Navier's most fundamental tool, and although Navier made use of the experimental results of others he did little empirical research himself. Finally, although Navier remained in the Corps des Ponts et Chaussées until his death, he carried out fewer engineering projects than Gauthey. Navier's time was occupied more with teaching, research, and publication.

Navier's ties to the world of practical engineering can be straight-forwardly delineated. At the close of 1807 he was assigned to the position of *ingénieur ordinaire* of the Corps des Ponts et Chaussées in the Department of the Seine in Paris. Between 1807 and 1820 he directed several engineering works around Paris, including the Pont de Choisy (1809–1811) and the Pont de la Cité (1819). He also drew up plans for a railway station at Choisy. In 1811 he was sent to Italy to draw up plans for several major engineering projects, most notably the reconstruction of the Horatius-Coclès bridge over the Tiber and the building of quays along the Tiber.[60]

In 1817 Navier was promoted to the rank of *ingénieur ordinaire 1ère classe*, and in 1822 to the rank of *ingénieur-en-chef*. The latter promotion came after a formal request, written in June of 1821, in which Navier reiterated his ties and his devotion to the profession of engineering. He explained that his uncle Gauthey, who had an unparalleled reputation and who had "planned and executed the greatest work" of the eighteenth century, had transmitted to Navier not his "pecuniary heritage" but his "heritage as engineer." Furthermore, wrote Navier, "being in some sense born an engineer, I dare to say that the administration must place me among those persons whose devotion and diligence it must most count upon."[61]

In subsequent years Navier was involved in designing and overseeing the construction of several bridges (including the Invalides suspension bridge over the Seine in Paris, which will be considered in detail in another chapter). He also consulted on a range of other projects, including railways. He remained a member of the Corps des Ponts et Chaussées until his death in 1836, having been promoted to *ingénieur-en-chef 1ère classe* in 1826 and to *inspecteur divisionnaire* after 1830.

Navier's ties to the world of mathematical science were not official, at least in the beginning. They were, however, a natural outgrowth of the closer links between the scientific and engineering communities that had developed in France during the previous decades. The rigorous mathematical training given to engineers and the structural relationship between the Ecole Polytechnique and the Ecole des Ponts et Chaussées (both located in Paris) meant that there was a deeper intellectual and institutional basis for interaction between Ponts et Chaussées engineers and the mathematical community.[62] Recall, for example, that the engineer Prony was a member of the Académie des Sciences, a professor at the Ecole Polytechnique, and the director of the Ecole des Ponts et Chaussées. Prony's closest colleagues therefore included not only other corps engineers but also mathematicians, among them Joseph Fourier, Augustin-Louis Cauchy, and Siméon-Denis Poisson.

In the case of younger engineers, close ties were often established first through teacher-student relationships. As we have seen, Navier's professors included Lacroix, Hachette, Prony, Poisson, and also (briefly) Fourier. Since Navier was assigned to an engineering post in Paris (already a sign of intellectual distinction, because these were coveted positions and were not given to just anyone), he was in an ideal position to keep up these contacts. He established a close relationship with Prony, who was prominent in both scientific and engineering circles. Prony, an *inspecteur général* in the Corps des Ponts et Chaussées, was one of Navier's superiors, and all the evidence available suggests that Navier was Prony's protégé. Several letters which Prony wrote on Navier's behalf testify to such a relationship. In one letter, Prony thanked the Corps' director for awarding Navier a promotion and indicated that, after surveying a bridge site with Navier in the village of Asnières (outside Navier's official service district), Prony wished to place Navier in charge of building the bridge, following a design drawn up by Navier and approved by the Corps' executive council. Prony also spent time with Navier in Rome when the latter was sent there in 1811. At the time, Prony was one of those in charge of the overall direction of engineering works in Italy. And when Navier died, it was Prony who wrote Navier's biographical memoir.[63]

A letter written by Navier in 1823 shows that he also remained in contact with his calculus professor, Lacroix.[64] And he became a particularly close friend and colleague of Fourier. When Fourier died, he bequeathed his papers to Navier, who subsequently prepared a posthumous edition of Fourier's *Analyse des équations déterminées* (Paris, 1831), with an historical introduction on Fourier's research in this area.[65]

Less formal, but probably equally important in fostering interactions between Parisian scientists and engineers, were the salons. The salons Prony and his wife hosted in their quarters in Paris's elegant Hôtel

Carnavelet (an elegant mansion that housed the Ecole des Ponts et Chaussées between 1814 and 1829) brought together the scientific, literary, and political elites of Paris and are reputed to have been particularly lively.[66] Two other well-known salons were that of J. L. N. F. Cuvier (a member of the Académie des Sciences) and that of the Count de Chabrol (Prefect of the Seine). Fourier may well have attended the latter—he had established a lasting friendship with Chabrol between 1798 and 1801, when the two of them accompanied Napoleon to Egypt. Since there was also a close working relationship between the prefect and the Ponts et Chaussées engineers in Paris, they probably interacted socially as well.[67] In addition, the director of the Corps des Ponts et Chaussées between 1815 and 1830, Louis Becquey, also had a salon, and he is known to have promoted intellectual exchanges between engineers in his corps and the scientific community. For example, he brought the engineer Augustin Fresnel back to Paris, from his post in Bretagne, so he would be in a better position to carry out his scientific research in optics (which was partly an outgrowth of his work to develop better optical systems for lighthouses). Fresnel was subsequently elected to the Académie des Sciences.[68]

In short, a variety of social and institutional mechanisms helped Navier gain access to, and encouraged him to become a part of, the French scientific community.[69] His growing ties to this community can be seen in the way his research interests evolved, in the way he internalized important priorities of the community, and in the honors the community bestowed on him. These honors included, most notably, his nomination to the Société Philomathique in 1819 (a major stepping stone to the Académie des Sciences),[70] his election to the mechanics section of the Académie des Sciences in 1824, and, in 1831, his appointment to what had been Cauchy's chair in analysis and mechanics at the Ecole Polytechnique. (Cauchy voluntarily went into exile after the 1830 Revolution.)

Navier's response to his candidature for a position in the Académie des Sciences reveals clearly the extent to which, by 1823, he had internalized priorities of the scientific community. These elections were always competitive, with two or more candidates vying for support.[71] An important prerequisite for election was an outstanding record of published research. In many sections of the academy, theoretical, mathematical research was looked on with great favor, and it was toward such research that Navier directed his attention. In a letter to

Sylvestre Lacroix in 1823, Navier wrote: "As to the election I forbid myself, as you can imagine, any remark about the person or the works of my competitors. . . . In confining myself thus to general considerations, I cannot help but hope that longer consideration will perhaps bring a person who has given great service to mathematics, on the belief that this species of knowledge is that which it is most important to encourage and to honor, and which will always constitute the most solid foundation for the glory of the Academy."[72]

Navier's research interests evolved from problems and topics rooted firmly in engineering toward an interest in problems that were potentially applicable to engineering but were primarily of concern within the scientific community. Navier's early publications, up to about 1819, were all substantially within the realm of engineering. These included most notably his edition of Gauthey's works on bridges and canals (for which he prepared extensive notes and additions), his own project for a railway station, and the preparation of thoroughly revised editions of two classic eighteenth-century French engineering treatises by Bernard Forest de Bélidor: *Science des ingénieurs* and *Architecture hydraulique*. In the case of the works of Gauthey and Bélidor, Navier's additions took the form of extensive notes and commentaries on the original text, which amounted sometimes to a new treatise being written alongside the old. These additions focused on matters of engineering theory, such as the stability of arches, soil mechanics, the theory of friction, and the theory of hydraulic prime movers. Navier transformed the analysis of these problems (as given in the original texts) by bringing more sophisticated theories and techniques from mathematics and theoretical mechanics to bear on them.

These new cognitive tools opened the possibility of a fundamental reshaping of the body of engineering knowledge. In the process of trying to achieve such a transformation, Navier moved further and further into the intellectual realm of mathematics and theoretical mechanics. He explained in his letter of request for promotion in 1821: " . . . the editing of the works of M. Gauthey . . . led me to devote myself to speculative studies. With much work and difficulty, I acquired [new] knowledge, some foreign to the building arts, some very important for engineers but which too often remains unknown to them. You have wanted, Mr. Director General, by one of these wise and noble visions that guides your administration, to make such knowledge useful for the general education [of engineers]. You have been able to judge yourself the efforts that I had to make to fulfill your vision, and the devotion with which I took it up."[73]

Increasingly, however, Navier's research memoirs began to embrace questions emanating strictly from the scientific community. For example, at a meeting of the Académie des Sciences in 1808, E. F. Chladni had given an experimental demonstration of the modes of vibration of flat plates. (These were made "visible" by covering the plates with sand.) Following up on this demonstration, the academy proposed a prize for an essay that would develop the mathematical theory of the phenomenon. The mathematician Sophie Germain was eventually awarded the prize in 1814, but with the explicit reservation that she had not fully solved the problem. At that point Poisson took it up, and he published his own analysis shortly thereafter. Navier became interested in the problem, and in 1820 he presented a memoir to the academy in which he used a different approach and different techniques (including the recently developed mathematical technique of Fourier analysis) to extend Poisson's results. Navier further generalized his own results the following year in a memoir that is generally acknowledged as a major step toward the development of a general mathematical theory of elasticity. Navier's work was then immediately taken up and extended by Cauchy, and it provided an impetus for further work by Poisson and others as well.[74] This episode shows the extent to which Navier, by 1821, had become involved in the research agenda of the mathematical community: Poisson, Fourier, and Germain were all mathematicians; only Cauchy had any direct involvement in engineering.[75]

Summary and Conclusion

A comparison of the careers of Gauthey and Navier shows that Navier's path of entry into the Corps des Ponts et Chaussées was structurally different from Gauthey's. Navier's formal training was more extensive and was oriented much more strongly toward the use of mathematics and theory to address engineering questions. Navier's initial posting in Paris, the center of French scientific culture, gave him both the opportunity and the impetus to become directly involved in the scientific community, especially since such involvement brought him rewards in his engineering career. Significant involvement in scientific research was also linked to Navier's teaching role at the Ecole des Ponts et Chaussées and the Ecole Polytechnique. In contrast, Gauthey came to work in Paris only at the end of his career, after many years of practical engineering work in the province of Bourgogne, and he never held a professorship. Navier responded to his qualitatively different environment and experience by embracing the analytical ideal to a much greater degree than Gauthey, and in the process of trying to realize this ideal he moved much closer to the research interests of the French scientific community.

These differences should not be allowed to obscure the important areas of continuity between the careers of Gauthey and his nephew, however. Both worked for the same engineering corps, and that corps continued to be hierarchically organized and to follow a military model rather than the kind of model suggested by the liberal professions. Furthermore, the members of the corps continued to see themselves as an intellectual and administrative elite, as representatives and servants of the state, and as important contributors to the tradition of monumental architecture. Navier's commitment to the analytical ideal was grafted onto a career pattern that continued to draw on many of the norms and values characteristic of corps engineers in the eighteenth century.[76]

5

Theorizing the Suspension Bridge[1]

Navier's work on suspension bridges must be understood in the context of his dual ties to the engineering and mathematical communities. His concern with suspension bridge technology came about as a result of an official request by the Ponts et Chaussées administration that he visit Britain in order to study and report on the progress of suspension bridge construction there. Available evidence indicates that British interest in this technology was initiated by knowledge of Finley's work in the United States. Finley's contributions had become known in Britain by the autumn of 1811. The earliest British proposal for a suspension bridge for vehicle traffic, prepared by Thomas Telford in 1814, was for an iron-wire suspension bridge with a span of 1,000 feet, to be built at Runcorn, between Liverpool and London.[2] This plan was never carried out, however. The first major suspension bridge to be completed in Britain was the Union Bridge (figure 5.1), designed by Samuel Brown[3] and constructed between August 1819 and July 1820.[4]

Taking up the request of the Corps des Ponts et Chaussées, Navier became one of the principal agents for the diffusion of suspension bridge technology to France.[5] He made two visits to Britain: one during the autumn of 1821 and one in the spring of 1823. While there, he read and collected available literature, talked with a number of engineers, and visited several suspension bridges completed or under construction.

The initial product that emerged from Navier's two visits—his response to this technology—was a 250-page report and treatise which he submitted to his superiors and published in 1823. The book had four parts: a summary report of about 25 pages, a descriptive and historical section of approximately 60 pages, a theoretical section of 115 pages, and a final section of roughly 35 pages in which Navier worked up two design projects—the first for a monumental suspension

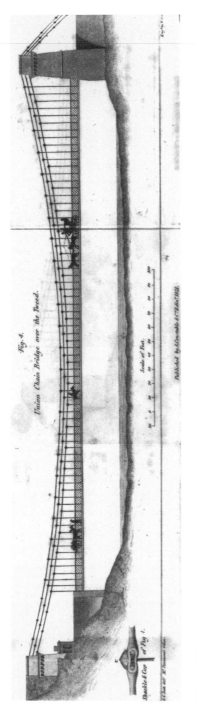

Figure 5.1
The Union Bridge, constructed by Samel Brown over the Tweed near Berwick in 1819–1820. Source: Robert Stevenson, "Description of bridges of suspension," *Edinburgh Philosophical Journal* 5 (1821), no. 10, plate VIII.

bridge to be built in Paris over the Seine at the site of the Hôtel des Invalides and the second for a suspension aqueduct. The remaining 15 pages consisted of appendixes in which Navier presented experimental data gathered in Britain concerning the strength of iron.[6]

A comparison of this report with the character of suspension bridge technology and knowledge in Britain at the time of Navier's visit shows that the general *problématique* around this technology—the kinds of questions, activities, and investigations connected with it—was fundamentally transformed in Navier's hands. From the many kinds of information that Navier obtained in Britain, he chose to ignore some and to pay careful attention to others. Drawing on this information, he created new knowledge. In order to understand and explain the structure of Navier's response to the suspension bridge—and how his cultural environment figured in this response—it is important first to understand the nature of the British response to the suspension bridge, and also to know what knowledge and information were available to Navier when he visited Britain. On this basis it will be possible to see how Navier "sifted" this knowledge and information so as to make it more relevant and meaningful in relation to his own environment—that is, in relation to the priorities of the Corps des Ponts et Chaussées, the French scientific community, and to the "analytical ideal."

Suspension Bridge Technology and Knowledge in Britain

Finley's innovation first became known in Great Britain in 1811, with the publication of Thomas Pope's *Treatise on Bridge Architecture,* which included a description of the Essex-Merrimack Bridge.[7] Pope's account engendered a short article on the Essex-Merrimack Bridge in *The Gentleman's Magazine* (81 (1811), September: 275).[8]

Among the first in Britain to take up Finley's idea were Samuel Brown (1774–1852) and Thomas Telford (1757–1834). Samuel Brown served from 1795 to 1812 in the British Navy, ending his naval career with the rank of Retired Captain. By 1808 he had become interested in promoting the use of iron rigging for ships, and by 1812 he had established a factory near London to manufacture iron cables for that purpose. His interest in suspension bridges was an extension of his previous technological work, in that much of the knowledge acquired in the former area was directly applicable to the latter. For example, before he became interested in suspension bridges, Brown carried out a series of experiments at his cable factory to determine the most efficient shape for iron links. He found that, weight for weight, ordinary links (such as those employed by Finley) were less strong than straight iron bars. In 1817, Brown incorporated this knowledge into a patent for suspension bridge cables composed of eyebar links.[9]

Brown became a leading builder of suspension bridges in Great Britain. He began by erecting, at his chain factory, a 100-foot trial suspension bridge capable of supporting loaded coaches.[10] Brown carried out experiments with the trial bridge to gain more insight into suspension bridge behavior and design. His earliest large-scale undertaking was the Union Bridge, mentioned above and shown in figure 5.1. The length of this bridge from tower to tower was around 436 feet, nearly twice the length of the longest span that had been constructed in the United States. This increase in length was due to major design changes. In addition to a more efficient type of chain (which cut by almost half the amount of iron needed to achieve a given strength), Brown developed sturdier anchorage and tower systems. Yet his roadway system did not have much in the way of stiffening. It did have longitudinal iron beams running the length of the deck on either side, but there were only two of these and they were only 3 inches deep. Brown attempted to achieve stability by means of a small sag/span ratio (nearly 1/17) which had the disadvantage of producing significantly higher stresses in the chains.[11] After completing the Union Bridge, Brown built a series of suspension bridges and two chain piers in the 1820s and the 1830s.[12]

According to one of his contemporaries, Thomas Telford began his career as a "common working mason."[13] By 1786 he had established himself as an architect, having acquired several wealthy patrons. In the 1790s he deliberately made a transition to civil engineering. Telford's interest in suspension bridges grew at least in part out of previous interest and experience in the construction of iron arch bridges. His first iron arch bridge was constructed around 1795.[14]

Telford's first proposal for a suspension bridge was for the 1,000-foot span over the Mersey River at Runcorn mentioned above. Such a span was quite daring at the time. Telford defended his proposal on two grounds. First, he pointed out that the practicability of suspension bridges had been demonstrated by their use in China and more recently by the construction in the United States of several designed to carry vehicular traffic. Telford specifically cited Pope's account of the Essex-Merrimack Bridge. Second, Telford explained that he had carried out more than 200 experiments at "Brunton's Patent Chain

Cable Manufactory," and that these experiments had confirmed the feasibility of long-span suspension bridges. They included tests of the strength of iron bars and experiments on sag/span ratios similar to those Finley had carried out. In Telford's view, experiments and good data—not mathematical theories—were the necessary basis for a new design such as the Runcorn project represented: "An undertaking of such immense magnitude, so perfectly original, and which, when completed, will perhaps be one of the most singular works of art that any age or nation ever produced, ought not to be attempted without the best data that could possibly be obtained, relative to the strengths of the proposed materials, under all the variety of strains to which they are likely to be exposed: and the following course of experiments were therefore made, as before stated, with this particular view."[15]

The Runcorn project did not win enough financial support to be carried out, but the time and effort that Telford devoted to it were not wasted: in 1818 that proposal served as a basis for a new plan to construct a suspension bridge over the Menai Straits. The Menai Bridge, built between 1819 and 1826, had a central span of 580 feet—well over twice the length of the longest American span—and was considered by contemporaries "one of the greatest undertakings ever executed."[16]

In order to understand the development of suspension bridge technology in Britain and its diffusion to France, however, it is not enough simply to enumerate who built what, and when, and how. We need to know more broadly how Brown, Telford, and other British engineers investigated and thought about suspension bridges. How did they determine cable tension? What kinds of research did they do on this technology, and what aspects of it did they deem most important? Several documents[17] from the period 1814–1821 provide answers to such questions. These documents show that British engineers were strongly preoccupied with the problems of cable design and the strength of iron but were very little concerned with creating a comprehensive mathematical theory of suspension bridges. They also show that British engineers were not even consistent in the choice of the mathematical techniques they used to determine the tension in a loaded cable. Indeed, Telford and Brown did not use mathematical techniques at all; they used empirical, experimental techniques to determine cable tension.

The civil engineer Robert Stevenson, in his 1821 article "Description of bridges of suspension," calculated cable tension as a simple force triangle—as though the cable were bearing a point load. With this assumption, the problem became one of determining the static equilibrium of a particle (rather than the equilibrium of a cable bearing a distributed load). Stevenson's assumption that the load was concentrated entirely at the midpoint of the cable, rather than distributed along the cable, made it possible to apply an elementary formula of geometry that derives the length of the hypotenuse of a right triangle from the length of one of its sides and the size of one of the angles bordering the hypotenuse. Applying this simple formula to Samuel Brown's Union Bridge, Stevenson determined that a load of 150 tons on the cables (the sum of the dead weight of the bridge and the weight of a crowd of people) produced a tension of 370 tons in the cables at the center.[18] For purposes of comparison, the standard formula for a cable loaded uniformly along the horizontal (which corresponds to a parabola and which more closely fits the case of a suspension bridge) gives a tension of 311 tons at the center point for the same bridge.

Another civil engineer—George Buchanan, who also lectured on mechanics at the School of Arts in Edinburgh—analyzed the Union Bridge in a letter to Samuel Brown in 1821.[19] Buchanan used a formula derived from the standard formula for a parabola, but he appears to have neglected to include a parameter for the load per unit span borne by the cable. Specifically, he assumed that the cable's minimum tension, T, was equal to $X^2/8Y$, where X represented the cable's span and Y its sag. The omission of a parameter for the load was probably an oversight on his part, because a paper he published several years later included a detailed derivation of the formula in which a parameter for load per unit span was included.[20] Without this parameter, Buchanan's result diverged widely from Stevenson's: he found a value of 900 tons for the tension at the center of the cables of the Union Bridge.[21] (Buchanan's modified formula, which was published in 1824 and which corresponds to the now-standard formula for a cable loaded uniformly along the horizontal, gives a value of 311 tons, taking into account the dead weight of the bridge—100 tons—and a distributed load of 50 tons.[22])

Still another method of estimating cable tension was published by Peter Barlow in his *Essay on the Strength and Stress of Timber* (London: J. Taylor, 1817). Barlow treated the curve of a suspension bridge as a catenary (i.e., a cable loaded uniformly along its length rather than along its span). This is a more difficult mathematical problem than assuming the cable curve to be a parabola, and solving it requires methods of approximation. From the point of view of engineering design, it is hard to justify this approach; suspension bridges carry most of their weight—namely the deck—distributed along the span of the cable rather than along its length. Thus, the cables of a suspension bridge are closer to parabolas than to catenaries. However, Barlow's main focus—his "unit of analysis"—was the cables themselves, to which the load of the deck would then be added. In other words, Barlow did not treat the cable-deck system as a unit (i.e., a single, integrated structure). Barlow applied his analysis to the case of Telford's Runcorn design in order to determine how much surplus strength the cables would have beyond the weight of the bridge itself and the traffic loads.

Although this issue of the catenary hypothesis may seem like a merely academic matter, it could have significant practical repercussions. George Buchanan pointed out in his 1824 paper that builders had used the catenary assumption to determine hanger lengths, with embarrassing consequences. Although Buchanan did not specify it, the instance he had in mind was almost certainly the construction of the Dryburgh chain bridge, detailed by Robert Stevenson in his 1821 article.[23] Originally designed in 1817 by John and William Smith (builders from Melrose, Scotland) as a cable-stayed bridge,[24] it was rebuilt as a chain suspension bridge after being destroyed in a storm in January 1818. During the reconstruction process, the hanger lengths were measured when the chains were merely hanging under their own weight (i.e., in a true catenary curve). When the roadway was added, the curve of the chains changed significantly, and the whole structure became distorted. The hangers connecting the roadway to the chains no longer hung vertically, and the roadway was "drawn off its level."[25] Stevenson noted that "between the center and either abutment, the roadway made two distinct curves, the [sags] of which measured about 7 inches."[26] To correct the problem, the hanger lengths between the center and ends of the bridge had to be altered. Evidence presented by Navier in his treatise suggests that a similar problem occurred during the construction of Samuel Brown's Union Bridge.[27]

Barlow's analysis, based on the catenary hypothesis, was extended in the 1821 article "On some properties of the catenarian curve with reference to bridges by suspension" by the mathematician Davies Gilbert.[28] Applying the method outlined in that article to the Menai Bridge, Gilbert first determined the theoretical maximum sag that its cables could have, given a span of 560 feet.[29] Next he simplified the mathematics of the problem by assuming that the sag of the cable would be very small in relation to its length. This assumption left him with a formula close to the standard formula for a parabola,[30] which he then applied to determine the tension in the cables of the Menai Bridge, assuming sags of 25 feet and 50 feet (for a span of 560 feet). The formula for cable tension which he derived expressed the value not in tons but, rather curiously, in feet. This value then had to be modified with a weight factor.[31]

Brown's and Telford's initial bridge designs did not rely on any of these mathematical methods. Buchanan's letter to Samuel Brown offers evidence that Brown used only experimental methods to determine the relation between load and tension in a suspension bridge cable. Buchanan commented on Brown's experimental results, and noted that they were "entirely conformable to the law of the catenary which I stated to you formerly."[32] He then suggested that Brown incorrectly assumed from his experiments that doubling the size of a bridge would not alter the cable tension as long as the span/sag ratio remained unchanged. Buchanan's theoretical analysis in the remainder of the letter was intended partly to prove that Brown's belief was incorrect and inconsistent with mathematical theory. Clearly, if Brown had been familiar with the mathematical methods used by Buchanan there would have been no reason for Buchanan to write this letter. It is also worth noting that this letter was written after the Union Bridge had been completed.

Thomas Telford's initial proposal for a suspension bridge at Runcorn is still extant, and it shows clearly that Telford also used strictly empirical methods to determine the relationship between load and tension in a suspension bridge cable. Telford used a setup similar to that employed earlier by Finley, and found that "an Iron Wire 1/10th of an inch diameter" would support 700 pounds when it was loaded axially (that is, when the weight was simply hung from one end of the wire). When the wire was supported at both ends, with a sag/span ratio of 1/50, Telford found that it would support a distributed load of 70 pounds.[33] With a sag/span ratio of 1/20, he found that the wire would support a distributed load of 233 pounds. When Telford later drew up his initial plan for the Menai Bridge (dated May 1818) he relied on his earlier experiments in proportioning the cables. He suggested that the necessary dimensions "can be satisfactorily ascertained by previous experiments; because with a given length and curvature, it is known that malleable iron of a good quality can support a certain weight more than its own, and therefore when the weight to be supported is known, a safe rule is obtained to determine the quantity of iron required."[34] In April 1919, Telford again cited this empirical method in evidence to the British House of Commons and again applied it to determine the amount of iron needed for the cables of the Menai Bridge. He explained that he had carried out more than 300 experiments, and he asserted, in reference to the problem of proportioning the cables for the bridge, "I do not go from any theory, I proceed upon experiments."[35]

Further evidence about knowledge and research on suspension bridges in Britain comes from the House of Commons hearings on the Menai Bridge, held in May 1818 and in April 1819. The Menai Bridge, a national undertaking, needed the approval of Parliament. Telford's proposal was therefore assessed by a special committee appointed by the House of Commons. The committee brought together some of Britain's leading engineers to review existing knowledge relevant to the problem of suspension bridge construction and to evaluate specific aspects of Telford's plan. Among the individuals who testified, in addition to Telford, were John Rennie, William Chapman, and Bryan Donkin, prominent civil engineers; Thomas Brunton, proprietor of the chain-cable factory where Telford had carried out many of his experiments; and Peter Barlow, professor of mathematics at Woolwich Academy.

The committee's reports, which record verbatim the testimonies of these experts, reveal how British engineers viewed the problem of suspension bridge construction and what questions about it they deemed most important. The reports show that they focused over-whelmingly on empirical evidence relating to practical concerns about the feasibility of Telford's plan. The topics that came up most frequently included the results, implications, and accuracy of recent experiments on the strength of iron; how best to fabricate iron cables for suspension bridges; what value to adopt as a safety factor for the strength of the cables; and whether the bridge towers could be made strong enough to withstand the downward force of the chains. The stability of suspension bridges under moving traffic loads and wind loads received some minor attention, as did the expansion and contraction of the iron cables due to temperature changes.[36]

These discussions gave only scant and superficial attention to theoretical calculations, however, and those that involved theory were concerned only with cable tension. Peter Barlow, the mathematical expert of the group, was asked if his mathematical calculations of cable tension

accorded with Telford's experiments, but the questions went no further. He was not asked to explain or justify the assumptions on which
his calculations were based. The strength of iron, the choice of a safety
factor, the stability of the bridge under wind and traffic loads, and the
effects of temperature changes were not treated as questions requiring
theoretical analysis.

In contrast to the scant attention given to theoretical calculations,
the discussions on how best to fabricate and test the iron cables for the
bridge were extensive and probing. Telford's initial plan for the Menai
Bridge involved a new cable design: iron bars were to be welded end
to end to produce continuous iron cables. Several experts were asked
detailed questions about the practicability of this scheme—for example, whether it would be possible to rotate such a long bar quickly
enough to achieve a good weld on all sides, and whether and how such
a bar could be tested before being used.[37]

The detailed discussions of the strength of iron reveal the existence
of a broader and more refined body of knowledge about iron than
that utilized by Finley. The discussions show, for example, that British
engineers had come to accept the need for routine testing of structural
materials such as iron chains. The discussions show also that British
engineers had formulated the concept of a yield point (i.e. a point at
which iron under tension begins to stretch sensibly and permanently
without actually breaking), and that they were coming to regard this
point as an upper limit for permissible loading.[38] Yet they did not agree
as to precisely what the yield point of wrought iron was: some felt that
it occurred at about half of the ultimate strength of the iron, while
others felt that permanent stretching occurred at roughly two-thirds
of the ultimate strength.[39]

Thus, the British response to the suspension bridge was dominated
by the great attention paid to questions regarding the strength of iron
and its use for suspension bridge cables. British engineers studied
these questions empirically and experimentally, and in doing so they
made extensive use of models and test bridges—both for general
experimentation and to help solve specific design problems. Besides
Brown's 100-foot trial bridge, Telford made and tested a 50-foot model
for the Runcorn proposal, and he used a $\frac{1}{4}$-size model of the Menai
Bridge to determine the lengths for the hangers. The historian Denis
Smith has found that virtually every would-be suspension bridge designer in Britain during the first half of the nineteenth century built
and tested models.[40]

British empirical research also included many experiments and tests on the strength of iron. Such tests were carried out not only by Brown and Telford but also by Buchanan, Rennie, Barlow, Brunton, William Chapman, Marc Isambard Brunel (who Navier visited), and others.[41] New cable designs were also proposed or patented by Brown, Telford, and Buchanan, and this tradition of attention to cable design continued at least through the first half of the nineteenth century. In the 1830s, for example, a British engineer, Charles Drewry developed and tested a design for a suspension bridge cable made from wooden links,[42] and in the 1840s another British engineer, James Dredge, developed a "taper chain" that became progressively thicker from the midpoint of a suspension bridge cable (where the tension is least) to the ends (where the tension is greatest). This chain design became the basis for a controversy that continued within British engineering circles for almost 10 years, giving rise to numerous tests and experiments.[43]

With respect to structural theory, some tentative steps toward the creation of a body of mathematical, theoretical knowledge concerning suspension bridges had been made in Britain by 1821, but the results of these efforts were still meager. Barlow's analysis occupied a little over two pages. Gilbert's paper was only slightly over five pages and offered no more than a means of estimating cable tension. Moreover, both of these papers were based on assumptions that ignored the effects of the deck on a cable's curve and equilibrium. Stevenson's method occupied a single paragraph in his paper, and the assumptions on which it was based were so wide of the mark that the results were little more than ballpark figures. The deliberations over the Menai Bridge project show that leading British engineers did not view mathematical theory as a crucial or even a creative tool for technological design. In short, there was no "analytical ideal" in evidence in the British suspension bridge community at this time, and British engineers did not attempt to integrate the results of their empirical experimental research on suspension bridges into a comprehensive mathematical theoretical framework.

Navier in Britain

When Navier visited Britain in the autumn of 1821 and then again in the spring of 1823, he studied most of the available literature on the subject of suspension bridges. He made a careful study of the first set of House of Commons committee reports and minutes on the Menai Bridge project, and he included long extracts from these discussions in his book. He apparently remained unaware of the second set of published minutes of this committee, printed in April 1819, which included extensive additional testimony from Thomas Telford, John Rennie, and Bryan Donkin. Navier did, however, cite a further report, published in 1823, in which recommendations by John Rennie, Peter Barlow, and Davies Gilbert for design changes to strengthen the Menai Bridge were discussed in detail.[44] Navier also read Pope's *Treatise on Bridge Architecture* (which included an extract from Finley's *Port Folio* article), Stevenson's article in the *Edinburgh Philosophical Journal* (which, in addition to a technique for calculating cable tension, described a number of suspension bridges and cable-stayed bridges erected in Britain and elsewhere), and Barlow's *Essay on the Strength and Stress of Timber* (which presented Barlow's catenary analysis, reported in detail some experiments on the strength of iron, and described Telford's 1814 design for a suspension bridge at Runcorn). Navier also read a number of articles describing particular suspension bridges, including an 1822 article by Samuel Brown describing his Trinity chain suspension pier at Newhaven (near Edinburgh).[45]

Navier spoke with two engineers who were directly involved in constructing suspension bridges: Samuel Brown and Marc Isambard Brunel. He also examined bridges they built. He made a detailed study of the design and dynamic behavior of Samuel Brown's Union Bridge, and he visited the Trinity suspension pier. During the spring of 1823 he inspected two of Brunel's suspension bridges (figures 5.2 and 5.3): one with a single span of approximately 130 feet and the other with two half-spans of about 130 feet, each flanking a central pier.[46] These bridges, completed and set up in Great Britain as of January 1823, were intended to be transported to a French colony, the Ile de Bourbon, near Mauritius.[47]

However, Navier evidently did not speak with Thomas Telford or view the site of the Menai Bridge (which was still under construction at the time of his visits). In general, Navier was careful to point out in his treatise when he had personally inspected a particular bridge site or spoken with a particular engineer. All his information about the Menai Bridge was secondhand. For example, he wrote: " . . . following unofficial information given to [me], it appears that the chains of the Menai Bridge are not constructed according to the original project presented by Telford, and that this engineer has adopted a system

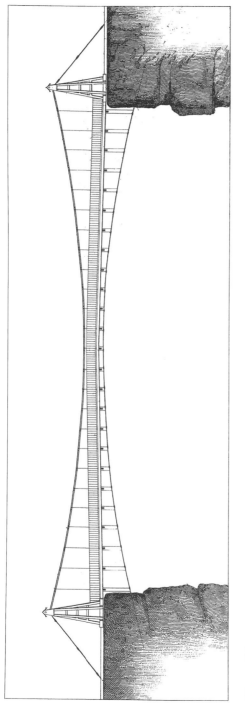

Figure 5.2
Navier's drawing of a suspension bridge constructed in 1823 by Marc Isambard Brunel for the River Mat at Ile de Bourbon. Source: Navier, *Rapport à M. Becquey et mémoire sur les ponts suspendus,* second edition (Paris: Carilian-Goeury, 1830). Courtesy of Special Collections Department, University of Virginia Library.

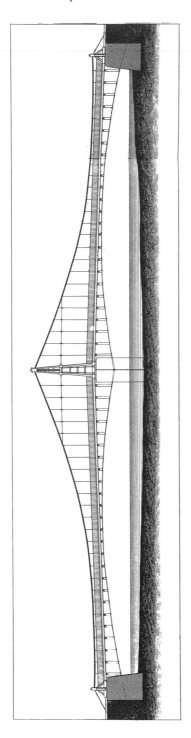

Figure 5.3
Navier's drawing of a suspension bridge constructed in 1823 by Marc Isambard Brunel for the River Sainte-Suzanne at Ile de Bourbon. Source: Navier, *Rapport à M. Becquey et mémoire sur les ponts suspendus*, second edition (Paris: Carilian-Goeury, 1830). Courtesy of Special Collections Department, University of Virginia Library.

analogous to that employed by captain Brown. . . . "[48] (Had Navier visited the site and spoken with Telford, he almost certainly would not have written this sentence in this way.) Other statements about Telford and the Menai Bridge lead to the same conclusion.[49]

What did Navier do with all the information he gathered, and how did he subsequently direct his own research into this technology? In order to answer this question, and to understand how Navier's cultural environment shaped his response to the suspension bridge, it is important to recognize what he did *not* do with the information he acquired. Unlike James Finley and Samuel Brown, Navier never attempted to patent a specific design system. And unlike Finley, Brown, Telford, and others, Navier did not build and test models, nor did he do extensive experiments on the strength of iron or experiments on the relation between load and tension in a suspension bridge cable. Nor did Navier investigate new methods for designing cables—a matter that was of central importance to British engineers. Rather than pursue the best method of fabricating suspension bridge cables, he created mathematical descriptions of their curves and movements.

Further, unlike Finley, Brown, and others, when Navier addressed economic issues in his treatise, he tended to do so in abstract terms divorced from any specific agenda for technological innovation. For example, at one point Navier considered the economic advantages of wooden chains relative to iron chains. (This was one of the few sections of his theory in which a cable became more than a mathematical curve.) Using data on the strength of wood relative to iron, and on its cost per unit volume relative to iron, Navier found that "the use of wood in the construction of chains would be 7 to 8 times more economical than that of iron."[50] Yet, unlike Drewry in Britain some years later, Navier did not address the issue of how a wooden chain might actually be constructed, nor did he try to build and test one. Navier also gave no precise guidelines about when wooden chains should or should not be used. He did recommend that they be used only for short spans (on the grounds that on long spans they would cause too much deflection, since wood stretches more than iron). Yet he did not specify more precisely what lengths he meant by "short" and "long."

In another section, Navier determined mathematically that a suspension bridge would be "most economical" when the height of the towers was equal to one-third the length of the span. This analysis rested on the assumption that the cost of constructing a bridge was simply a function of the weight (and hence the cost) of the materials.

Navier ignored, however, the roles played by time and manpower requirements in determining construction costs, and how their relative importance might be altered through technological innovation and design.[51]

In short, the context for understanding and investigating suspension bridges was fundamentally altered in Navier's hands. Navier did draw extensively on information obtained in Britain: information about the strength of iron, about its behavior under loading and its tendency to stretch before rupturing, about the movements of suspension bridges due to wind and traffic, about the curve and equilibrium of suspension bridge cables, and about the interactions between the cables and the towers supporting them. All these issues were discussed in the British literature Navier had read. Yet Navier brought these topics into an entirely different intellectual framework. Whereas in Britain they had been discussed separately and had mostly been analyzed empirically, without recourse to mathematical theory, Navier mathematized them all and merged them into a single, comprehensive theory of suspension bridges. Thus, the first way in which Navier's research on suspension bridges was socially shaped was the very fact that he chose *not* to continue along the path of empirical research and technological innovation that he had encountered in Britain, but rather to go the route of mathematical theory.

Navier's Theory of Suspension Bridges

Navier's theory of suspension bridges testifies to the importance which the "analytical ideal" held for him. The unified mathematical framework in which he integrated the scattered bits of empirical information he had gathered in Britain consisted of a basic mathematical model of the suspension bridge, which was then successively modified to take into account the weight of the cables and hangers, surplus loads (particularly concentrated loads), the kind of tower system adopted, the use of multiple spans, the elasticity of the cables and hangers, the expansion and contraction of the cables due to temperature variations, and vibrations and oscillations resulting from traffic and wind.

Navier wanted this theory to serve as an exemplar of what the analytical ideal could achieve; he wanted to show that mathematical theory was a useful, necessary, and creative guide to constructive practice. Accordingly, before presenting the details of his theory, he discussed the challenge of mathematization in more general terms.

Technological structures, he admitted, were generally too complex to be fully described in mathematical terms. Indeed, to analyze them at all demanded "a particular art which consists of replacing the very questions to be resolved by other questions that differ as little as possible and to which mathematics may apply."[52] In other words, mathematical theories in technology had to rest on hypotheses that in some way idealized or simplified the systems they were attempting to describe. Navier contended that a mathematical theory could nevertheless yield valuable information about the behavior of a system if used with care. But unfortunately, he added, too often not enough attention was paid "to the differences which existed between the phenomena and the hypotheses that were submitted to analysis, and the results of the computations were accorded too much confidence and too much authority."[53]

Navier's comment reflected his experience with the works of Bernard Forest de Bélidor, which he had spent some years editing. One of the explicit aims of Bélidor's treatises had been to convince doubters of "the necessity of theory" in engineering practice.[54] Bélidor's vision remained largely unfulfilled at the time, however, and Navier's editions of Bélidor's works tried to show that much if not most of the latter's theorizing was of little practical value. For example, the chapter on friction in Bélidor's *Architecture Hydraulique* consisted of 105 numbered sections (218–322), of which Navier cited 85 as largely false or inapplicable. Navier found the theories in the chapter on hydraulics equally problematic, noting of sections 491–517 that "all the notions given here by the author, the rules and the calculations that he presents are absolutely erroneous, and the reader can do no better than to skip this part entirely."[55] Regarding most of the remaining sections of the chapter, Navier counseled the reader "not to stop at all," since whatever was not false was useless for practice anyway, and he pointed out in particular that "the results in sections 524, 526, 540, 542–547, 550–554, are absolutely faulty and illusory."[56] And Navier dismissed most of the other theoretical material in Bélidor's treatises in similar fashion.

The proper antidote for this state of affairs, according to Navier, was not to abandon mathematical theorizing but to give more careful attention to how necessarily idealized theories could be made to yield useful information with minimum distortion and error. Navier outlined two basic ways in which this might be done. First, he suggested that results be formulated so as to specify the theoretical limits within which certain phenomena would operate, rather than their exact value under given circumstances. Second, he suggested that mathematical analysis be used to determine the relationships between important parameters (that is, how these parameters varied relative to one another) rather than their values in absolute terms, on the ground that the former were more accurate and just as useful to engineers. Navier made frequent use of both of these techniques in his theory. Another technique, which he did not discuss explicitly, but which was actually the principal conception underlying his model, may be termed "selective complexification." What Navier did was develop an abstract and idealized mathematical model of the technology he was investigating and then add hypotheses, one at a time, so as to gain more accurate information about selected features of the system's behavior.

Navier's Theoretical Model

In physical terms, Navier's abstract and idealized model portrayed the suspension bridge as a perfectly flexible, inextensible, massless cable, supported at both ends and loaded uniformly along the horizontal so as to form a parabola. (Thus, as he explicitly noted, Navier did not follow Barlow's catenary model.) In mathematical terms, Navier's model amounted to a set of interrelated formulas for the cable's curve, tension, angle of inclination, sag, and length. The crucial parameters on which the equations depended were the cable's sag/span ratio and its load per unit of span.[57]

Two general points can be made about these equations. First, Navier did not use the British system of calculus ("fluxions"), as Gilbert had; he used the continental European system. Second, in developing the equations, Navier drew on an extensive body of mathematical knowledge that had been developed since the seventeenth century, some of which directly concerned the equilibrium of loaded cables.

Navier credited John Bernoulli with being (in the early eighteenth century) the first to study this problem, but the concepts and techniques Navier used were products of a longer research tradition. By the late seventeenth century, Christian Huyghens, Gottfried Wilhelm Leibniz, Brook Taylor, and others had begun to investigate the different curves produced by cables loaded and supported in various ways. James Bernoulli formulated the concept of tension around 1700. He also established the practice of resolving forces into their x and y components, and of analyzing the x and y forces acting on a small element of a cable (techniques Navier relied on in deriving his equilibrium equations). The use of a uniform Cartesian coordinate system in the solution of problems involving loaded cables made its appearance in the early 1740s. By the middle of the eighteenth century, this diverse body of research had been generalized and systematized by Leonhard Euler, John Bernoulli, and others, and the solutions for the case of a cable loaded uniformly along the horizontal (i.e., a parabola) were established. Thus, Navier's basic equilibrium equations did not represent something fundamentally new; they were applications of ideas and methods that had become standardized in French mechanics texts. In principle, these ideas and methods could have been used by Finley, Brown, and Telford.[58]

The remainder of Navier's theory explored how the equilibrium conditions represented by this core set of equations would be influenced by taking into account the list of specific factors cited above, considered on a case-by-case basis. Each case involved adding different complexifying hypotheses to the original model, so as to yield information about a particular aspect of the behavior of a suspension bridge. Let us look briefly at Navier's treatment of each of these cases.

Finding the Cables' Curve

The idealized equilibrium equations were based on the hypothesis that the dead load borne by the cables was distributed uniformly along the horizontal. But the dead load of a suspension bridge consists of the weight of the deck plus the weight of the hangers and cables. Although the deck load is distributed pretty much uniformly along the horizontal, the weight of the cables is distributed uniformly along their curve, and the weight of the hangers increases from the center of the span (where the hangers are short) to the towers (where the hangers are long). Thus, the curve of a suspension bridge cable deviates somewhat from a parabola.

To develop an equation for this curve, Navier assumed that the load—expressed as a function of the cable's span, $w(x)$—was no longer a simple constant (as it would be if the curve were really a parabola), but was rather the sum of a constant and two variable terms:

$$w(x) = w_r + w_c(x) + w_h(x).$$

The constant term w_r referred to the deck's load; the second and third terms accounted for the loads due to cables and hangers, respectively.

Navier derived an expression for the second term based on the general differential equation for the length of a curve. To derive an expression for the third term, he first imagined the hangers not as individual rods, but rather as a continuous sheet between cable and roadway with a total weight of W_h. He then had only to construct $w_h(x)$ so that its integral, evaluated over the cable's span, would be equal to W_h.

Navier used the above equation to deduce new formulas for the cable's curve, length, tension, sag, and angle of inclination. These formulas showed that the cables of a suspension bridge would be longer than parabolas of equal span, that they would be more steeply curved at the supports, and that their sags would be less. The price of achieving greater accuracy, however, was a much greater degree of mathematical complexity: the new equations were considerably longer and more intricate than those for a parabola. It is perhaps not surprising, therefore, that although Navier used the new equations to work out an algorithm to calculate hanger lengths, he did not use them to derive the rest of his theory: the calculations would have been dauntingly complex.[59]

Concentrated Loads

Suspension bridges must bear not only their own weight, but also surplus loads, which may be either uniformly distributed (as with snow loads) or concentrated (as with traffic loads). In practice, concentrated loads on a suspension bridge are often moving loads. For example, when examining Brown's Union Bridge, Navier carefully observed what happened as vehicles crossed the bridge. He noticed that "the outline of the deck [was] continually modified throughout the duration of the crossing, the place where the vehicle was situated sinking while the other parts rose."[60] Yet Navier realized that he could isolate each point in the vehicle's passage across the bridge and treat it as a problem of static equilibrium, using the same mathematical techniques he had used to determine the basic equilibrium equations.

Using this approach, Navier analyzed three cases (figure 5.4): (1) the case where a surplus load P was distributed over a small arbitrary section of the span, (2) the case where P was distributed over a small section centered at the span's midpoint, and (3) the case where P occupied only a point precisely at the midpoint. Case 3, on which I will focus, represented the situation in which a given load would produce the greatest effect. In other words, the solution of case 3 gave

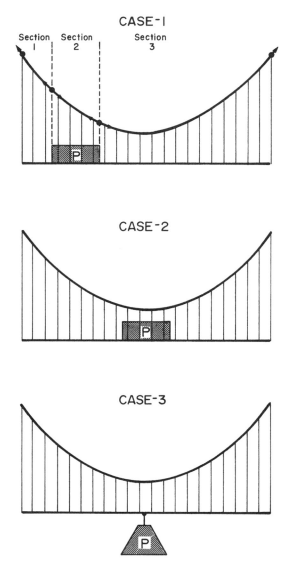

Figure 5.4
Navier's analysis of concentrated loads.

the theoretical upper limit for the effects of a given concentrated load on a parabolic cable. It must be emphasized, however, that Navier's formulation of the problem was such that he analyzed the effects of concentrated loads as though they acted directly on the cable rather than on the roadway. In his own schematics, the roadway was not even pictured.[61]

The eventual aim of this analysis was to understand the effects of concentrated loads on cable sag and tension relative to bridge size. Toward this end, Navier derived formulas for the cable's altered tension (T') and sag (Y') due to a surplus load P. He then computed the differences, $T' - T = \Delta T$ and $Y' - Y = \Delta Y$, where T and Y represented the values for the tension and sag in the absence of extra loading. The resulting expression indicated that the tension increase due to a given concentrated load would be roughly proportional to the cable's span/sag ratio, X/Y. In other words, ΔT would be greater for a bridge where X/Y was large than for a bridge where X/Y was small. This result implied that designers should proportion suspension bridges to make X/Y small in order to minimize tension increases due to live loads. Yet the expression for ΔY implied that designers should proportion suspension bridges to make X/Y large, for it showed that the greater was X/Y, the less would be the deformation produced by a given concentrated load. The expressions for ΔT and ΔY taken together thus pointed to a dilemma. Designers could not hope to minimize both values simultaneously. Rather, they would have to choose X/Y so as to steer a middle course between maximum stability (X/Y large) and minimum tension (X/Y small)—in other words, so as to optimize the relationship between stability and tension.

The resulting expressions for ΔT and ΔY also showed that neither would be any greater for a large bridge than for a small one, if the sag/span ratio was held constant. This result, which Navier termed "very remarkable," showed that the larger the bridge, and the heavier, the less would be the relative influence of traffic loads and, hence, the greater would be the bridge's stability.[62] Finley had also reached this conclusion, although he had never attempted to mathematize it. British engineers prior to Navier may have shared Finley's intuition as well, but the idea is not mentioned in any of their extant writings.

The Equilibrium of Tower-Cable Systems

Navier's idealized mathematical model assumed a suspension bridge to be merely a cable suspended from fixed points. Yet the cables of a suspension bridge usually pass over towers and then continue down to anchorages, so that a cable's parabolic central section is flanked by two anchor sections (which may be separate cables). There are, moreover, different types of tower-cable systems, each with its own behavior. Finley used curved links at the towers so that the cables could not move back and forth over the towers. Samuel Brown supported the cables at the towers of the Union Bridge on rollers; this arrangement allowed some movement to ensure that cable tension would be equal on both sides of the towers.[63] For the Menai Bridge, Telford used cable saddles mounted on rollers.[64] Brunel used another method to achieve an analogous result: he hung an iron link within the framework of each tower and suspended the cables from these links, so that they could pivot back and forth to a certain degree.[65]

These various design solutions show that Finley and certain British engineers thought about tower-cable interactions, yet they did not attempt to classify or analyze mathematically the range of possible systems. Tower-cable interactions were also specifically discussed at the hearings for the Menai Bridge, yet the main concern was to ensure that a tower could be made strong enough to resist the downward force of the cable.[66] Navier's approach was to analyze these systems mathematically. First he categorized the basic possible types of tower-cable systems according to their structural characteristics,[67] then he developed a set of equilibrium equations for each type of system.

Navier identified two main systems (figure 5.5). In type 1, a slim vertical tower (having little resistance to toppling) supported a central cable and a separate anchor cable. In type 2, a massive, rigid tower supported a continuous cable that was free to slide back and forth over the tower (that is, the anchor cable was the continuation of the main cable).[68] The equilibrium conditions of type 1 were straightforward. It was necessary for the horizontal tension components in the main cable and the anchor cable to balance one another; otherwise there would be a net horizontal force that could overturn the tower. The vertical tension components necessarily reinforced one another, resulting in a net downward force on the tower. Type 2 was more complex because the equilibrium conditions depended on the amount of friction between the cable and its support. To show the extent of the range of possible behavior for type 2, Navier analyzed the two extreme cases: one where the cable could slide back and forth with no friction and one where the cable was assumed to lie directly on the tower so that friction was maximized. In the frictionless case, the magnitude of the cable's tension would be the same on either side of the tower, since it

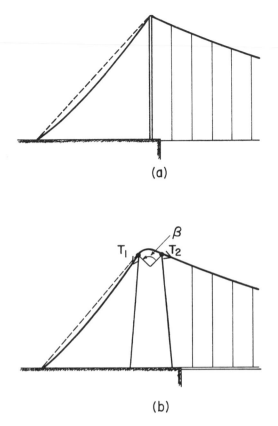

Figure 5.5
Tower-cable systems analyzed by Navier.

was free to move as if on a pulley. Yet the angles at which the tension was directed might differ because of differences in the cable's inclination on either side of the tower; the result would be a net horizontal force on the tower. In the case of maximum friction, the tension in the cable's central section would not be fully transmitted to the anchor section, so the magnitude of the tension would be different on either side of the tower. By assuming the tower top to be an arc of a circle, Navier calculated the tension difference by applying a formula for belt friction derived first by Euler and subsequently by Coulomb.[69]

Neither of the two extreme cases of tower-cable interaction was likely to be met with in practice. As Navier pointed out, supporting cables on rollers diminished but did not eliminate frictional forces. Navier's analysis of tower-cable interactions, like his previous analysis of concentrated loads, was thus intended to reveal the theoretical limits of behavior of a system rather than its average or most probable behavior.

Having studied the basic equilibria of the two types of tower systems, Navier then looked at how each would be affected by the addition of a uniformly distributed load (e.g., snow) on the main span. He explained that for type 1 the effect of a surplus load on the main cable would be to increase the cable's tension and cause some displacement of the tower (since it had no resistance to toppling). Through a clever and intricate line of reasoning, Navier was able to derive a formula that gave the tower's displacement as a function of the additional load. This formula revealed that the magnitude of the displacement did not depend on the height of the tower but did depend significantly on the distance between the tower and the anchorage. Specifically, the displacement was proportional to the cube of this distance.[70] This result suggested that designers should attempt to minimize the distance between towers and anchorages.

Type 2 proved less amenable to mathematization. Navier pointed out that the additional load could cause the cable to slide toward the middle of the span or could even cause the tower to tip. What would actually occur depended on the amount of friction between the tower and the cable and on the design of the tower. Thus, no general mathematical theory of this type of tower system could be given; each specific bridge design required a mathematical analysis of its own.

Multiple Spans

Up to this point Navier's analysis of tower-cable interactions extended only to single spans, but he was also interested in how live loads would affect the equilibria of multiple-span suspension bridges and of the multiple-span chain piers designed by Brown. Specifically, Navier wanted to know how a surplus load on one span would affect a structure's overall equilibrium, particularly as the number of spans increased. Like that of a tower-cable system, however, the behavior of a multiple-span bridge depends on the specific technological arrangements. Navier analyzed the most unfavorable possibility: that where the cable was assumed to be continuous and able to slide over rigid, intermediate towers. Once again, he considered the extreme cases of negligible and maximum friction.

Navier pointed out that, in qualitative terms, a surplus load on one span would cause the cables to slide over the towers, increasing the

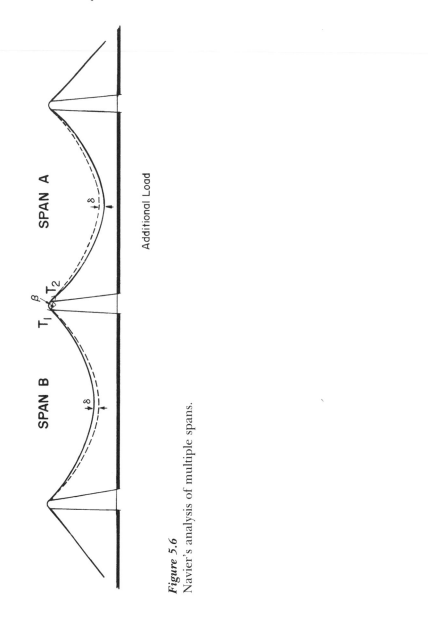

Figure 5.6
Navier's analysis of multiple spans.

sag of the loaded span and decreasing the sags of the other spans until the tension was equalized throughout (figure 5.6). Mathematically, Navier sought a precise formula for the increase in sag, ΔY, as a function of the surplus load. He was able to derive this by using the belt-friction equation modified by a series of substitutions based on the idealized equilibrium equations. (He also had to adopt some simplifying assumptions in order to make the calculations tractable.) He adapted the formula to account for the cases of maximum and negligible friction. The analysis indicated that even a moderate surplus load on one span would produce significant deflection and, moreover, that the amount of this deflection would increase with the number of spans. In other words, Navier proved that the relative stability of a multiple-span bridge with continuous cables would decrease as the number of spans increased.[71]

The other major conclusion Navier drew from his analysis of multiple spans was that intermediate towers were significantly less stable than end towers. For example, if an end tower began to overturn, friction between tower and cable would cause the already taut anchor cable to become even more taut, its small sag would become even smaller, and this would help to stabilize the tower and prevent it from overturning. Yet an intermediate tower would be flanked on both sides by cables with substantial sag. Thus, Navier argued, if an intermediate tower began to overturn, the cables would not be able to provide much resistance to overturning, even with the influence of friction. Navier's reasoning here was not only qualitative. Earlier, in his treatment of tower-cable interactions, Navier had carried out a mathematical analysis of the situation in which an end tower was about to overturn, and he had calculated exactly to what extent the anchor cable would contribute to the tower's stability. Subsequently he showed that this same analysis could not be applied to the case of an intermediate tower. On the basis of his conclusions, Navier recommended that designers take special care to proportion intermediate towers so as to have extra stability.[72]

Elasticity[73]

Thus far, Navier's analysis had depended on the hypothesis of inextensible cables. But iron cables stretch when loaded. Navier wanted to determine how much a cable would stretch relative to the proportions of a bridge, and how this stretching would affect the cable's sag and tension. Here Navier sought to mathematize a phenomenon that Finley and British engineers had largely ignored. Finley, as we have seen, paid attention only to the rupture strength of iron; he ignored the phenomenon of stretching before rupture. British engineers did note in their experiments that iron began to stretch perceptibly before rupturing. Yet, as Navier pointed out in his treatise, their concern was only with the phenomenon of *permanent* stretching (now referred to as *plastic deformation*), which occurs only after the yield point of the iron has been reached. Navier was concerned with the slight stretching that occurs when iron is loaded below its yield point. This kind of stretching is not permanent: the iron returns to its original length when the load is removed.[74]

Attention to the latter kind of stretching was at the heart of the theory of elasticity, which was the product of a research tradition extending back at least to 1678. In that year, Robert Hooke had pointed out in his paper "Of spring" that an iron wire stretches proportionally to the applied load—an observation that is now referred to as Hooke's Law. In 1776, Charles-Augustin Coulomb determined that every substance has its own elastic constant (which governs how much each substance will stretch under a given load). Thomas Young formulated the term *modulus of elasticity* for this constant in 1807.[75]

Nevertheless, in the early nineteenth century most engineers did not see much relevance in the theory of elasticity. Navier, in editing the second volume of Gauthey's *Traité de la Construction des Ponts* in 1813, had argued that determination of the modulus of elasticity for iron had no practical relevance. He wrote: "It is thus above all resistance to rupture that it is necessary to consider for iron in order to arrive at the limit for the weight with which it may be loaded: from which it follows that research on the numerical value of the moment [i.e. modulus] of elasticity is little more than an object of curiosity, and offers little of interest for applications."[76]

Between 1813 and 1823, with his growing involvement in the French scientific community and his direct involvement in mathematical research on the theory of elasticity, Navier's views on this subject changed profoundly. He came to see a fundamental connection between the study of elasticity and the study of strength of materials, and he came to view both as having direct relevance to engineering practice. He brought these two perspectives to bear on his research on suspension bridges.[77]

Navier applied these new perspectives in evaluating loading limits for iron cables to ensure that they would not be loaded beyond their elastic limit. More important in the present context, however, he applied these new perspectives to determine how the elasticity of iron would affect the equilibrium of suspension bridge cables. To solve this problem—to determine how much iron cables would stretch under various loads—Navier needed a value for the modulus of elasticity of iron. This value, which had seemed irrelevant to him in 1813, was provided by a series of some seventy experiments on the stretching of iron bars under small loads below the yield point. The experiments had been carried out some years earlier by A. Duleau, who was also an engineer in the Corps des Ponts et Chaussées.[78] The value Duleau found was 20,000 kilograms per square millimeter.[79] This value could not be determined from the experiments of Telford, Brunel, and other British engineers, because it required data relating stress (amount of loading) to strain (amount of stretching) *below* the yield point. The experiments of British engineers noted only the point at which significant permanent elongation occurred.

In his treatment of cable elasticity, Navier analyzed how much iron suspension bridge cables would stretch due to surplus loads that were either distributed uniformly along the horizontal or concentrated at the middle of the span. By applying Hooke's Law in conjunction with his set of ideal equilibrium equations and Duleau's value for the elastic constant of iron, Navier arrived at simple, approximate formulas for each case, which gave the cable's change in length (ΔL) and sag (ΔY) as functions of the surplus load. From the sag formulas he further derived proportionalities which indicated how ΔY depended on bridge size, assuming a constant sag/span ratio. For case 1 he found that ΔY was proportional to the length of the span. For case 2 he found that ΔY was independent of the length of the span, which meant that traffic loads would have relatively less effect on a large bridge than on a small one. Thus, Navier concluded, even when cable elasticity was taken into account, the hypothesis that long suspension spans were the most stable remained valid.[80]

Temperature Variations

Like the analysis of elasticity, Navier's investigation of the effect of temperature variations on the equilibrium of suspension bridge cables required empirical data about iron—in this case, about how it was affected by heat. Navier cited three experimental studies on the expansion of iron due to temperature increases, carried out, respectively, by Smeaton, by Laplace and Lavoisier, and by Dulong and Petit. The results of the studies varied slightly, and Navier took their average: he assumed that a temperature rise of 1°C would produce a 0.122 percent elongation in an iron rod. He did not carry out any experiments of his own to corroborate that value.[81]

As in the investigation of elasticity, the goal behind Navier's examination of temperature variations was to determine their influence on cable sag and hence on roadway deflection. Navier may have come across evidence of this problem in Britain. The issue was not discussed in the first set of hearings on the Menai Bridge, which Navier cited, but it was discussed in the hearings the following year (1819), and particularly in a set of written comments and observations submitted by Telford. Telford made a very rough estimate of how much a 700-foot iron cable would expand for a temperature increase of 90°F. According to the committee's report, Telford and Rennie estimated "a rise or fall to the extent of four or five inches."[82] Yet Telford and Rennie agreed that this "would not derange the bridge," and Telford suggested that the expansion and contraction effects could be minimized if the cables were insulated.[83] There must still have remained some concern, however; a decision was made in 1823 to increase the height of the Menai Bridge's towers by 2 feet, so that the center of the roadway would be higher and would not sag below the horizontal as a result of temperature changes. (After the bridge was completed, the roadway was found to shift 11 inches at mid-span between summer and winter.[84])

Navier analyzed the problem of temperature changes as it applied to the two basic types of tower-cable systems. Here we will focus only on the case where a rigid tower supported a continuous cable. Guided by a careful, qualitative assessment of the range of effects that a temperature change would produce, Navier derived a formula for ΔY in which three terms accounted, respectively, for the absolute change in length of the main cable, the change in height of the towers, and the effective change in length of the anchor cable. The third term comprised four subterms, which concerned the absolute change in length of anchor cables due to temperature changes, their change in length due to elasticity effects brought on by temperature-induced expansion or contraction, the change in tower height, and the change in sag of the anchor cables. As if this degree of complexity were not enough,

Navier had to compute certain terms of the overall formula differently depending on whether the temperature rose or fell, because the friction effects differed in the two cases. Nevertheless, each term of the equation was still based on a significant degree of idealization of the system. Indeed, Navier's entire analysis hypothesized an ideal world where the sun never rose or set—that is, where differential heating never occurred.[85] But of course bridges are affected by the sun's movement: different sections expand and contract during the course of a day, depending on whether they are in sun or shade. This very practical problem was not amenable to a generalized theoretical analysis, however, and Navier never raised it in his treatise; we will see, however, that differential heating became a problem when he attempted to build a suspension bridge.

Vibrations

The most sophisticated part of Navier's theory from a mathematical standpoint was his treatment of cable vibrations due to impact loading. In a suspension bridge, vibrations from impact loading first affect the roadway and from there are transmitted through the hangers to the cables. Navier witnessed such vibrations when he examined Samuel Brown's Union Bridge:

> When vehicles cross [the bridge], different types of effects are produced: the deck sags; because the presence of the vehicle alters the distribution of the load supported by the chains, these chains must take a new shape . . . which conforms to the new state of equilibrium being established. The result of this situation is that the contour of the deck is continually modified throughout the duration of the crossing. . . .
>
> Apart from these changes of contour . . . which must be considered as a static effect resulting from the flexibility of the chains and the deck, dynamic effects are produced, some of which result equally from the flexibility of the construction, and others which are due to the elasticity of the materials. The little jolts [on the roadway] stemming from the movement of vehicles and horses . . . are transmitted by the hangers to the chains, which, during the passage of vehicles, always display a very small horizontal swaying . . . such that these chains . . . are almost always in movement.[86]

Such vibrations were also described by Robert Stevenson in the article on suspension bridges cited by Navier. Referring to the Dryburgh Bridge, Stevenson called attention to "the sudden impulses, or jerking motion of the load."[87] British engineers did not attempt to analyze this phenomenon mathematically as did Navier, however.

Navier based his analysis on the assumption that a load, *P,* would fall directly onto a cable precisely at the middle of the span. In other words, roadway and hangers (and also towers and anchor cables) were abstracted out of the model. Navier justified this idealization on the grounds that his analysis would establish the theoretical upper limits for the effects of impact loading. A further advantage of this approach for Navier was that the problem became analogous to the vibrating string problem, a mathematical classic that had been investigated by a host of prestigious researchers, including Euler, d'Alembert, Lagrange, the Bernoullis, and others, up to and including Fourier.[88]

Navier considered two types of vibrations: transverse oscillations (such as those produced when a violin string is plucked) and longitudinal vibrations (i.e., elastic expansions and contractions of the cables). To solve the problem of transverse oscillations, he worked out two partial differential equations that gave the cable's displacement as a function of time and position (along the *x* axis). One equation held for all points along the cable; the other held strictly for the midpoint, which carried the load *P.* It was one thing to derive these equations and quite another to solve them. A general methodology for solving them had eluded eighteenth-century mathematicians; it was worked out only in Navier's generation, largely through the efforts of Fourier, whose technique Navier explicitly followed. This technique enabled Navier to determine the desired solution by computing the constants of its trigonometric-series expansion.[89] The resulting function occupied two full lines and comprised multiple sine and cosine expressions.

Owing to its mathematical complexity, the solution function Navier derived gave no immediate physical insight into the character of suspension bridge vibrations. Navier had to simplify and restructure the expression carefully in order to find out roughly how the frequency and the amplitude of the oscillations were related to bridge size. He discovered that both the frequency and the amplitude would decrease as the overall size of the bridge increased, with the sag/span ratio assumed to be constant. He found, in other words, that large bridges would be relatively less affected by vibrations than small ones.

Navier used analogous procedures to analyze longitudinal cable vibrations, except that here the partial differential equations were based on Hooke's Law. He also extended the analysis to consider hanger vibrations. He found that, for both cables and hangers, the frequency and amplitude of elastic vibrations would be less for large bridges than small ones (if not in absolute, then in relative terms), with the sag/span ration assumed to be constant. The implication, once again, was that suspension bridges would become inherently more stable as their size and weight increased.[90]

Wind-Induced Oscillations

The potential seriousness of wind-induced oscillations was brought to Navier's attention several times during his trip to Britain. For one thing, Brunel's bridges, which Navier examined, were specifically designed to be able to withstand hurricanes. They had extra sets of chain cables below each deck to hold it in place. (The inverted cables were like mirror images of the main cables.) Navier explained that these inverted chains were "intended solely to restrain the deck against the wind's action during storms to which these bridges will be exposed."[91] Another French builder, Marc Seguin, who visited Brunel the same year as Navier, made the point even more clearly. He explained that Brunel's decision to build one of the bridges in the form of two half-spans flanking a central pier was itself an arrangement intended to impede wind-induced oscillations:

This able engineer believes that this arrangement is above all useful to prevent the bridges from entering into vibration [during windstorms] and to oppose the *cumulation of motion* that follows. . . . Not content with this precaution, Mister Brunel also added inverse arcs, made of iron chains. . . . These arcs are fixed to the deck . . . which enables it to resist a force *acting upwards, tending to lift it*.[92] (my italics)

Navier reported that the deck of Samuel Brown's Trinity suspension pier was likewise secured against wind. When Navier visited the structure in 1821 it had no wind bracing, but he explained that it was added the following year:

It seems, however, according to the report of a person who visited the construction during the summer of 1822, that more violent gusts of wind during the preceding winter made it necessary to install the wind bracing described earlier, and even to hold the deck in place by lines attached to anchors placed at the bottom of the sea.[93]

Navier also learned during his trip about the collapse of the Dryburgh cable-stayed bridge in a windstorm in 1818. He reported the incident in his *Mémoire sur les Ponts Suspendus:*

. . . on January 15, about six months after the completion of the bridge, a very violent gale occurred, and the vibratory movement became so great, that the longest inclined stays were . . . ruptured, the platform carried away, and the structure entirely destroyed.[94]

Navier noted that when the structure was rebuilt it was provided with wind bracing in the form of chains fixed to the deck and anchored on either bank of the river.[95]

Despite the fact that Navier realized that wind-induced oscillations could build up enough to destroy a bridge, and that Brunel, Brown, and other builders were taking technological measures to keep suspension bridge decks from lifting and undulating during windstorms, Navier's treatment of this problem was much more cursory than his treatment of transverse and longitudinal vibrations (a mere four pages, as opposed to thirty for the section on vibrations). His analysis was based on assumptions that made the problem easy to treat mathematically but did not do justice even to the limited empirical data to which he had access.[96]

Navier's assumptions misrepresented his empirical data concerning wind effects. He hypothesized that wind would cause the cables to oscillate laterally like swings, and he therefore analyzed the oscillations according to the laws of pendular motion. He further hypothesized that the oscillations would originate in the cables, but that they would always be quickly damped out because of the deck's lateral rigidity. Observations reported by Stevenson, however, showed very clearly that a bridge deck could be lifted vertically by wind. (See figure 5.7, which dates from 1836, for a visualization of such a phenomenon and its destructive effect.) And suspension bridge cables have very little capacity to dampen vertical or torsional deck oscillations. For example, when discussing the collapse of the Dryburgh Bridge, Stevenson reported that the movements of the roadway were "such as would have pitched or thrown a person walking along it into the river."[97] And referring to the period of the rebuilding of the Dryburgh Bridge after it was destroyed in a storm, Stevenson reported:

A high wind having occurred before the side-rails were erected, one end of the platform was *lifted above the [normal] level of the roadway,* and the undulating motion produced on this occasion is described as resembling a wave of the sea; an effect which pervaded the whole extent of the bridge, and went off with a jerking motion at the farther end. But after the side-rails were attached, this *vertical motion* was checked, and is now found to be greatly reduced.[98] (my italics)

Navier recapitulated this description virtually word for word in his own treatise.[99] Finally, Navier hypothesized that a windstream would

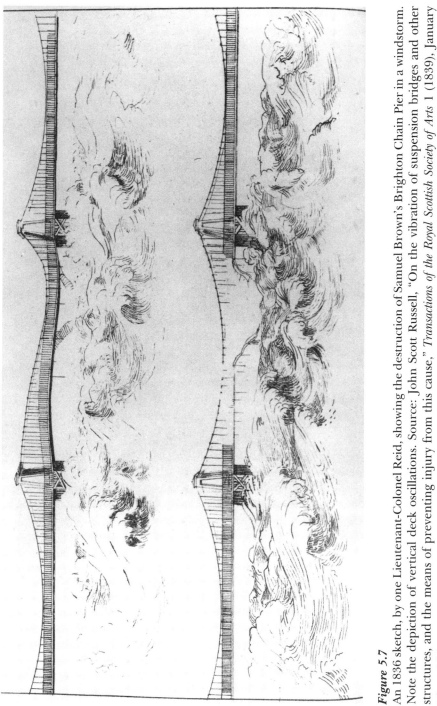

Figure 5.7
An 1836 sketch, by one Lieutenant-Colonel Reid, showing the destruction of Samuel Brown's Brighton Chain Pier in a windstorm. Note the depiction of vertical deck oscillations. Source: John Scott Russell, "On the vibration of suspension bridges and other structures, and the means of preventing injury from this cause," *Transactions of the Royal Scottish Society of Arts* 1 (1839), January 16, p. 304.

remain perfectly uniform and horizontal, although he knew this to be an overidealized assumption.

To say that Navier oversimplified and misrepresented the phenomenon of wind-induced oscillations is not to imply that he could have "solved" the problem if he had merely proceeded from another set of hypotheses. In fact Navier was at the tip of a very large iceberg, and more than a century later the problem was still inadequately understood. This lack of understanding led most spectacularly to the collapse of the 2,800-foot Tacoma Narrows suspension bridge, in Washington State, in 1940. The engineers who drew up the official report on this collapse the following year could only reach tentative conclusions as to the process by which oscillations were induced and built up in the structure: "The vertical oscillations of the Tacoma Narrows Bridge were probably induced by the turbulent character of the wind. Their amplitude may have been influenced by the aerodynamic characteristics of the suspended structure. . . . At the higher wind velocities torsional oscillations, when once induced, had the tendency to increase their amplitude."[100] Nine years after this report appeared, Freidrich Bleich, in collaboration with three other engineers, published a 400-page theory of wind-induced vibrations in suspension bridges, again as part of the investigation of the failure of the Tacoma Narrows Bridge. They likewise acknowledged a sense of uncertainty about their research: "One of the principle difficulties lies in the fact that the early research appears to lack orientation. In certain instances there is evidence that we are not quite sure just what we are searching for."[101] Further on, they were more specific: "Several mathematical or semi-empirical explanations of the interaction of wind stream and structure have been presented, but it is doubtful that any of them at this time can be considered adequate to explain and predict the behavior of a suspension bridge under all conditions of wind action. There is strong reason to believe, however, that much truth is contained in some of them."[102]

Despite the uncertainties expressed by these engineers writing in 1941 and 1950, they had at their disposal considerably more knowledge and information than was available to Navier. First and foremost, they had a visual and statistical record of the oscillations and the ultimate failure of the Tacoma Narrows Bridge. Motion pictures were taken of the bridge before and during its failure. Extensive data on the frequency, amplitude, and pattern of the oscillations, together with measurements of the wind's velocity and direction, had been collected daily over a period of several months up to and including the day the bridge failed.[103]

In addition, researchers in 1941 and 1950 had extensive data from tests done in wind tunnels (an instrument which was developed only in the second half of the nineteenth century). Tests on models of the Tacoma Narrows Bridge were carried out at the University of Washington and at the California Institute of Technology Aeronautical Laboratory, both before and after the collapse, and their importance for developing insight into the causes and mechanisms of the oscillations cannot be overestimated. Indeed, there was not a chapter of Bleich's book that did not refer to knowledge and data gathered from these tests. Among other things, the tests helped to convince researchers that vortex shedding had a major role in producing oscillations, and they provided data on the relationship of this phenomenon to deck design.

Finally, the engineers who analyzed the failure of the Tacoma Narrows Bridge had the advantage of being able to draw on many theoretical concepts and techniques that were developed after Navier's time. Of particular importance were ideas and knowledge drawn from aerodynamics, such as the concept of lift; the concepts and theories of eddy formation, vortex shedding, and of negative damping; and the mathematical theory of flutter (originally developed to explain the combined vertical and torsional vibrations that occur in airplane wings but applicable to the problem of suspension bridge oscillations). The relevance of aerodynamics for understanding the behavior of the Tacoma Narrows Bridge is shown by the fact that the three-member Board of Engineers appointed to report on the bridge failure included the renowned aerodynamicist Theodore von Kármán.[104]

Although we could not expect Navier to have given a complete mathematical theory of wind effects on suspension bridges, it is still rather surprising, in view of the detail and care with which he developed the rest of his theory, that he treated the wind problem so superficially—particularly in his mathematical modeling. For example, his analysis assumed that a cable would be given a *single* push by a gust of wind, and then he merely calculated theoretically how rapidly the ensuing oscillations would die out. In his theory Navier never entertained the possibility that oscillations could build up if the cables were given *multiple* pushes—that is, if *vis viva* (what Navier called *force vive*[105]) were continually added to the system.[106] Yet, as we have seen, Navier had empirical evidence of such a possibility: aside from the fact that

Navier's own description (in his treatise) of the collapse of the Dryburgh Bridge depicted a buildup of oscillations by wind, Brunel had consciously sought to prevent a "cumulation of motion" in his design, and Stevenson (cited at length by Navier) had specifically referred to the "effects of gusts of wind, *often and violently repeated, which destroy the equilibrium of the parts of a bridge of suspension*" (my italics).[107]

In short, Navier downplayed the significance of wind-induced oscillations and ignored the theoretical implications of existing empirical evidence and design practice. No doubt he did so in part because he felt unable to handle the problem mathematically:

Known research on the laws of impact [*lois du choc*] and fluid resistance do not offer the means to understand the action of wind on suspension bridges with the exactitude that one would desire. It is moreover evident that we cannot undertake to submit to exact consideration the tumultuous action of irregular gusts of wind, with variable speeds and directions. The accidents resulting from this action can only be understood and prevented on the basis of insights provided by observation and experience.[108]

This caveat is not altogether convincing, however. Navier could have done experiments to study the problem. There were, moreover, physical concepts and mathematical techniques which he could have attempted to apply to the wind problem, such as the concept of *vis viva* (*force vive*). His own research on cable vibrations was potentially applicable. Alternatively, he could simply have emphasized the need for adequate wind bracing and the need to make suspension bridge decks stiff enough to impede deflections and vertical oscillations. But Navier did none of these things. In the section on wind effects, Navier even counseled against the need for deck stiffening, ostensibly guided by the prior results of his theory:

The roadways may be regarded as perfectly flexible in the vertical sense, unless special precautions have been taken to make them rigid; [these are] useful precautions in light constructions destined for pedestrians, but become superfluous in more important bridges.[109]

When the superficiality of Navier's wind analysis is contrasted with the depth, sophistication, and attention to detail of his analyses on cable vibrations and temperature variations, it becomes evident that he did not seriously apply himself to the wind problem. He did not devote to this problem anything like the attention he was devoting at that time to his research on the theory of elasticity and the theory of

fluid dynamics.[110] Rather, Navier allowed himself in this instance to
follow the route he had earlier criticized: the route of "not paying
enough attention to the differences which exist between the phenomena and the hypotheses submitted to analysis."[111]

Conclusions

Having reviewed the specifics of Navier's theory, we can now draw
some more detailed conclusions about the ways in which his research
was socially shaped (beyond the mere fact that he chose to develop a
mathematical theory at all). First, Navier's theory was socially shaped
in the sense that, to create it, he employed a set of mathematical tools
and techniques which were developed, taught, used, or promoted
within his own institutional environment. These included basic mathematical tools, such as calculus, which in principle were known to every
graduate of the Ecole Polytechnique (but which were not known to
Finley, Telford, and Brown, for example). They also included more
specialized mathematical techniques and equations, such as the equation for belt friction and the general equation for the length of a curve
and the equation for the movement of a pendulum. Finally, they
included advanced and specialized mathematical knowledge and techniques that were not widely known or used at that time outside of a
small research community, of which Navier was a part. Prominent
among the latter were the theory of elasticity and Fourier's technique
for solving partial differential equations by means of trigonometric
series expansions.

It might be argued that the mere use of a particular set of mathematical tools and techniques does not reflect any social shaping. But
it does reflect such shaping in the sense that people most readily use
the tools that are available, accepted, and relevant within their own
environment. For example, in principle Finley could have read and
used Lagrange's *Mécanique Analytique* in his research on suspension
bridges, but it would have been unusual had he done so: that tool was
not readily available to him, and it had no accepted place or particular
relevance in early-nineteenth-century Fayette County. Of course, many
creative individuals master tools and techniques that are foreign to
their environments; yet it is extremely rare for an individual to reject
all the tools provided within his or her own environment. Navier did
adopt a new technique when he helped to introduce suspension bridge
construction into the Corps des Ponts et Chaussées, yet he attempted

to make this new and foreign technique more acceptable and relevant by evaluating it, by reinterpreting it, and (as will become evident in the next chapter) by restructuring it on the basis of mathematical tools and techniques that were already accepted and respected within his community.

Navier's use of a particular set of mathematical tools reflects social shaping at another level as well. The Continental system of calculus and the style of Navier's mathematical analysis were very different from those used by Gilbert and other British researchers. As Joan Richards has observed: "The English clung to Newtonian fluxional notation, which was geometrical and also more constraining than the algebraical Leibnizian *dy/dx* symbols used on the Continent. Throughout the eighteenth century, English and Continental calculus developed along these different lines and in virtual isolation from each other."[112] Richards has argued convincingly that these differences, in the case of France and England, were linked to the very different institutional structures in which mathematics was promoted and diffused in the two countries—structures that reflected very different philosophies concerning the nature of mathematics and its role in society.

Navier's theory was also socially shaped in the sense that it embodied many decisions about what to analyze and what to ignore that reflected goals and priorities stemming from or rooted in his environment. Why did Navier devote considerable attention to the problem of cable vibrations but completely ignore the question of the behavior of the deck, which was at least as important for builders? Why did he choose to analyze in detail the problem of cable slippage over the towers (taking into account friction, elasticity, and temperature variations) but, for the most part, to ignore the issue of cable design?

The evidence provided in the preceding sections has shown that Navier's decisions about what to analyze and in what depth were not simply dictated by "the technology," or "the evidence," or even just by the needs of builders. This is not to imply that these things were irrelevant. The way Navier mathematized his observations of vehicles crossing the Union Bridge and the way in which he classified different tower-cable systems show that his decisions *were* made partly on the basis of empirical evidence and technological precedent. But they were also made partly on the basis of who he wanted to impress, and what mathematical issues were currently of concern to the Parisian scientists with whom he was increasingly interacting.

Navier's decision to devote much attention to cable vibrations due to impact loading from traffic but only cursory attention to wind-induced oscillations provides an example. The difficulty of the wind problem cannot fully account for this decision, nor can it be accounted for by the empirical evidence available to Navier. As we have seen, Navier had clear evidence of the potential seriousness of wind-induced oscillations: direct evidence from the Dryburgh Bridge and indirect evidence from Brunel's bridges and from Brown's Trinity chain pier. (Brunel's two bridges and the Trinity chain pier had provisions to keep the roadways in place during windstorms.) Navier's empirical evidence concerning cable vibrations, in contrast, did not suggest such danger. Referring to the vibrations of the Union Bridge due to traffic, Navier commented that they "in any event do not appear to jeopardize the solidity of the construction."[113]

Navier's decision to devote nearly thirty pages of difficult mathematics to the problem of cable vibrations but only four pages to the problem of wind becomes more understandable when we recall certain facts about his growing stature within the French scientific community. As we saw in chapter 4, Navier, who had been elected to the Société Philomathique in 1819, was hoping in the early 1820s to achieve the next step: election to the Académie des Sciences. (Recall the 1823 letter to Lacroix in which Navier stated his belief that the candidate "who has given great service to mathematics" should be the one elected to the Academy.) By 1820–21 Navier had begun to do important research on the mathematical theory of elasticity, which was one of the main interests of Parisian mathematicians. It should also be noted that Fourier became *secrétaire perpétuel* of the Academy of Sciences in 1822, which was the most powerful position in that organization.[114] And Fourier's research on heat, published in *Mémoires de l'Académie des Sciences* in 1819 and 1820 and then as a book in 1822, introduced his new technique of using trigonometric series to solve partial differential equations. What better way to make an impact on Fourier and on mathematicians concerned with the theory of elasticity than to apply their theories and mathematical techniques to solve a new and difficult problem? In 1823 Navier even extracted a key portion of his cable-vibration analysis and published it separately in the *Bulletin de la Société Philomathique,* where it would be more visible to the mathematics community.

Navier's analysis of cable vibrations was especially praised by his scientific colleagues,[115] but it was not used much by bridge builders.

The renowned suspension bridge designer David Steinman, in his 1922 book *A Practical Treatise on Suspension Bridge,* ignored the theory of cable vibrations. Nevertheless, Steinman did present other elements of Navier's theory. In a chapter on unstiffened suspension bridges, Steinman presented the same set of basic equilibrium equations found in Navier's work and then went on to show, following Navier, how the equilibrium conditions would be affected by temperature variations, elasticity, surplus loads (distributed and concentrated), and by movements of the towers or cables. But Steinman did not include any analysis of cable vibrations.[116]

The idea that at least parts of Navier's theory were intended as much for mathematicians as for bridge builders and practicing engineers gains further credence when we reflect more carefully on one of the hypotheses underlying his analyses of cable vibrations and deflections due to concentrated loads: the hypothesis of a completely yielding deck. Mathematically, Navier treated a bridge deck as a straight line with uniformly distributed mass and no stiffness. He chose *not* to analyze it as a technological variable that could display different degrees of stiffness.[117] (In the same manner, a bridge tower can embody different degrees of resistance to toppling, and Navier did incorporate this technological possibility into his mathematical model.)

The hypothesis of a deck without stiffness was necessary for the analysis of cable vibrations, but it did not serve the needs of builders. As we have seen, Navier justified his hypothesis partly on the grounds that it would reveal the maximum possible degree of vibration and deflection that could be produced by a given load. From the point of view of builders this justification could be seen as having little validity because it did not provide knowledge about the relative effects of different technical options. Designers want to know in advance roughly how each technical possibility will affect a system's behavior. Conversely, they want to know what technical solution can best (or most economically) achieve a desired effect. In this sense, Navier's dictum that technological theorists should only attempt to find the maximum limits of behavior of a system or structure reflected the mindset of a mathematician more than that of a builder.

Navier's idealization of the deck as having weight but no stiffness led him to ignore the question of how to analyze the equilibrium of a suspension bridge with stiffening, which became one of the central issues in suspension bridge design and theory. The technological direction taken by builders was increasingly to design suspension bridges with *stiffened* decks, so as to prevent deflection. The need for deck stiffness, already indicated by Finley and by Stevenson, was picked up by other builders. For example, Marc Seguin, a manufacturer and inventor in Privas who became interested in suspension bridges, wrote in 1824 that "the most important characteristic [*la première qualité*] of suspension bridge decks is to have the greatest rigidity possible."[118] And the British engineer Charles Stewart Drewry noted in a treatise on suspension bridges published in 1832 that "the object . . . in building a suspension bridge is, either to make it so light that its own vibration shall not hurt it; or if, as in nine cases out of ten, that cannot be done, then to make it so heavy and stiff, in proportion to the load it will have to carry, that the load shall not cause it to vibrate much."[119] Recognition of the need to eliminate vibrations and deflections led by the middle of the nineteenth century to suspension bridge decks being constructed as stiffening trusses.[120] Concomitantly, the main direction of theoretical development was precisely to analyze the equilibrium of stiffened structures, to which Navier's theory did not apply.

Two major lines of theoretical development occurred. First, beginning with work by W. J. M. Rankine in 1858, a theory was elaborated (referred to as the Elastic Theory) which assumed that a cable and a deck would suffer no deformation under concentrated loads. Second, beginning in the later nineteenth century and continuing at least up to the middle of the twentieth century, a more exact version of the Elastic Theory was developed (referred to as the Deflection Theory), which took into account the small deformations that would in reality occur.[121]

The results of the earlier Elastic Theory were presented in a comprehensive and definitive manner by Steinman in his 1922 treatise. This analysis occupied about 75 percent of his book. In contrast, only one short chapter was devoted to unstiffened bridges, the analysis of which had been the sole focus of Navier's treatise.[122] Steinman regarded unstiffened suspension bridges as technologically unimportant; his chapter on the theory of these structures (following Navier's analytical model) opened with the statement that "the unstiffened suspension bridge is not used for important structures."[123] Further on, Steinman counseled that "unstiffened suspension bridges should not be used except for footbridges."[124]

This gap between the thinking of Steinman and Navier cannot be ascribed simply to differing views of what constituted an "important" structure. Navier used his theory to derive a formula for the maximum permissible span of a suspension bridge. He calculated, for a sag/span ratio of 1/15, a maximum span of approximately 3,000 feet. The spans that had actually been constructed by 1922, however, did not reach even 2,000 feet. The gap between Steinman and Navier was due, rather, to the conclusions Navier drew from his theory. In Navier's view, the theory showed that deck stiffening became *less* necessary as the size of the bridge increased, and indeed that stiffening was necessary *only* for footbridges.[125]

Navier wanted his theory to be an essential and a creative guide to engineering practice. As this brief survey of the stiffening truss reveals, the theory did not achieve either of these goals to the extent Navier wished or thought possible. It did not point at all to the line of technological development that led to the stiffening truss and its analysis. On the contrary, Navier counseled against the need for deck stiffening on all but light pedestrian bridges.

Many builders of Navier's era and later felt his treatise to be needlessly long, complex, and abstract. It was never translated into English. It did not supersede more empirical traditions of research and design of suspension bridges (which continued to flourish, as is shown by the example of the stiffening truss). Another example of the continuing importance of empirical design traditions concerns the use of radiating cable stays in conjunction with parabolic cables, an arrangement intended to help limit deck vibrations and deflections. On the basis of theoretical considerations, Navier strongly counseled against the use of radiating stays in conjunction with parabolic cables. According to him, theory showed that parabolic cables and inclined stays were incompatible in their structural behavior, and that the use of stays in conjunction with parabolic cables would mean that the loads could not be borne properly by the cables but would be transferred to the stays.[126] Nevertheless, builders continued to use these two systems together—including, most notably, John Roebling, for his Niagara and Brooklyn bridges.

The foregoing observations are not intended as a criticism of Navier's theory; they are meant, rather, to emphasize the utopian nature of his vision of what theory could offer. Navier's work did represent a significant contribution to knowledge. Perhaps most notable, it took into account a much broader range of variables than previous theories had (and here I am referring to technological theories in general, not merely those concerning bridges). Yet the "analytical ideal" to which Navier was so committed, and which his theory was intended to realize, was a vision of how the world should be rather than a reflection of how it is.

Thus, a further important way that Navier's theory was socially shaped was in its embodiment and acceptance of this analytical ideal. As we have seen, the level of Navier's commitment to this ideal was heightened by his dual career path, through which he came increasingly to take up ideas, problems, and tools from the French mathematical community while still working professionally as an engineer. Navier's strong commitment to the analytical ideal shaped his response to the suspension bridge. One senses, in reading Navier's theory, that he wanted it to be a timeless statement of immutable laws. One senses that, in his view, including a stiffened deck in the analysis would merely have served to obscure the "true laws" of suspension bridge behavior: "The characteristic property of these bridges is that they constitute a flexible system in which the equilibrium is stable: that is to say that they take up . . . all the changes of shape produced by whatever cause, and following these changes, the structure, left to itself, spontaneously returns to its original shape."[127] Many builders, however, saw the suspension bridge differently. Finley, Telford, Brown, Seguin, Roebling, and other builders saw it first and foremost as a technological system that permitted greater distances to be spanned between piers than other bridging systems, or that allowed a given distance to be spanned at lower cost. Although they recognized the tendency of these structures to deflect and vibrate, they nevertheless regarded the question of flexibility more as a problem to be controlled technologically than as an inherent, natural characteristic that had to be given free reign. Seguin put it this way: "One of the greatest drawbacks of suspension bridges being the wobbling that is caused by any moving body of any significant mass, all possible means must be employed to give them rigidity."[128]

Navier's acceptance of the analytical ideal may also account in part for his reticence with regard to the wind problem, because the two did not fit together easily. The wind problem involved messy and irregular phenomena and it did not correspond neatly with any existing mathematical-physical framework. The wind problem was also linked with a sense of danger—of powerful, unruly forces that thwart human control. And what could be more alien to the analytical ideal?

Finally, we must recognize that the analytical ideal, as manifested within the French institutional environment, was an ideology with social roots and functions. This ideology was associated with the learned, rationalist, and socially elitist traditions of the French engineering corps. And, as incorporated into the structure and goals of the Ecole Polytechnique, it was used to justify social privileges—in particular, the privilege by which *polytechniciens* alone had access to the engineering corps and to the security, prestige, and power which these positions offered. It would not be accurate to say that Navier cynically viewed his theory as a way to preserve power and privilege. But he did seek, through his theory, to preserve the corps' honor and place within French society by demonstrating the worthiness of an ideal he felt to be central to the corps' mission. Referring to deficiencies in earlier attempts at technological theory, Navier commented: "Such errors gave rise to that common allegation that theory does not accord at all with practice; an opinion so false, and which can find no place in a corps as enlightened as that of the engineers of Ponts et Chaussées."[129] By ostensibly proving that theory did accord with practice, Navier hoped to show that the corps' shepherding of this tradition was indeed essential to the progress of French engineering.

6

The Pont des Invalides

Although Navier's research on suspension bridges reflected his growing ties to the French mathematical community, he did not abandon his identity as a practicing engineer and builder. Not content simply to theorize, he wanted also to design and build suspension bridges, and this interest was embodied in two designs which he presented and analyzed in his treatise. One design was for a suspension aqueduct/canal with a span of 320 feet; the other was for a monumental suspension bridge, with a span of 560 feet, to be built in Paris over the Seine.

From a technological perspective, the canal was the more innovative of the two projects. It represented an entirely new application of the suspension principle, and it involved new technological arrangements. It was also a utilitarian project that manifested a spare, "technological"[1] aesthetic. Nevertheless, the design of this structure reflected Navier's idealism. A neat application of theory, it epitomized the principle that increasing the weight of a suspension bridge would increase its stability. Indeed, there was probably no way to achieve greater weight in both dead and live loads than to replace a suspension bridge roadway by a large cast-iron canal filled with water. The structure was 18 feet wide and $6\frac{1}{2}$ feet deep, and its weight per linear foot when filled was 4.63 tons.[2] This was around 16.5 times the dead weight per linear foot of Samuel Brown's Union Bridge,[3] and 5.3 times that proposed by Telford for the Menai Bridge.[4] Another way in which the canal was a neat application of theory was that it did away with the problem of cable deflections produced by concentrated loads. The forces due to these loads—caused by boats passing through the suspended canal—would effectively be distributed by the water over the canal's entire span.[5] Moreover, each dimension of the structure was calculated theoretically, and many elements were designed so as to accord with theoretical principles. Navier explained: "One sees that every part of these

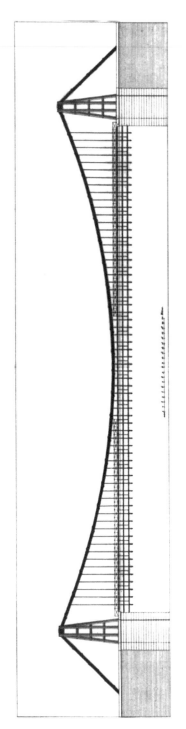

Figure 6.1
A suspension aqueduct designed and drawn by Navier. Source: Navier, *Rapport à M. Becquey et mémoire sur les ponts suspendus*, second edition (Paris: Carilian-Goeury, 1830). Courtesy of Special Collections Department, University of Virginia Library.

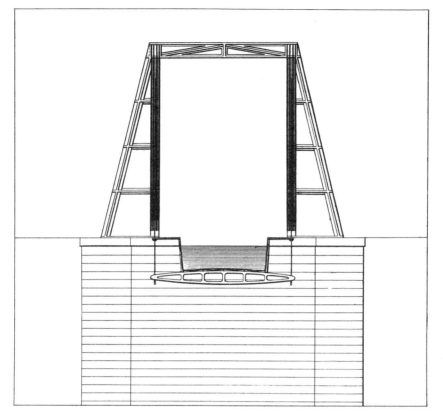

Figure 6.2
Cross-section of a suspension aqueduct designed and drawn by Navier, show-ing cast-iron towers. Source: Navier, *Rapport à M. Becquey et mémoire sur les ponts suspendus,* second edition (Paris: Carilian-Goeury, 1830). Courtesy of Special Collections Department, University of Virginia Library.

structures, precisely because of the simplicity of the design, is propor-tioned according to geometric laws in a way that is exempt from arbitrariness."[6] For example, the transverse cast-iron beams which were suspended from the main cables and which supported the canal were shaped as elongated ovals. They were "solids of equal resistance," designed so that their resistance to bending was just as great in the middle as at the ends. The idealism of the canal design can be seen not only in its theoretically derived structural elements, but also in the fact that Navier was so confident in the correctness of his calculations that he adopted only minimal safety factors (a few percentage points) beyond the stipulation that the iron not be exposed to stresses beyond 14 kilograms per square millimeter.[7]

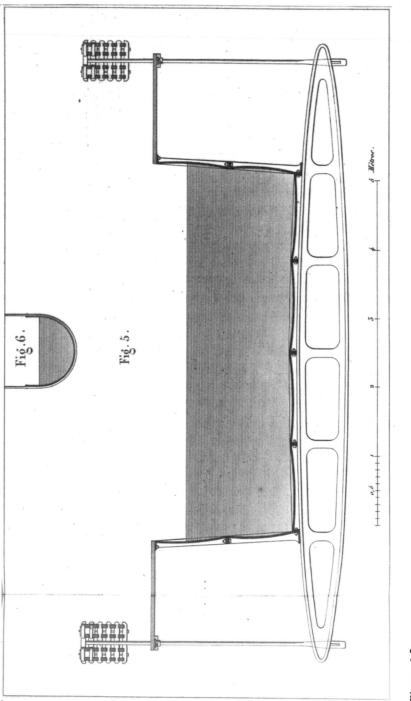

Figure 6.3
Suspension aqueduct designed and drawn by Navier. Cross-section showing detail of aqueduct design. Source: Navier, *Rapport à M. Becquey et mémoire sur les ponts suspendus*, second edition (Paris: Carilian-Goeury, 1830). Courtesy of Special Collections Department, University of Virginia Library.

The canal design was no more than an idea on paper, however. It was not meant for any particular site, and Navier never tried to have such a canal built. His heart lay with the monumental design rather than the utilitarian one. His sympathies are revealed by the fact that the monumental bridge design was accorded more than 26 pages in the treatise, and was described first; the canal design was accorded only a little over eight pages. More important, Navier immediately requested to build the monumental bridge—he presented the proposal to his superiors months before his treatise on suspension bridges was published.[8]

A Monumental Design

Navier's wish to build a monumental suspension bridge, and the ideas that guided his design, reveal goals and priorities very different from those that guided James Finley. Whereas Finley's plan was intended for mass application with little reliance on the skills of professional engineers, Navier's bridge was deliberately intended to serve as a monument to the glory of France and the Corps des Ponts et Chaussées. Whereas Finley patented his design system, the very concept of a patent was inimical to Navier's aims, since he wanted to build a unique design that would not be replicated. (Navier did not take out a patent in his entire career.)

Navier's attitude was shaped by the tradition of monumental architecture with which the Corps des Ponts et Chaussées was closely associated and which students encountered at both the Ecole Polytechnique and the Ecole des Ponts et Chaussées. A major public work, such as a bridge in the nation's capital, was expected to be a unique work of art that conveyed a sense of majesty and harmonized architecturally and decoratively with its surroundings. This attitude led many Corps engineers to manifest a certain disdain for what they regarded as the more petty objectives and the limited vision of entrepreneurs, who tended to seek out economical designs rather than majestic ones. As the director of the Corps put it in 1831: "When it is a question of constructing a bridge in the heart of the capital and in the vicinity of other magnificent monuments, it is indispensable that the new construction be in harmony with its environment. There are special considerations of taste and of decoration which would make it difficult to leave the work to the entrepreneur, who, as we all know, always seeks that combination which costs him the least."[9]

As a backdrop for his monumental suspension bridge, Navier chose a site which he regarded as "one of the most beautiful which exist in the world."[10] The bridge was to link the "magnificent esplanade" of the Hôtel des Invalides (an edifice built during the reign of Louis XIV to house 7,000 invalid soldiers) with the "lovely promenades" of the Champs-Elysées, at the site where the Pont Alexandre III now stands.[11] Navier's bridge was thus to provide a link between two of the city's most elegant districts.[12] Contemporaries argued that it would have been better, from a practical standpoint, to choose a site closer to the center of the city, around the area known as the Marais. A bridge there would have considerably shortened the route to and from work for some 4,000 people.[13] But this bridge was not intended for the practical needs of workers; it was intended for the "enjoyment of [those] who know how to appreciate works of art."[14] Indeed, Navier explicitly disregarded any criterion of utility: "There exists no urgent necessity to construct a bridge to the Champs-Elysées; there is no obligation to build a suspension bridge in Paris. But if it is desired that one be built, let it be made into a monument; let the character of grandeur be given to this work that the style of construction admits of; let its disposition be calculated with the idea of forming an edifice approved by artists, agreeable to the public, honorable to the administration."[15]

The Pont des Invalides was to consist of a single 170-meter span, giving the idea of "a difficulty overcome, of a great endeavor due to the progress of the arts."[16] A single large span would also add, in the words of François Eustache (*ingénieur en chef* for the Department of the Seine), "interest and movement to all the magnificence of this part of the capital." The towers on each bank were to consist of individual columns with an Egyptian palm motif, linked together at the top by a hollow, rectangular cast-iron beam (figure 6.6). Navier avoided the use of more massive, triumphal arches, so as not to block the view of the Hôtel des Invalides. Each anchorage, to be located 32 meters behind its respective column, would appear as a rectangular stone pedestal supporting a reclining lion. The decorative style of the bridge was meant to complement that of the Invalides.[17]

One aim of Navier's plan was to demonstrate that iron structures could be made compatible with the goals of monumental architecture: "A construction in iron, if it achieves grandeur and simplicity of form, can, just as much as an edifice in stone, deserve to be called monumental."[18] Two fundamental principles of beauty in the arts, Navier explained, were "the arrangement of objects by *masses*" and "*accord*"

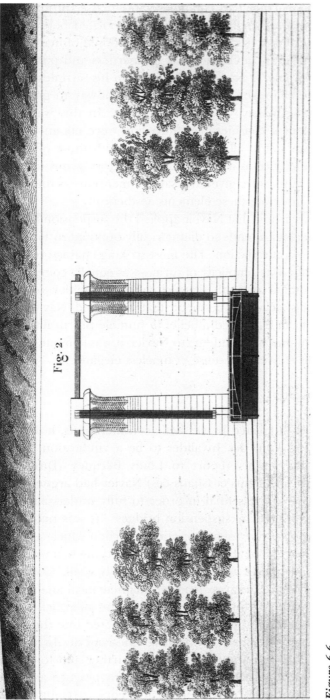

Figure 6.6
A cross-section of the Invalides Bridge showing the tower design. Source: Navier, *Rapport à M. Becquey et mémoire sur les ponts suspendus*, second edition (Paris: Carilian-Goeury, 1830).

between masses. A structure that combined large masses of masonry with a "great number of small, isolated pieces" violated both rules. Indeed, Navier asserted, a suspension bridge composed of many slim iron chains could take on "a character of meagerness and paltriness that would be intolerable."[19] Navier avoided this in his own design by interconnecting the chains on each side of the roadway to give the appearance of single, massive rectangular cables. "In this way," he concluded, "masses of a considerable dimension were obtained, the volume of which was not at all disproportionate with those of other parts, and with the edifice as a whole." The cast-iron beams uniting the towers on each bank were given the same dimensions as the cable arrays, in order to integrate these elements aesthetically.

It may seem rather ironic that Navier applied the suspension bridge to aesthetic and ideological ends so diametrically opposed to Finley's, but this involved no contradiction. The most striking characteristic of the suspension bridge—its simplicity of form—was equally compatible with a pragmatic, utilitarian aesthetic, such as guided Finley, or with the kind of formalist, classical aesthetic that motivated Navier. To Finley, the simple form expressed a belief in humble practicality and a disdain for presumptuous grandeur; to Navier, it was comparable to a Platonic ideal and conveyed a sense of timeless elegance.

A Vindication of Theory

Navier was not only motivated by aesthetic considerations, however; he also intended the Pont des Invalides to be a vindication of his theoretical approach. In his report to Louis Becquey (Directeur Général of the Corps des Ponts et Chaussées) Navier had argued that mathematical analysis was essential in order to fully understand the practical potentials offered by suspension bridges: "It was necessary . . . to make a thorough study of a type of construction which seemed to offer great advantages and about which almost nothing had yet been learned through time and experience; but this study would not have been possible without the progress made in mathematical analysis in recent times, and without the institutions by means of which those charged with the direction of public works are initiated into the most advanced ideas of mathematics."[20] By means of the Pont des Invalides, Navier sought to carry this argument one step further and to give a material proof of the applicability of his theory to practice and, by extension, of the relevance of mathematical theory to engineering practice generally. Navier opposed empirical approaches to structural design because in his estimation they led to overbuilding. He believed that only on the basis of theory could a precise correlation between the design and the function of structural elements be achieved.

Accordingly, in Navier's plan for the Pont des Invalides, every possible element—cables, hangers, columns, floor beams, etc.—was designed and proportioned on the basis of theoretical analysis. For example, in his theory Navier had developed equations to determine the forces acting on the towers of a suspension bridge, tending either to crush or overturn them. He had shown that the overturning forces depended on the amount of friction between cables and towers, and he had applied the belt-friction equation to this problem. This equation assumed the top of a tower to be an arc of a circle, and consequently Navier actually designed the tower tops of the Pont des Invalides that way, so that the equation would remain valid in practice. Navier applied the equation to determine the upper limit of the overturning force that could develop in the Pont des Invalides, and through this calculation he found the minimum dimensions that would ensure stability of the columns. He also determined the vertical crushing forces that would act on the columns, and to ensure that these could be withstood he designed the columns with an internal network of iron reinforcing bars.[21]

Navier also calculated the dimensions of the iron chains theoretically, using 14 kilograms per square millimeter as a stress limit for the cables so as to ensure that they would not be stressed beyond the yield point of wrought iron. James Finley had figured the maximum tension in the cables by multiplying the projected deck weight by 5 (assuming a 1/7 sag/span ratio). Navier's method was more intricate. First, he calculated the combined weight of the hangers and the deck, including the side rails. To this he added a value for the maximum live load the bridge would be required to bear, based on the assumption of the deck's being entirely covered with people standing shoulder to shoulder like "troops marshalled for battle."[22] (He had calculated that this weight would be greater than if the bridge were entirely covered with cattle, cavalry, or the most heavily loaded wagons.) Then, using his formula for cable tension, he calculated the maximum tension that the total load would produce in the cables, and from this he determined the total cross-sectional area of iron needed to resist this tension.

Navier's method of estimating live loads shows, once again, his habit of seeking the theoretical limit for the value of a given parameter. It presents a significant contrast with more pragmatic builders, such as Telford. In the hearings on the Menai Bridge, one of the questions posed to Telford concerned the weight that would result if a drove of cattle were to cover the bridge "as full as might be." Telford's answer showed that he thought the question misguided, because cowherds never allowed so many cattle to get so close together: "The weight, as far as I can calculate, of covering it with oxen will be about three hundred tons, it depends on the weight of the cattle; but in the usual way of driving cattle, there is never that quantity together, that is, supposing the whole bridge covered from end to end and no void space left, which is a very unusual thing. In driving cattle, nobody ever thinks of driving 200 head all in a heap."[23]

Navier used theory not only to proportion various structural elements of the Pont des Invalides but also to guide aspects of the overall design. In particular, he followed precepts derived from his analysis in order to work out a design to maximize the structure's stability. To begin with, the fact that the bridge was composed of a single, long span reflected not only aesthetic considerations but also Navier's finding that such spans were less prone to vibration and deformation. Second, Navier chose a small sag/span ratio of 1/15, which theory had indicated would contribute to stability. Third, the decision to interconnect the chains on each side of the roadway had advantages in terms of stability as well as aesthetics: this arrangement made the cables less flexible and hence less prone to vibration or deformation. Finally, guided by the finding that heavy bridges were relatively more stable than light ones, Navier designed a bridge whose dead weight per linear foot exceeded that of other suspension bridges. He proposed to use transverse beams of cast iron, iron side rails, and oak joists and flooring (which are heavier than pine), resulting in a dead weight of roughly 1.3 ton per linear foot. The dead weight of Telford's initial design for the Menai Bridge was 0.87 ton per linear foot,[24] Finley's design system applied to a 300-foot span 30 feet in width provided for a dead weight of roughly 0.47 ton per linear foot,[25] and Brown's Union Bridge had a dead weight of only 0.28 ton per linear foot.[26]

Significantly, however, Navier did not attempt to make a very rigid deck (i.e., one with substantial vertical stiffness). Following his view that vertical stiffness was unnecessary for longer spans, he proposed a design that, although more rigid than Brown's Union Bridge, was less rigid than the deck design proposed by Finley. Finley had strongly emphasized the need for vertical stiffness, and this design principle was embodied, as we have seen, in a deck composed of three layers: a bottom layer of transverse beams, a middle layer of longitudinal joists running the entire length of the bridge, and a top layer of planking. Finley had insisted that the longitudinal joists needed to be at least 12 inches deep, and that four or more such joists were needed, depending on the width of the bridge. His narrowest bridge, which, following his principle, needed four joists, had a width of $12\frac{1}{2}$ feet.[27] The Lehigh Gap chain bridge, which was not built by Finley but which largely followed his plan, included seven longitudinal joists for a roadway width of 16 feet.[28]

Since Finley never designed a suspension span beyond 250 feet, his design provided a minimum vertical stiffness ratio (the ratio between depth of stiffening and length of span)[29] of 1/250. In practice, Finley used the guard rails at the sides of the roadway as a further means of providing vertical stiffness; they were deliberately designed as stiffening trusses. For a 200-foot span pictured in his *Port Folio* article, these heavy truss parapets are shown to be 5 feet deep. If this depth is added to the depth of the longitudinal joists, the vertical stiffness for the model bridge pictured in his article becomes 1/33.[30]

Samuel Brown's Union Bridge stands out as the extreme of lack of vertical stiffness. The roadway planking was laid down longitudinally, directly upon the transverse beams suspended from the hangers. The only vertical stiffening in the deck, which was 18 feet wide, consisted of two longitudinal iron beams, 3 inches deep, which ran along each side of the roadway. The parapets, or guard rails, added little or no additional rigidity to the deck, since they were not designed as stiffening trusses.[31] The deck of the Union Bridge provided a vertical stiffness ratio of only about 1/1,500.

Navier's design fell between these two poles. Unlike Brown, Navier did employ a layer of longitudinal joists between the transverse beams and the roadway planking. But these wooden joists were only $7\frac{1}{2}$ inches deep, for a roadway length of 500 feet, giving a vertical stiffness ratio of 1/800. There were seven of these joists for a roadway width of approximately 31 feet. The parapets at the sides of the bridge, although not as deep as Finley's, also provided some stiffness. They were made of iron, with diagonal bracing, and were about 3 feet tall. If the parapets are included in the stiffness ratio of Navier's bridge, the value becomes 1/125. However, it should be noted that the considerable width of Navier's bridge limited the effectiveness of the parapets in adding vertical stiffness.[32]

Although Navier's deck design did not provide much rigidity, he did attempt to limit deflections and movements of the cables by means of a small sag/span ratio as well as by means of his tower and anchorage designs. First, the cables were supported at the towers upon fixed saddles. Navier recognized the advantage of a mobile saddle system to accommodate variations in cable tension. Yet his theory showed that a more rigid arrangement, with the force of friction harnessed to resist movement, would provide greater stability, and he believed that such an arrangement would prove more durable too.[33] Second, the anchorages were of a new design invented by Navier and described in his treatise on suspension bridges. The most common practice had been to extend each anchor cable into the ground without altering its direction, but Navier viewed this method as wasteful. The cable could be assumed to be secure only if the weight of earth above the anchoring point exceeded the cable's tension. Yet the cable was directed slantwise rather than vertically into the earth, so it had to extend quite far before sufficient depth was reached. What Navier proposed was to make the cable change direction at or just below ground level and then descend vertically to the required depth (see figure 6.4). The cable was to be supported at the point of curvature by a half-arch or buttress built underground. This arrangement saved space because the anchorages could be located nearer to the towers, an important consideration for the Pont des Invalides.

In principle, Navier's anchorage design had the added advantage of making the bridge less susceptible to deflection, because it effectively decreased the length of the anchor cables. Navier had shown theoretically that, the shorter the cables were, the less movement would be produced by traffic loads, temperature variations, and the like.[34]

Navier also used his theory to predict how the Pont des Invalides would behave when built. He made extensive calculations to determine how much the roadway would deflect as a result of the cables' elasticity. He found that the dead weight of the bridge would cause the cables initially to stretch by 7.64 centimeters, which in turn would bring about a sag increase of 20.2 centimeters. The addition of a uniformly distributed live load—the maximum permissible—would, for its part, cause the sag of the cables to increase by 13.8 or 14.3 centimeters, depending on whether or not it was assumed that the towers would deflect. Navier's calculations took into account not only the elongation of the main cables but also that of the anchor cables. He concluded that these variations would not be great enough to cause any damage to the towers.[35]

With respect to temperature changes, Navier found that the maximum probable temperature variation (25°C) would cause the cables to slide at the tower tops by no more than 8 or 9 mm and would produce a variation in cable sag of 18 or 19 centimeters.[36]

Next, Navier determined how much the cables of the Pont des Invalides would sag under traffic loads. He calculated that the largest probable load—19,600 kilograms, equal to the weight of two large, heavily loaded wagons, and assumed to be concentrated at a single point at the middle of the span—would alter the cables' sag by 18–20 centimeters. When elasticity effects were also taken into account, the increase in sag would be 23 centimeters.[37]

Finally, Navier analyzed the nature and extent of the vibrations that would be produced by moving traffic. In particular, he used his mathematical results on cable vibrations to compare theoretically the behavior of Samuel Brown's Union Bridge with his own proposed bridge. This approach was necessary, according to Navier, because his formulas did not provide absolute numerical values but rather had to be used comparatively. He found that the maximum amplitude of vibrations on the Pont des Invalides would be 1/22 that of the vibrations on the Union Bridge, and that the elastic extensions and contractions of the cables would be around 1/30 as much as on the Union Bridge.[38]

In terms of the practical aspects of the design—the particular methods and hardware to be employed in constructing the bridge—Navier mainly followed existing techniques. For example, a recurrent problem was how to adjust the chains precisely to the correct length. Small differences in their lengths could tilt the roadway to one side or cause loads to be distributed unequally among the chains. Marc Isambard Brunel had earlier solved this problem by means of short coupling links equipped with expandable bolts (figure 6.7). Each bolt was formed of two semi-cylindrical halves into which special wedges could be driven. Navier adopted this system essentially without change. He also employed adjustable hangers invented by Brunel. A nut on the lower end of each hanger could be screwed up or down, thereby altering the hanger's effective length.[39]

Official Approval of the Pont des Invalides Project

Navier's theory and his design for the Pont des Invalides were both favorably received by the Corps des Ponts et Chaussées. The Corps' director, Becquey, was impressed enough by Navier's theory to have copies of it distributed to other Corps engineers.[40] The design,

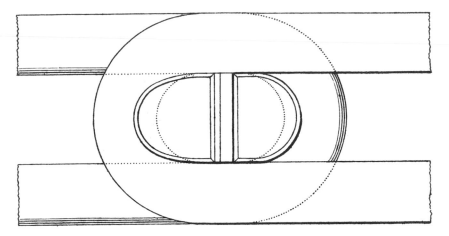

Figure 6.7
An expandable bolt devised by Marc Isambard Brunel and used by Navier in the Invalides Bridge. Source: Navier, *Rapport à M. Becquey et mémoire sur les ponts suspendus*, second edition (Paris: Carilian-Goeury, 1830). Courtesy of Special Collections Department, University of Virginia Library.

illustrated by Navier himself in a beautiful scale drawing, was circulated at Becquey's salon gatherings on several occasions, which was considered to be a special token of success.[41] With this positive response, Navier decided to seek formal approval to construct the bridge. He submitted the project to François Eustache in April of 1823. Eustache reviewed the project, reported favorably on it, and transmitted the project along with his own report to the departmental prefect, who in turn submitted it to Becquey. Becquey then appointed a panel of experts to draw up a comprehensive evaluation of the project. This report was presented at a meeting of the Ponts et Chaussées executive council.[42]

The committee of experts that reported on Navier's project included several leading Corps engineers, notably Prony, Sganzin, and Bruyère. They gave high praise to the project's beauty, to the choice of location, and to the way it was derived from mathematical theory: " . . . the details of this project have been studied in the most complete and luminous manner. Theory everywhere illuminates practice: all is foreseen, all is calculated, and the graphic part has great beauty."[43] The elegance and technical perfection of the drawings, which Navier's colleagues particularly admired, helped to stimulate interest in the project.

Although the committee members recognized that the Pont des Invalides was "not indispensable" (in other words, did not fill a pressing economic need), they approved the plan with several changes. They stipulated that the anchor cables' point of curvature be made lower and that the anchorage works include additional layers of masonry. They also established specifications for the chain links and coupling bolts and specified that all links and hangers had to be tested under a load of 18 kilograms per square millimeter with no evidence of permanent deformation.[44]

At least two of the changes that were requested, however, countered measures that Navier had deliberately taken to increase the structure's stability. First, the committee requested that the parapets be "suitably designed [*convenablement disposés*] so that the undulations of the roadway will not cause them to rupture."[45] Did the committee mean simply that the parapets should be made strong enough to resist rupture? Or was it thinking along the lines of Valentine Shockey, the blacksmith who made hinged railings for a Finley-type chain bridge? The first possibility, although the more logical, is placed in doubt because the committee did not specify dimensions for the parapets; if they wanted the parapets to be made thicker, they almost certainly would have specified dimensions. In any case, the comment reveals that the committee did not ascribe any stiffening role to the parapets. The second change in this category concerned the materials from which the deck was to be fabricated. Navier had proposed oak for the deck joists and planking, and cast iron for the transverse beams. At the request of the commission, however, the joists, the planking, and the transverse beams were all to be made of pine. Pine was lighter than oak and cast iron, which meant that the deck would become lighter and hence, according to Navier's theory, less stable. Pine also had less resistance to stretching than oak and cast iron, which meant that pine joists would deflect more under a given load than oak ones. Yet the members of the committee felt that the greater elasticity of pine was an advantage: they argued that transverse beams of pine, because of their greater elasticity relative to cast-iron beams, would better help to absorb shocks due to impact loading.[46]

The members of the committee not only oversaw the technical details of the project; they also decided how it was to be financed and built. They felt that "the site, the novelty of the system, and the importance of the projected bridge" demanded "great precautions in its execution."[47] Yet they knew that, because the of the capital requirements and because the enterprise was not an urgent necessity, Becquey wanted it to be financed through the private sector. (This was part of Becquey's stated policy of expanding public works by means of "prudently directed" private enterprise.[48]) Consequently, they recommended that the enterprise be made "the object of an absolute concession": the bridge was to be privately financed and built in return for rights to collect all tolls for 55 years.[49]

Navier felt that the project, because of its magnitude and importance, should be funded and undertaken by the state. He sensed that there were fundamental contradictions between the character of the project and the aims of the entrepreneur.[50] And in fact several potential difficulties with the arrangement proposed by the committee are evident. For one thing, as is evident in the case of Seguin, entrepreneurs have real reason to prefer economical designs; but monumental designs are not generally economical—particularly those that embody new, untried technologies. The projected cost of the Pont des Invalides was on the order of a million francs; although this was significantly less expensive than a stone arch bridge of the same overall length, it was not inexpensive relative to other suspension bridges. Seguin's suspension bridge over the Rhône between Tain and Tournon, which was constructed at the same time as the Pont des Invalides and which had the same overall length, cost only 200,000 francs. Furthermore, entrepreneurs prefer undertakings that are vital economically, because that is where they have the best chance to earn a good rate of return on their capital; yet it is evident that the Pont des Invalides did not fall into this category (although one of Navier's studies indicated that the bridge would bring an adequate profit to investors[51]). Finally, from the entrepreneur's perspective, it is very important to minimize construction time in order to achieve a rapid return on invested capital; yet major new undertakings usually take substantially longer to complete than expected. The Menai Bridge provides a good example: although the construction time was originally set at 3 years, it actually took 5 years to complete; and although the cost was estimated at 60,000 pounds, the actual cost proved to be double that amount.[52]

Constructing the Bridge

If Navier ever expressed his reservations to his superiors in the Corps des Ponts et Chausées, they were ignored. In July of 1823, the Corps' executive council adopted both the technical and financial recommendations of the special committee. The plans were then returned to Navier, who was to make the necessary revisions and draw up a code of specifications. Navier completed this work in March of 1824 and resubmitted the project to Eustache. After final approval from Becquey and from the Minister of the Interior, the project was opened to bidding on April 7, 1824. In May the concession was granted to Allain Desjardins, and in August construction began.[53]

The contractual relationship between Desjardins and the state was designed to ensure the latter's complete authority over every aspect of the design and construction of the bridge; the entrepreneur was put into the passive role of merely carrying the work out. Thus, the contract stipulated that the bridge was to be built precisely in accordance with a detailed set of plans and specifications provided by the Corps. No change could be made without the prior written consent of the Corps' director. Nor was there any clause authorizing the contractor even to suggest modifications, despite the fact that such a clause was reportedly a standard feature in this type of contract—all of which suggests that the administration did not want the contractor to take any initiative in this regard. Finally, it was stated that the entrepreneur was obliged "to submit to the administration's authority and supervision" over the building of the bridge and subsequently over its maintenance.[54]

An agreement was reached between Desjardins and the Corps des Ponts et Chaussées whereby Navier took on the overall direction of the works in the interest of Desjardins. In addition, the administration appointed two Corps engineers to monitor the construction: Navier's superior Eustache and Charles Stapfer, an *ingénieur-en-ordinaire*. Stapfer was assigned to inspect the work daily, to keep a detailed journal of the construction process, and to verify and enforce compliance with the approved plan.[55]

It has been suggested that successful completion of the Pont des Invalides was jeopardized by a poor system of inspection and material testing,[56] but this was not the case. Following the example of British builders, Navier took great care to test all the links and hangers. He devised new equipment for this purpose. One of the two types of link testing machines employed in Britain overestimated the force on a link; the other underestimated it. Navier designed a new machine that he claimed was more accurate than either. It was also equipped with a device to measure the elongation of the links under loading. Navier also invented a machine to test curved links, which were employed at the point where the anchor cables changed direction. Finally, he devised a method to test the resistance of the hangers to impact loading. A hanger was suspended from its upper end, and a nut was threaded onto its lower end. A cylindrical weight encircling the hanger (like a bead on a string) was allowed to drop onto the nut below. With the weight of the cylinder and the height of its fall known, it was easy to calculate the force on the hanger.[57]

Though Navier followed British precedent when it came to testing the links, he did not follow British methods of determining the lengths of cables and hangers. Here he returned to theory. Charles Stapfer was assigned the task of calculating the necessary lengths for the hangers and for the links of the main cables, both of which varied over the length of the bridge. To do this, Stapfer used modified equilibrium equations, derived by Navier, that took into account the weight of the cables and hangers in addition to the weight of the deck. Stapfer's calculations followed a 14-page mathematical procedure; the equations took into account not only the weight of the cables and hangers but also the effects of elasticity and the potential sag of the deck. Using this complex mathematical procedure, Stapfer calculated 47 hanger lengths and 47 link lengths, each to four decimal places.[58]

Stapfer's method of determining the lengths of hangers and links shows the confidence Navier placed in his mathematical theory. Yet although computing each length to four decimal places certainly gave the impression that "all was foreseen and calculated," the day-to-day experience of building the bridge revealed the limits of Navier's analytical ideal. For example, Stapfer's careful calculations did not obviate the use of adjustable nuts and expandable bolts to regulate hanger and link lengths empirically.

The limits of calculation became particularly obvious when it came time to join the separate chains (Navier called them "partial chains") to make the heavy rectangular cables specified in his design. The task proved quite difficult, and Navier's description of the experience, recounted in the second edition of his treatise after the abandonment of the Pont des Invalides, has a different and more pragmatic flavor than his mathematical analysis and his original design proposals. The trouble began when these combined cables, left hanging under their own weight, were exposed to sunlight:

. . . immediately one saw the effects resulting from the action of the sun's rays, and the temperature inequalities among the various partial chains. The upper tier, hit by the sun's rays, expanded more than the middle tier. At the same time, the outside column of partial chains deflected more than the inner columns which were in the shade. The result was a sort of sloping of the group of chains. The cross-section of the cable was no longer a rectangle with its sides aligned horizontally and vertically.[59]

Navier explained that the tilt of the rectangular array amounted to as much as 6 centimeters. Furthermore, the direction of the tilt changed over the course of the day. In the morning the cables tilted toward the east, in the evening toward the west. The cables were so sensitive to sun that passing clouds would bring noticeable shifts in the way they hung. This problem aggravated the difficulty of correctly regulating the lengths of the partial chains:

Several days after the chains were left to hang on their own, it was recognized that, despite the extreme precision we endeavored to achieve in the lengths of the links and in putting them into place, it was necessary to make use of double (i.e. expandable) bolts and of wedges which were placed in the portions close to the towers, in order to regulate suitably the lengths of the partial chains such that the four chains of each tier would hang at exactly the right level and the three tiers of chains would all be exactly parallel to one another. The continual and unequal variations of the lengths of the chains made this operation almost impossible when there was direct sun.[60]

The problem was ultimately resolved by placing small canopies above the cables to shade them from the sun until the necessary adjustments could be made.

The limits of calculation were also revealed in the behavior of the towers and anchorages under the weight of the cables and the deck. Once the roadway was hung from the chains, small cracks appeared at the base of each tower column, extending around one-eighth of the column's circumference. The behavior of the anchorages was still more problematic. In early July of 1826, when the cables were first allowed to hang on their own, small vertical fissures appeared in the anchorages in the area where the anchor cables changed direction.[61] Initially this aroused no concern, because some settling had been expected. Navier wrote a letter to Becquey informing him that the cables were fully in place, hanging on their own, and that the anchorages showed no signs of trouble. He concluded that "this success would seem to leave no uncertainty about the success of the structure."[62] On July 24, Navier wrote to Becquey that "the masonry continues to behave in the in the manner which could be expected."[63]

By the end of August, however, when the roadway was completed except for the railings, the fissures had widened to about 5 centimeters for each anchorage. By this time it was recognized that the sections in question would have to be strengthened. Then, on the night of September 6, an accident occurred. A water main that passed close to the underground anchorage buttresses on the Champs-Elysées side of the bridge ruptured and flooded the area around the buttresses. Some of the supporting earthwork ceased to bear against the anchorages. As a result, the two fissures on the Champs-Elysées side suddenly widened (to 11 and 17 centimeters, respectively), and the buttresses suffered some upturning and displacement in the area above *b–c* in figure 6.8. This movement, in turn, caused the towers on the Champs-Elysées side to tilt toward the river. One tilted more than 4 inches, with consequent damage to the stonework at the base. The anchorages and columns on the Invalides side, however, showed no further signs of dislocation.[64]

The accident accelerated what would otherwise have been a more gradual dislocation of part of the anchorage, showing in dramatic fashion that the anchorage was not strong enough. But what conclu-

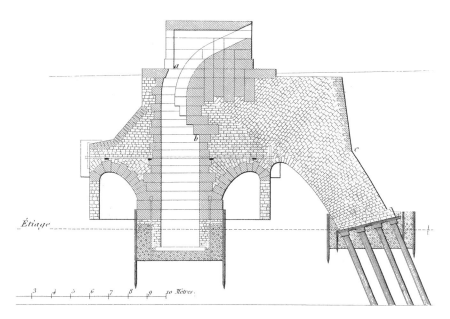

Figure 6.8
An anchorage of the Invalides Bridge as built, incorporating modifications requested by the executive council of the Corps des Ponts et Chaussées. Source: Navier, *Rapport à M. Becquey et mémoire sur les ponts suspendus,* second edition (Paris: Carilian-Goeury, 1830). Courtesy of Special Collections Department, University of Virginia Library.

sions should be drawn from this episode? Was the problem unavoidable, or should it be ascribed to miscalculation or negligence on Navier's part? It is clear is that the basic idea of having the cables curve so as to descend vertically to the anchoring points was workable, because it was subsequently adapted to many other suspension bridges (including the Brooklyn Bridge).[65] The problem with the Pont des Invalides lay in Navier's specific implementation of this idea.

A visual examination of Navier's anchorage design and a comparison with later, successful designs suggest that his anchorages needed more support at and above ground level, particularly in the area below the point where the cables exited from the anchorage tunnels into the open air. In view of the tremendous pull of the chains (each rectangular cable theoretically bore a tension between 1,200,000 and 1,760,000 pounds), the lack of any stonework or earthwork in these areas seems, in hindsight, a poor decision.[66]

The aim here is not to assign blame, but rather to understand how Navier conceptualized the anchorage and why he did not see any need for extra stonework or earthwork. Navier's treatise analyzed the problem of suspension bridge anchorages in general terms. That discussion shows that his conceptualization focused on theoretically calculating the resultant of the forces on the anchor cable where it changed direction beneath the ground, and on constructing a buttress (H–I in figure 6.9) to resist that resultant force. Navier recognized that all of the earth to the right of area G–N–C (figure 6.9) actually helped to resist the resultant at N. Nevertheless, the essence of his thinking was that the force exerted by the cable on the earth between G and C could be abstracted into a single resultant, acting in a particular direction at a certain point. Both his analysis and the schematic he used to illustrate it suggested that the resultant at N could be resisted by a properly positioned underground buttress with a sufficiently solid foundation. That Navier conceptualized the anchorage in this way can be seen from the close resemblance between his schematic (figure 6.9) and his initial anchorage design for the Pont des Invalides (figure 6.4).[67]

The nature of this conceptualization reveals, once again, Navier's strong belief in the efficacy of mathematical theory to guide constructive practice. He evidently felt confident that he could determine the resultant force accurately enough to position a comparatively slim buttress at just the right point to provide the necessary resistance. Yet he gave no indication of how to ensure that in practice such buttresses would be placed exactly where theory said they should be. Moreover,

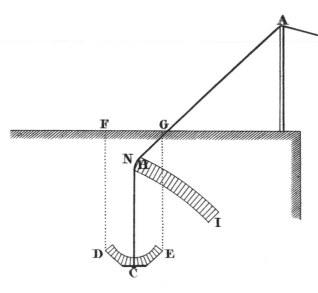

Figure 6.9
Navier's schematic illustrating the general conception underlying his anchorage design. Compare this with figure 6.4, which shows the Invalides anchorage as originally design by Navier. Source: Navier, *Rapport à M. Becquey et mémoire sur les ponts suspendus,* second edition (Paris: Carilian-Goeury, 1830). Courtesy of Special Collections Department, University of Virginia Library.

he never mentioned the possible effect of powerful shear and lifting forces on such buttresses if the cables began to shift. Navier may have felt such a possibility to be precluded by the use of curved eyebars at the point of curvature of the anchor cables.

Furthermore, although Navier's anchorage design was new and untried, there is no evidence that he attempted to test the idea empirically before applying it on a large scale. Nor, in the absence of any empirical test (which might have been too expensive), did he feel it necessary to overbuild the anchorage substantially in order to compensate for lack of empirical knowledge of how it would function. Yet Navier as much as admitted that theory could not entirely predict the behavior of such an anchorage. In particular, he noted that the resistance of earth could be calculated accurately only in the case of *vertical* forces opposed by a given weight of earth or stone. For forces in the horizontal direction, resistance depended significantly on the cohesion of the earth, "the evaluation of which is subject to great incertitude."[68] Yet Navier's anchorage had to resist tremendous tension in the horizontal direction.

Navier's manner of conceptualizing the anchorage blinded him to other perspectives that might have led him to proportion or structure the anchorage differently. Navier's superiors did not entirely share his conception, however. When they evaluated his original proposal, they demanded that the anchorages be modified to include many more layers of masonry. The structures built were therefore much bulkier than Navier had initially intended (compare figures 6.4 and 6.8), although they were still not strong enough.

From Technical Failure to Social Collapse

The accident to the Pont des Invalides generated a wave of legal and political conflict, but at first the executive council of the Corps des Ponts et Chaussées saw only a straightforward repair problem. After inspecting the site in the wake of the accident, the executive council concluded that the damage to the anchorages could be repaired and the anchorages strengthened without undue expense. Estimating that an additional 100–200 cubic meters of masonry would be needed, which would increase the total cost by only 1–2 percent, they directed that steps be taken to begin the work. Navier immediately drew up a proposal to reconstruct the anchorages.[69] The repairs were never made, however. Growing controversy during the fall and winter of 1826–27 led Becquey to cancel the project in the spring of 1827. The Pont des Invalides was then dismantled.

Becquey's change of policy cannot be ascribed simply to the technical defect of the anchorages. Not only do such technical problems usually have fairly straightforward solutions but they are more common than we tend to imagine. Relatively little has been written about the history of the maintenance and repair of engineering structures, but this history is at least as long and complex as the history of their initial construction, and equally important. For example, the Menai bridge was built without any significant stiffening or protection against wind despite the fact that its location exposed it to severe gales. A month after its completion, in 1826, it was damaged in a windstorm; several cross-beams supporting the roadway broke as did a number of hangers. When repairs were made the deck was strengthened, but it was again ripped apart by a hurricane in 1839.[70] The Brooklyn Bridge likewise had a significant technical problem involving a hinge in the stiffening truss at the middle of the span.[71] Yet neither of these projects was abandoned because of its initial technical defects, and indeed both are classed today among the world's great engineering monuments. In the case of the Pont des Invalides, not only was it technically possible to repair the bridge but the projected cost of the repairs was said to be less than the cost of dismantling the structure![72]

What finally led to the abandonment of the Pont des Invalides was a series of problems linked to the organizational structure for financing the project and to opposition from the Paris city council and outspoken members of the public. First, there were financial difficulties. The contractor Desjardins was unable to raise new capital to make repairs, because the accident had caused panic among investors. Moreover, a dispute emerged between the contractor and the government as to who should pay for the repairs. The Corps took the position that the contractor should pay, since he had undertaken to build the bridge "at his own risk and peril."[73] But Desjardins objected on two grounds. First, he argued, he could not be held responsible for the accident because he had fulfilled all his obligations. He had built the structure exactly according to specifications, and he could prove that all the work had been carried out properly. (The proof lay in the fact that a Corps engineer, Charles Stapfer, had daily overseen and approved all materials and workmanship during the entire period of construction; the evidence was in Stapfer's daily log book.[74]) Second, Desjardins argued, the phrase "at his own risk and peril" could not be taken to include the risk that the design might be flawed. Rather, the risks he rightly had to bear were the risks that there might be increases in the price of materials or labor, that the tolls collected might not provide a sufficient return on the invested capital, and so on. How, Desjardins asked, could the state expect him to pay for altering a design he had been legally obliged to follow—a design he was not even supposed to question? He added that the correctness of his view was shown by the fact that the Corps had earlier requested certain design changes and had agreed to pay for them since they were not originally specified in the contract. In sum, Desjardins felt that he was being unfairly called on to pay for the Corps' own mistakes, and he was vexed enough to threaten to take the matter to court. Events also prompted Desjardins and the other stockholders to reevaluate the project, and they evidently concluded that the original site chosen for the bridge was not the most desirable from an economic standpoint.[75]

Not only did the Pont des Invalides lose the support of its contractor and its investors; it also lost most of what support it had from the public. In fact, there had been strenuous opposition to the project from the beginning, especially from within Paris's municipal council. Many disliked the location chosen for the bridge. It was argued that the beauty of the Champs-Elysées would be marred by the creation of a major thoroughfare across it, and some people felt that the bridge would have a negative impact on the esplanade of the Hôtel des Invalides, which would become dirty and clogged with traffic. Finally, some believed that the bridge columns would mask the view of the Invalides (an accusation Navier denied).[76]

Of course, virtually every major engineering project is at some point opposed by certain groups; yet when such undertakings respond to real economic and social needs, they usually win considerable public support. The Brooklyn Bridge provides a case in point.[77] By 1850 at least ten percent of the population of Brooklyn commuted every day to Manhattan to work. Their only means of transportation was the Brooklyn Ferry, which was often delayed or cancelled because of bad weather. Thus, when the cost of building the Brooklyn Bridge ran so high that the entrepreneurs who had initiated the project could no longer afford to invest in it, public support made it possible for an arrangement to be made whereby the cities of Brooklyn and New York jointly took control of the bridge and provided the funds to complete it.[78] In contrast, the Pont des Invalides was imposed on Parisians by the Corps des Ponts et Chaussées with little concern for what they wanted or needed; reservations expressed by the city council before the project was approved were simply ignored. Hence, it is hardly surprising that the council later took advantage of the accident to call the entire enterprise into question and to declare itself in favor of the removal of the bridge.[79]

The Invalides accident also became a focal point for the expression of general hostility to the Corps des Ponts et Chaussées. One group that was particularly outspoken in this regard was the growing community of entrepreneurs and private-sector engineers. Civil engineers in particular often felt either thwarted or belittled by Corps engineers, and they resented the fact that a small elite had control over broad areas of technological development.[80]

Opponents of the Corps' mathematical approach likewise used the Invalides accident as an opportunity to scorn theoreticians and their calculations. One technical writer referred to "the eminent men of science" who were "seeing their calculations fail in Paris."[81] Balzac, an outspoken critic of the Corps, expressed a similar view in his novel *Le Curé du Village:* "All France knew of the disaster which happened in the heart of Paris to the first suspension bridge built by an engineer, a member of the Academy of Sciences; a melancholy collapse caused by blunders such as none of the ancient engineers . . . would have made."[82] A British writer reported facetiously that Navier, when called on to explain the accident, had responded "C'était seulement une petite distraction dans mes calculs" ("It was only a small oversight in my calculations").[83]

These strands of conflict and opposition were able to take root in part because of the timing of the accident. As a Paris newspaper explained, "If the accident . . . occurred in the spring, it would hardly merit public attention."[84] But it occurred in the fall, with the result that the repairs became more difficult and expensive and could not be immediately undertaken. To make the repairs the chains and deck had to be supported on scaffolding so as not to exert any pressure on the anchorages. In the spring this would have posed no problem, but in the fall and the winter it did: the scaffolding could not be protected against the high water or ice that might form on the river during those months. It was therefore necessary to remove the deck and the chains for the entire winter. The resulting delay created a context in which conflict and opposition could escalate and become organized.

The growth of organized conflict and opposition—in the form of lawsuits, city council resolutions, sarcastic articles, and other negative publicity—transformed the Invalides accident into a major source of embarrassment that jeopardized the reputation of the entire Corps. Under the circumstances, Becquey and the Corps' executive council perhaps feared that, if completed, the Pont des Invalides would become an object of public scorn or a symbol of the Corps' fallibility. Abandoning the project was a way to save face. (The Corps' earlier decision requiring the contractor Desjardins to pay for the repair of the bridge may also have been taken partly in order to save face and to avoid any admission of fallibility.) More concrete, abandoning the project was a way to avoid a damaging lawsuit, which the Corps probably could not have won. At least Navier suggested as much in a letter published in *La Pandore,* when he stated that the contractor had clearly held to the specifications of the contract: "It has been officially admitted in two meetings of the executive council of the Corps des Ponts et Chaussées, chaired by the director general, that the accident which suspended work on the bridge could in no way be attributed to any defect in the execution of the works."[85]

Resolution of the Conflict

Concomitantly with the abandonment of the bridge, the Corps des Ponts et Chaussées reached a new agreement with the contractor Desjardins and project's investors. Both contractor and investors were compensated for their losses. In addition, Desjardins was permitted to build three suspension bridges in Paris in lieu of the original one and to use the materials from the Pont des Invalides. Desjardins was also allowed to provide his own designs for the three bridges, the only stipulation being that the completed structures had to pass load tests supervised by the Corps.[86]

The three bridges that were eventually built presented a striking contrast to the Pont des Invalides. They were designed to make a profit, not to achieve glory, or to beautify the city, or to outdo the British, or to show how theories could be applied in practice. One of them, the Pont des Champs-Elysées (located about 200 meters downstream from the original site of the Pont des Invalides), was described by the editor of the *Journal du génie civil* as "truly a villainous thing"[87] and as a poor imitation of Samuel Brown's Union Bridge. The Pont des Champs-Elysées consisted of a span of 72 meters flanked by two half-spans, one of which, through some miscalculation, was made 63 centimeters shorter than the other. The bridge, moreover, lacked stability; it was said to vibrate considerably when a carriage or even a single horseman crossed over it. Nevertheless, it stood for 25 years until the Corps decided it should be replaced by a stone arch.[88]

Whereas Desjardins seems to have come through the affair relatively unscathed, the same cannot be said for Navier. Having to witness the demolition of a structure that embodied much of what he had worked for in his professional career was a devastating blow. Navier also had to bear personally the brunt of the public's criticism, even though it was implicitly directed toward the Corps as a whole. He became a scapegoat—not only for the public but also for the Corps, which punished him by withholding promotion. Between October 1830 and August 1831, Navier wrote four letters to the Corps' director Simon Bérard (appointed in 1830), asking to be promoted.[89] He complained that he had been passed over in favor of a younger engineer with less time in service and poorer credentials. A marginal comment on one of the letters, presumably written by Bérard, explained that in awarding promotions, "one must consider not merely works undertaken, but also those that are successfully completed."[90] A few years later, in 1836, Navier died at the age of 42. His colleagues believed the Invalides debacle to have been a factor in his early death. As Prony cryptically explained in his biographical notice of Navier:

When a great monument . . . comes thus suddenly to disappear at the moment when it is ready to fulfill its very useful purpose, as the number of those who can know and appreciate the true causes of this disappearance is always infinitely small in proportion to the number of those whose beliefs, although founded upon false appearances, constitute nonetheless what is called *public opinion,* there necessarily results in this opinion, a disfavor that . . . exerts a very fatal influence on the mental condition and even the physical constitution of the one who is its object.[91]

Conclusion

Although the immediate motives for abandoning the Pont des Invalides were public opposition and the financial dispute between the administration and the contractor, an important underlying cause was the reluctance of the Corps des Ponts et Chaussées to adapt to a changing world. In particular, the process of industrialization that was taking hold in France during this period was accompanied both by a rapid development of new transportation systems—bridges, canals, railroads, etc.—and by the growth of a community of entrepreneurs and of engineers who worked in the private sector. Yet the Corps was reluctant to share power with this new group (and, in many instances, even to cooperate with it).

The Corps often adopted policies that made it difficult or impossible for entrepreneurs to work efficiently or even to work at all. Complaints to this effect were a constant theme in French technical literature of the nineteenth century; some of them, recounted in detail, seem like nightmares of red tape and bureaucratic interference and sabotage.[92] This conflict had long-term effects. For example, it is well documented that the policies and decisions of the Corps' administration were at least partly if not largely responsible for the comparatively slow growth of the French railway network before 1850.[93] Similarly, contemporaries argued that the Corps hindered the development of long-span bridges later in the nineteenth century. Every time a proposal for one was submitted, they complained, it was "rejected on all sides by the Conseil des Ponts et Chaussées, even when the cost was less than that of a bridge with several spans."[94]

Contradictory policies followed by Becquey contributed significantly to the failure of the Pont des Invalides. Becquey saw a need to stimulate private enterprise, but he was insensitive to the broader implications this policy entailed. He wanted to choose what projects entrepreneurs would be allowed to undertake, but his choices reflected the priorities of the Corps des Ponts et Chaussées rather than those of entrepreneurs. He decided to have the Pont des Invalides contracted out because it was primarily monumental, but that is the kind of project that private enterprise is often unable to carry through. Moreover, for Becquey to assume that private enterprise could be directed at will by the state was to overlook one of its fundamental characteristics: its success depends on the willingness of individuals to invest, and that cannot be dictated administratively. After the accident, the contractor's and investors' attitudes toward the Invalides project changed dramatically, and Becquey's declarations could not alter that fact.[95]

Another dimension of the Corps' inflexibility was its stubborn adherence to an ideology and to policies that were increasingly regarded by the public as reactionary. Public opinion generally favored democratization of the Corps and a shift to a more pragmatic style of engineering, such as existed in Britain and America.[96] Instead, the Corps continued to nourish mathematical theories and elitist attitudes. These circumstances inevitably shaped the public's attitude toward the Pont des Invalides. Many saw the bridge as nothing more than a material embodiment of the Corps' world view, a "sterile monument"[97] that looked backward to an aristocratic past.

Conclusion to Part I

The preceding chapters have detailed the responses of Finley and Navier to the suspension bridge, which differed in terms of both research and design. The lag of roughly ten years between the time Finley stopped working on suspension bridges and the time Navier took up the problem obviously accounts for some of the differences in the ideas and methods of the two builders. Navier borrowed new technical ideas from British builders—for example, expandable bolts from Brunel. He also benefited from British research on the strength of iron, which called attention to the phenomenon of permanent stretching at a stress significantly lower than that which would produce rupture.

Nevertheless, the time gap alone cannot account for all of the differences in the ways these two individuals approached this technology. Most important, it cannot account for the fact that Navier used mathematics extensively and Finley did not. Navier did not "borrow" his mathematical analysis from the British: Brown, Telford, and other British builders hardly used mathematics at all, and the published research results of Gilbert and other British theoreticians were meager. Moreover, the basic mathematics used by Navier to study the equilibrium of loaded cables was already codified and published before Finley began his research. Of course, Navier's analysis went beyond the statics of loaded cables to consider other phenomena, such as cable elasticity and vibration. And some of this research did draw on mathematical theories and techniques developed after Finley's 1808 patent, such as the technique of Fourier analysis. Yet the fact remains that the portions of Navier's theory which related to questions posed by Finley were based on mathematical techniques and results that were in principle available to Finley.

A fuller understanding of Finley's and Navier's different responses to the suspension bridge has to go beyond the mere existence of a time gap; it must take into account the very different environments in which the two individuals worked. Environments shape strategies. Individuals take stock of their environments in order to decide which strategies are most likely to bring success. This does not mean that environments are one-dimensional or all-knowable. An individual responds according to his or her perceptions of what are the most significant elements and characteristics of the environment, and these perceptions may in turn be shaped by a host of factors (including personal psychology). But the environment places limits on the range of strategies that can be made to succeed.

How far would Navier have gotten if he had tried to do in Fayette County what he sought to do in Paris? How would he have mobilized the equivalent of a million francs to have a monumental suspension bridge built there? How would he have convinced the farmers who served as county commissioners to let him build such a bridge, or even to contribute to the cost of it? Where would he have found an engineer assistant who could compute hanger and link lengths to four decimal places, following a complex 14-page mathematical procedure? How many in Fayette County would have agreed that such a bridge should be built partly as a proof of the applicability of mathematical theory to constructive practice?

Leaving aside this image of a monumental Pont de Fayette, the preceding chapters have shown that Finley and Navier did, in practice, attempt to structure their research and design efforts in relation to their environments. For the purposes of comparison, it is helpful to think of these environments in terms of three dimensions: an intellectual dimension, an economic dimension, and an organizational dimension.

With regard to the intellectual dimension, Finley adopted an empirico-inductive research methodology advocated by the Common Sense philosophy that was popular within his intellectual community. By following this methodology, Finley helped to ensure that his research findings would be understood and accepted by his peers. He would have had much less of a chance to convince a rural blacksmith or carpenter or county commissioner to build one of his bridges if he had given a proof of its technical adequacy based on the mathematics of Euler, Bernoulli, and Lagrange, which the majority of them had not read and did not care about.

In contrast, Navier followed a theoretico-deductive methodology that was respected within his intellectual community. By so doing, Navier helped to ensure that his research findings would be considered valuable and insightful by certain groups of his peers (and his superiors) in the scientific community and in the Corps des Ponts et Chaussées. If Navier had gone the route of Finley, Brown, and Telford, and had studied the equilibrium of loaded cables empirically, without reference to the mathematical contributions of his colleagues and predecessors, his work would probably have been scorned by the likes of Becquey, Prony, and Fourier.

Both Finley and Navier sought a deeper understanding of the character and behavior of suspension bridges, but their distinct research methodologies and intellectual perspectives led them to employ different tools to achieve this understanding. Finley used models, and experimental apparatus comprising cables, pulleys, and weights. He also sought understanding through encyclopedia articles on bridges and on the strength of iron, and through the construction (in 1801) of a 70-foot trial span over a stream close to Uniontown. Navier's tools were mainly mathematical: calculus, Fourier series, elasticity theory, mathematical statics, and a variety of specific equations such as the equation for belt friction. Navier also sought understanding through inspection of existing British bridges, through discussions with the builders of some of those bridges, and through published literature, not only on suspension bridges, but also on mathematics, theoretical mechanics, and elasticity theory.

Finley's research was closely tied to his design goals: he wanted knowledge that would enable him to come up with a workable yet simple and generic design. He therefore paid careful attention to such matters as the procedures smiths should follow to fabricate the links of the chain cables. Navier's research was not focused as much on design and hardware; it was wider ranging, it was often more abstract, and it was explicitly linked to research agendas that were of scientific rather than technological interest. Navier's analysis was also more comprehensive than Finley's. Finley could say nothing specific or quantitative about the effects of friction, cable elasticity, or temperature variations on the equilibrium of suspension bridges; Navier analyzed these issues in detail. Yet Finley had no special reason to seek out such knowledge. It made sense for him to seek out design solutions to minimize effects of vibration, expansion, and contraction, rather than to study the extent of the manifestations of these phenomena in extreme cases.

Navier's intellectual environment included an acceptance of monumentality as a worthwhile goal in structural engineering. This was not the only goal that was considered valid by his peers, but it was an important one: within the Corps des Ponts et Chaussées, monumental projects were the ones that received greatest respect and acclaim. It was quite reasonable for Navier to attempt to integrate the suspension bridge into this tradition. Finley, on the other hand, deliberately avoided monumentality, because it conflicted with the strict religious outlook that was prevalent within his environment, which equated monumentality with wastefulness, "vanity," "contamination" of the mind, and "idle elegance and show." (These were expressions used by Finley to describe the nature and consequences of monumentality and to justify the non-monumental character of his design.) Finley also perceived that monumentality would not be a practical selling point in the many communities where bridges were needed but where money was scarce. In Navier's environment of the Corps des Ponts et Chaussées, however, monumentality was a good selling point.

Concerning the economic dimension of the environment, Finley adapted the suspension bridge to succeed in a competitive market against other bridge designs—particularly wooden arches and trusses. Since price was an important criterion, and since Finley had to "sell" many bridges to earn a reasonable profit through royalties, he needed to do everything possible to minimize the construction price of his bridges. Navier's environment did not oblige him to compete in the marketplace at all. He was a salaried employee, and in principle it made no difference to his income whether he built one bridge or a hundred. Navier probably had more to gain by building one expensive, monumental bridge than a host of cheap ones. To ensure that he could build the monumental one, however, Navier had to convince his superiors that such a bridge was a worthwhile project. By presenting his project as a sophisticated application of theory that would establish a new technology in France on a monumental scale, Navier was striking chords that he knew his superiors—especially Becquey and Prony—would respond to.

Finley and Navier adapted the technical elements of their bridge designs to these different economic contexts. Whereas Finley's design choices show that he tried always to minimize costs, Navier's design choices reveal that keeping costs low was not an overwhelming priority for him. In his treatise, Navier had pointed out that the cost of a bridge was roughly proportional to the weight of the materials in it. Yet he did not attempt to minimize the weight of materials in the Pont des Invalides, but rather to maximize it (relatively speaking), so as to improve stability, despite the added cost. Finley sought to minimize weight in order to save money. Navier proposed to use heavy oak and iron for his bridge deck; Finley advised builders to use light pine. Finley sought stability through stiffness, Navier through weight. The latter was the more expensive option, at least in these instances.

Finley's use of short, multiple spans was also a money-saving choice. He saw the possibility of long-span suspension bridges, but he chose not to take up that possibility in his design. Shorter spans meant shorter, cheaper towers that could be made of wood. Shorter spans also made it possible to rely on the skills of local craftsmen, because shorter spans were easier to construct (unless the intermediate piers created special difficulties). Navier chose the long-span option to accord with his aim of monumentality, yet this choice resulted in a comparatively more expensive structure. Navier's use of compound cables to achieve a more monumental look also increased costs. The compound cables were more expensive because they were more difficult to construct. Recall the problems Navier had trying to adjust the cables in the sun—the job required that special canopies be built to shade the cables. Finally, Navier's design had technically unnecessary decorative features that were undoubtedly expensive, such as the reclining lions at the anchorages.

Finally, Finley and Navier adapted their designs to be built within different organizational structures and networks. Finley made use of the patent system, aiming to collect royalties. Navier worked within the administrative structures of the Corps des Ponts et Chaussées. These organizational structures involved different tradeoffs and compromises with regard to technical control over the building process, financial control (whether and how to finance the building process), and control over the supply of labor.

Because Finley used the patent system, permitting anyone to implement his design as long as they paid royalties, he often had no immediate involvement in the building of a bridge based on his design. This was an advantage in terms of his own career. Being a judge and a farmer, with local commitments in Uniontown, he probably did not wish to travel around the country building bridges. Yet the consequent lack of technical control became a significant problem for Finley, to the point of jeopardizing the reputation of his design system.

Using the patent system also meant that Finley generally could not exert direct control over financing. This organizational context gave him no authority to decide if a bridge should be financed by a local government, or through the creation of a corporation, or by some other means, or not at all. All Finley could do was advertise his bridge design, and otherwise encourage prospective builders, and then demand a royalty payment if a bridge were built. Finley also could not exert direct control over the supply of labor for the building process, and therefore he could not ensure that those doing the building would have specialized technical qualifications (apart from e.g. a general knowledge of carpentry).

Finley adapted his bridge design to this organizational context in several ways. First, he created a generic design, to which he added a step-by-step plan to guide builders through the process of adapting the general design to a specific location. Second, he attempted simultaneously to maximize the ease of construction and to minimize the need for technical expertise (both of which reduced costs), by creating a design that could be built by local craftsmen without any need for mathematics beyond arithmetic and without any need to differentiate between load and tension in a cable. The key here was the 1/7 sag/span ratio. Builders could likewise determine hanger and link lengths using a "board fence" and a simple empirical technique.

Navier's tradeoffs were different. Working within the organizational context of the Corps des Ponts et Chaussées, Navier first had to accept design changes ordered by his superiors. Yet this context guaranteed that, once a design was agreed upon by the Corps, there would be strict technical control over the building process. This control was achieved by means of a legally binding set of technical and administrative specifications (the *cahier des charges*), by daily inspection of all work and materials by an assistant engineer, and by the agreement that made Navier the technical director of the project. Navier's context also guaranteed that he would have access to specified materials (e.g., stone of a certain type) and to a labor supply with the technical expertise he wanted. He could count on having an assistant with enough mathematical and technical knowledge to understand and apply his mathematical analysis, and to oversee materials, labor, and construction methods. Like Finley, however, Navier could not exert significant financial control over the project. Here he had to submit to the authority of the Corps' executive council, despite the fact that their decision to have the Pont des Invalides financed by private capital directly opposed his own wishes. The only influence Navier managed to exert on this score was to carry out a study to demonstrate that the bridge would bring an adequate return to investors. (Certainly this was an important way to attract capital, however.)

Navier adapted his design to the organizational context. His design presupposed the availability of highly skilled labor and the ability to control the technical details of the construction process. It is evident that Finley, in his organizational context, could never have hoped to succeed with the kind of anchorage used by Navier. Navier's anchorage required that the underground buttresses be placed where theory said they should be—the design presupposed that someone conversant with the theory would monitor the construction process to ensure that the buttresses were correctly built and positioned. Navier's design also required careful and systematic testing of links, and mathematical calculation of hanger and link lengths. The latter presupposed the availability of highly skilled labor and expert monitoring of the construction process.

Both Finley and Navier innovated, but their innovative work was not a transcendent quest for novelty; it was fundamentally structured by their respective environments. Finley created a new structural form—the suspension bridge with a level roadway—with the idea that it would provide a cheap means of spanning the many streams and rivers of a largely undeveloped country. He innovated to achieve a design that could be technically and economically successful in regions with little money, comparatively few inhabitants, and laborers having only general technical skills. Navier created a new mathematical theory that built on the mathematical contributions of his colleagues and his predecessors. He also innovated in adapting the suspension bridge to the classical French tradition of monumental architecture, and in working out a design that followed guidelines and precepts deduced from his theory.

In the long term, some of Navier's and Finley's innovations had considerable impact beyond their initial environments. Finley's idea of a level-roadway suspension bridge opened a realm of possibilities which builders have explored ever since. Navier's theoretical work— manifested not only in his theory of the suspension bridge but also in his teaching—brought more sophistication to structural theory and helped to make statics, mechanics, and the theory of elasticity more powerful tools in the hands of engineers.

In the short run, however, the innovative work of Finley and Navier did not prove entirely successful. Both designers found their efforts thwarted by forces beyond their control. The initial success of Finley's system—because it occurred within a competitive market environment—stimulated further competition. Other designers began to beat Finley at his own game with new wooden arch and truss designs that cut into the market for chain bridges. At the same time, his lack of control over the building process led to accidents which may have damaged public confidence in his design system. Navier was thwarted by forces both within and beyond the Corps des Ponts et Chaussées. Constructing the Pont des Invalides within that organizational context protected his project from outside technical and economic competition. Yet it also meant that the project was dependent on the Corps' continued protection. And when the failure of the anchorage led to financial disagreements, to public opposition, to embarrassment to the Corps, to a hostile resolution from the Paris municipal council, and to the threat of a lawsuit, the Corps withdrew its support and the Pont des Invalides was removed.

II

Social Determinants of Engineering Practice: A Comparative View of France and America in the Nineteenth Century

The analysis of Finley's and Navier's work in part I revealed a fundamental difference in how the two men approached the study of suspension bridge technology. Navier took a theoretico-deductive approach and focused on deducing general laws and specific design proportions mathematically; Finley adopted an empirico-inductive approach and focused on discovering the laws of suspension bridges through systematic experimentation and on working out design details using this knowledge and information derived from the use of models and prototypes. Navier's tool kit consisted of mathematical methods and equations; Finley's consisted of physical apparatus—models, cables, pulleys, weights, and a prototype bridge. These two approaches constituted different forms of technological practice in the realms of both research and design.

I will argue in part II that these differing approaches were part of broader patterns that persisted in France and America throughout the nineteenth century. Finley's and Navier's approaches reflected distinct structural trends in the way technological research was done in the two countries—trends that shaped many technological endeavors other than bridge design. France had a strong and continuous tradition of research in the mathematical engineering sciences during the nineteenth century, but a rather weak tradition of experimental research oriented toward industrial and design problems; America had a strong and continuous tradition of experimental research oriented toward industrial and design problems, but a rather weak tradition of research in the mathematical engineering sciences.

The extent of the divergence was considerable. During the nineteenth century, French technologists were among the leading contributors to the development of mathematical engineering science. They made fundamental contributions to statics and mechanics, to the theory of elasticity and strength of materials, to soil mechanics, to hydrodynamics, to thermodynamics, to the theory of machines, to the theory of friction, to the analysis of structures, and to the analysis of vibration. Indeed, many of their ideas and methods have become basic intellectual equipment for engineers the world over. In contrast, American technologists made few contributions to mathematical engineering science in the nineteenth century. When they did use mathematical theory, it was most often imported. For example, the mathematical theories or formulas presented in the widely used engineering treatises of Dennis Hart Mahan and Charles Storrow were European (mainly French) in origin.[1] Later in the nineteenth century, Americans also drew extensively on the work of British and German theorists, including W. J. M. Rankine and Julius Weisbach.

Histories detailing the development of specific branches of mathematical engineering theory in the nineteenth century rarely mention Americans. Stephen Timoshenko, in his classic *History of Strength of Materials,* does not mention any American mathematical theorists other than Squire Whipple and Herman Haupt (both of whom developed theories of trusses), whereas he discusses the work of dozens of French researchers.[2] Of the (mathematical) theoretical works discussed by T. M. Charlton in *History of Theory of Structures in the Nineteenth Century,* five were written by American researchers and 49 by French researchers[3]; this ratio is much higher than the ratio of the populations of the two countries at the time.[4]

In industrial research, the situation was different. In the United States, dozens and dozens of firms and institutions began to carry out industrial research and testing on a more or less consistent basis during the nineteenth century: chemical companies such as Dow, Dupont, Kodak; railroad companies and steel companies such as the Pennsylvania Railroad and Carnegie Steel; companies involved in mechanical or electrical engineering, such as the Sellers company (machinery), the Proprietor of Locks and Canals in Lowell, Massachusetts (hydraulic turbines), Sperry Gyroscope, Edison Electric, and Bell Telephone; government bodies such as the Army, the Navy, and the National Bureau of Standards; private institutions and professional societies such as the Franklin Institute; and colleges and universities. Overall, American industrial research was remarkable for its thoroughness, precision, and creative character, and it became the basis for technological innovations of international importance in every imaginable field.[5]

France had little tradition of industrial research until well into the twentieth century. Striking evidence for this conclusion is presented in a study by Terry Shinn, who analyzed the research activities of 34 French firms between 1880 and 1940.[6] The firms Shinn analyzed were distributed over four sectors: chemical, metallurgical, electrical, and mechanical engineering. Shinn found that, out of 34 companies, only nine possessed laboratories for industrial research or product testing. French firms tended to acquire new technology by purchasing licenses and patents from other companies or inventors, and many of the patents were of American origin. Moreover, some of the French companies that did not sponsor industrial research were quite prominent, such as CGE (Compagnie Générale d'Electricité), Renault, St.-Gobain, and Pathé. Support for industrial research by the French government, by private organizations, by professional societies, and by universities was also comparatively limited. For example, the Conservatoire des Arts et Métiers, which had established laboratories and undertaken significant experimental research in the first half of the nineteenth century, had, according to one American observer, stagnated by the end of the century.[7]

It is not possible at present to quantify the investments in industrial research in France and America in the nineteenth century, or even the numbers of contributors. But there is at least some validity in looking at patent statistics; one would expect a relatively stronger tradition of industrial and design-oriented research to lead to more patents per capita.[8] What we find is that, except during the period 1840–1865, more patents were taken out in America than in France. During the period 1790–1815, the overall total of patents taken out in America was twice that in France, although France's population was several times that of the United States. American patents continued to outnumber French patents until 1836.[9]

In 1836 the American patent system, which since 1793 had been a simple registry system like that in France, was changed by law to require preliminary examination to verify the novelty and patentability of each invention.[10] France retained a simple registration system, without any guarantee of the novelty or workability of the patented invention. Moreover, a change in the French patent law in 1844 made it easier to obtain a patent. The change in the American system immediately slowed the growth of patenting in the United States, while the change in the French system led to an increase in patenting. As a result, until 1865 more patents were taken out per capita in France than in the United States. The tremendous impact of the change in the U.S. law can be seen in the fact that between 1842 and 1858 there were nearly 43,000 requests for patents but fewer than 22,000 patents were granted. Under the old system, all 43,000 requests would have been accepted.[11]

Beginning in 1865, however, and continuing until the end of the nineteenth century, more patents were taken out per capita in America than in France. In the period 1865–1875, the average number of patents taken out per year in America was at least twice the number taken out in France, although the populations of the two countries were roughly equal. From 1875 to 1890, the average number of patents taken out per year in America was between two and three times the number in France, although the American population remained less than twice the French population. And throughout the period 1865–1890 the French still held to a simple registry system, while the Americans still required preliminary examination to verify novelty and patentability.[12]

In the following chapters, I will explore the differences in orientation between French and American technologists. My basic aims are (1) to show that these differences were rooted partly in the different social and institutional contexts in which technologists learned and worked in France and America and (2) to show that they shaped both ideology and technological practice in domains beyond bridge building, and at a group level rather than merely at an individual level.

In order to achieve the first aim, chapters 7 and 8 will look, respectively, at the French and American technological communities in the nineteenth century, in order to understand more about their social and institutional organization. These chapters also include an overview of the growth of a system of formal technical education in each country. Formal technical education is often viewed primarily as an indicator of the rise of professionalism, but the concern here is rather to understand how French and American technical education was structured, both in terms of its social organization and in terms of its intellectual focus. For example, were students mainly taught mathematics, or did they spend most of their time in the laboratory? Technical schools' curricula embody ideas about what students need to learn in order to become "good" technologists. Likewise, technical schools may teach students to favor certain approaches over others, and we will see that this was indeed the case in French and American technical schools. In looking at technical education, we will also briefly consider the significance of apprenticeship, self-study, and on-the-job training.

In order to achieve the second aim of showing that the differences between French and American approaches extended beyond bridge building, and beyond the individual level, chapter 9 will look at rhetoric, ideology, and practice among French and American technologists. How did French and American technologists conceptualize the relations among mathematical theory, experimental research, and technological design? Did they tend to advocate and to follow different strategies in the creation and use of knowledge and in the design process? These questions will be addressed in general terms, but they will also be addressed in relation to a particular technology: the steam injector. This was a different kind of technology than the suspension bridge, which dated from a later period. The steam injector was a device, composed of a series of nozzles, that used steam from a boiler to feed water back into the same boiler. It was invented in France by Henri Giffard, and was patented in 1858. By comparing the bodies of research that were built around this technology in France and America, and by seeing the evolution of the technology in the two countries, we will see—at a group level and for a later period than represented by the work of Finley and Navier—how the different approaches of French and American technologists led to distinct patterns of technological practice and evolution.

Several points about the hypotheses underlying this analysis must be made clear at the outset.

First, this study is not intended to suggest that mathematical theory was an intrinsically "French" approach or that experimental research was an intrinsically "American" approach. These are simply two basic ways to generate knowledge. The fact that French technologists gave preference to one of these approaches, while American technologists gave preference to the other, thus has nothing to do with "national character" in any ethnic or biological sense. It does, however, have to do with the fact that different laws, institutions, and social networks created distinct environments at a national level. To the extent that laws, institutions, and networks function at a national level, it makes sense to speak of national differences. Looking at national trends in technology is thus quite justified. The existence of patterns at a national level does not preclude the possibility of other patterns' existing at a regional or municipal or individual level. The national level is just one of the levels at which structural patterns can be studied.[13]

Second, this study is not intended to suggest that American technologists completely ignored mathematics or that French technologists did not do any experiments. The issue is one of probabilities rather than absolutes. In absolute numbers, there were relatively few French technologists who devoted a great deal of time to mathematical theory; proportionately they were a minority. Moreover, as will become evident in the following chapters, the stratification of the French technological community must be taken into account in understanding the importance of mathematics for French engineers. The mathematical approach was particularly associated with technologists who worked in the state engineering corps, although I will show that the presence of this group led other technologists in France toward an interest in mathematical theory. My argument, therefore, is not that an absolute majority of French technologists worked on mathematical theory; it is that French technologists, as a group, devoted more time and energy to mathematical theory than American technologists, and that American technologists, as a group, devoted more time and energy to experimental research than French technologists. The extent to which this claim holds true evidently depends in part on the field of technology considered. One might expect mathematical research to have less relevance in the realm of mining technology, even in France; alternately, one might expect it to have more relevance in the realm of electrical technology, even in America. The present study does not attempt to achieve a sectoral analysis of this type. Its value is more heuristic, in that it does not presume to be a complete analysis of French and American technological approaches or to be uniformly applicable to every technological sector.

The third point concerns terminology. In comparing French and American technologists, I specifically mean those who did research, designed machines and structures, or innovated in some way. They may have been formally trained, as was Navier, or they may have been self-educated, as was Finley. Often the term *engineer* is applied only to individuals who have had a formal, academic technical education, but in the present context the term is used analogously to my use of the term *technologist*. Thus, Finley and Navier were both engineers in the sense that both were engaged in "bridge engineering": researching, designing, and building bridges. Similarly, Thomas Edison, although he had no formal academic training, is referred to as an engineer because he did what is commonly referred to as electrical engineering: he researched, designed, and built electrical apparatus.

A further clarification about terminology concerns the meaning of the terms *theoretical* and *practical*. The closer one looks at technologists' research practices and ideologies, the more it becomes evident that these terms may have many different meanings. A theory can be non-mathematical, and "practical" does not necessarily imply an absence of mathematical theory. As we saw in part I of this study, the analytical ideal held by Prony, Navier, and others in France had the explicit goal of making mathematical theory practical in the sense of making it a useful and creative tool for design and innovation. On the other hand, Finley sought a theoretical basis for proportioning a suspension bridge—a basis derived from "natural laws"—yet he attempted to discover these laws not through mathematics but by means of experiments. In studying the issue of French and American ideologies regarding the relation between "theory" and "practice," attention must be given to how technologists in each country interpreted these terms. The analysis in the following chapters thus does not intend to suggest a simple dichotomy between French "theory" and American "practice." Rather, it considers, on the one hand, how technologists in each country interpreted these terms, and, on the other hand, the extent to which they used either mathematical or experimental approaches in their research, whether or not that research was "basic" or "applied."

The French Technologists

Two features of France's society especially shaped that nation's technological community in the early nineteenth century: social stratification and a high degree of state control over technological activity. The latter was manifested in the power of the state engineering corps, which was not matched by the U.S. Army Engineering Corps or the Engineer Corps of the U.S. Navy. Besides Navier's Corps des Ponts et Chaussées, there were eleven other corps by 1837, among which were the Corps des Mines, the artillery corps, the corps of military engineering (fortifications), and the corps for naval construction, maritime engineering, munitions manufacture, state manufactures, geography, and tobacco. These corps played many important roles in French technological development. For example, anyone who wanted to build or use a stationary steam engine in France had to get prior permission from the Corps des Mines and, until 1865, had to follow detailed design and safety standards established by the engineers of this corps. Engineers of the Corps des Mines also inspected, regulated, and kept track of mining and metallurgical establishments throughout the country.[1] Engineers of the Corps des Ponts et Chaussées did the same for railroads, bridges, canals, and ports, and indeed they designed and oversaw the building of many of these structures.[2]

The corps varied in size, but they all mainly employed graduates of the Ecole Polytechnique. (For most of the corps this was a legal requirement; the Corps des Ponts et Chaussées employed only *polytechniciens* until late into the nineteenth century.) Statistics on admissions to this school therefore give a rough upper limit of the total number of engineers serving in these corps. Between 1805 and 1883, a total of 11,757 students were admitted to the Ecole Polytechnique—an average of slightly under 150 per year. All but 10–15 percent of these entered the state and military corps.[3]

Data on the Corps des Ponts et Chaussées help to provide a sense of the size of the state corps. Over the course of the nineteenth century, the number of engineers in this corps remained between 450 and 650, including students. In 1804 the total number of engineers and students in the Corps des Ponts et Chaussées was 622; in 1851 the total was 617; in 1891 the total was 455.[4] Recruitment of *polytechniciens* to the Ecole des Ponts et Chaussées averaged a little under 22 per year between 1806 and 1850; subsequent recruitments to the Corps des Ponts et Chaussées during the same period averaged between 20 and 21 per year.[5]

The military engineering corps was comparable in size to the Ponts et Chaussées; the artillery corps (more a military than an engineering corps) was significantly larger.[6] The Corps des Mines and the others were smaller than Corps des Ponts et Chaussées, however. In 1810, the Corps des Mines comprised a total of 68 engineers, including students (*aspirants*).[7] This corps grew slowly during the nineteenth century; available statistics indicate that it drew no more than one-third as many students as the Corps des Ponts et Chaussées.[8]

Another feature of French society that shaped the technological community was social stratification. Although the French Revolution had considerably diminished the power of the *ancien régime* aristocracy, the system of social values and relations which the latter had engendered had an enduring influence in nineteenth-century France.[9] Among the values that persisted, one that particularly affected the technological community was a disdain for manual labor. French artisans and mechanics commented on this throughout the nineteenth century. A characteristic example was the following complaint written by a clockmaker, Alfred Beillard, in 1888: "Unfortunately, aristocratic prejudices are so deeply rooted in our morals and among our public authorities, that it will be necessary to battle for a long time to come in order to make their deadly influence disappear. Instead of sincerely encouraging workers, there has always been, why not say it, a hint of disdain for those who resort to [manual] labor in order to live."[10] Social stratification promoted credentialism, and it fostered the growth of a hierarchical three-tier system of technical education in which each tier tended to draw students mainly from a particular social level; disdain for manual labor shaped the technical curricula.

Technical Education

Formal technical training already played a significant role in France in the first decades of the nineteenth century.[11] In the United States only the U.S. Military Academy at West Point (established in 1802) and the Rensselaer School (established in 1825 and later known as the Rensselaer Institute and then the Rensselaer Polytechnic Institute[12]) offered anything approaching a comprehensive technological training during the first three decades of the nineteenth century; however, a range of schools in France offered such training. First, the Ecole Polytechnique fed several *écoles d'application.* Among the latter were the Ecole des Ponts et Chaussées, the Ecole des Mines (Paris), and the Ecole du Génie Militaire. There were, in addition, other schools that were not linked to the Ecole Polytechnique. Among these were the Ecole Centrale (established 1829), the Ecole des Mineurs at St.-Etienne (1816),[13] and the Ecoles d'Arts et Métiers at Châlons and at Angers (ca. 1803–04).

A number of French technical schools had substantial enrollments. Between 1800 and 1848 more than 6,000 students were admitted to the Ecole Polytechnique, an average of about 120 per year.[14] After its establishment in 1829, the Ecole Centrale admitted more than 1,000 students from 1829 to 1847, about 60 per year. In later years, its enrollment grew: the graduating class of 1850 was 67, and the class of 1865 was 135.[15] As for the Ecoles d'Arts et Métiers, maximum annual recruitment was set by law in 1817 at 100 students per year for the school at Châlons, and at 50 per year at Angers; in 1832 the numbers for both schools were set at 100 per year, and this limit remained in force throughout the nineteenth century. In 1843 a third arts and trades school was established at Aix-en-Provence, with a yearly recruitment limit likewise set at 100. As of 1851, there were about 4,000 living graduates of the Ecoles d'Arts et Métiers.[16]

Although this system of technical schools continued intact throughout the nineteenth century, other institutions were established during the latter decades of the century. Prominent among these were the Ecole Supérieure de Télégraphie (1878), the Ecole de Physique et de Chimie Industrielle (1885), and the Ecole Supérieure d'Electricité (1892).[17]

The Ecole Polytechnique and the affiliated *écoles d'application* constituted the top tier of the hierarchy of French technical schools.[18] Their high prestige earned for them the official title of *grandes écoles*. The Ecole Polytechnique kept a military organization throughout the nineteenth century (except for a brief period during the Restoration, from 1815 to 1830). *Polytechniciens* wore military uniforms and were trained to become officers. The Ecole Polytechnique's recruitment policies and entrance requirements ensured that its students and the students of the affiliated *écoles d'application* would be drawn largely from the upper echelons of society. The entrance examination was so difficult that lengthy and expensive preparatory training (generally lasting one or two years beyond the *baccalauréat*) was a necessity for prospective candidates. The written part of the examination occupied approximately 24 hours, distributed over four days. To do well, candidates had to have a thorough mastery of algebra, geometry, and trigonometry, a grounding in mechanics, physics, and chemistry, and knowledge of a foreign language.[19]

Up to 1880, the proportion of *polytechniciens* drawn from the *classes populaires*—workers, peasants, artisans, and shopkeepers—never exceeded 4 percent, although these groups represented about 90 percent of the population. Rather, these schools drew the majority of their students (60–70 percent) from the highest levels of society: the families of large landowners, high-ranking officers and administrators, industrialists, wealthy merchants, and members of the liberal professions. From 1880 to 1900 recruitment broadened somewhat, and the percentage of students from the *classes populaires* increased to almost 20 percent.[20]

The recruitment policies of the Ecole Polytechnique made it inevitable that the *écoles d'application* would draw their students from the same social strata, because most of the latter drew students only from among the graduates of the Ecole Polytechnique. The Ecole des Ponts et Chaussées provides a good example. Between 1804 and 1851 it drew less than 7 percent of its students from the *classes populaires*. More than 32 percent came from the families of large landowners, wealthy bankers, merchants, or industrialists, and another 21.6 percent from families of high-ranking administrators, officers, or members of the higher levels of the liberal professions. Nineteen percent came from the families of middle- and lower-ranking officers, administrators, and members of the liberal professions.[21]

The second tier of French technical schools was represented by the Ecole Centrale des Arts et Manufactures, founded privately in 1829 to train engineers for industry.[22] Tuition costs were high, but the Ecole

Centrale had a more industrial orientation than the Ecole Polytechnique, and the entrance exam was not as stiff. What this meant in terms of recruitment in the period up to 1850 was that, whereas around 50 percent of *polytechniciens* were from the highest ranks of society (the families of large landowners or high-ranking civil servants), only around 30 percent of the *centraliens* were. A significantly larger proportion of *centraliens* than *polytechniciens,* however, were sons of industrialists, merchants, or middle- and low-level civil servants: around 50 percent for the Ecole Centrale, as opposed to about 30 percent for the Ecole Polytechnique.[23]

The lowest tier was represented by the Ecoles d'Arts et Métiers. These schools were nominally intended to train artisans, mechanics, and foremen, although in fact they were a major source of mechanical engineers in France in the nineteenth century, and an important source of civil engineers. Nevertheless, graduates were not accorded the title *ingénieur* until the twentieth century. Candidates for admission had to have completed at least a year of apprenticeship. It is therefore not surprising that very few students of these schools (who were called *gadzarts*) came from families in the higher ranks of society.

Data compiled by Charles R. Day in his comprehensive history of the Ecoles d'Arts et Métiers shows that, in the period 1830–1890, 27 percent of students came from families of artisans and small shopkeepers; 27 percent had fathers who were technicians, draftsmen, foremen, or workers; and 22 percent had fathers who were manufacturers, managers, or "engineers"—however, the "engineers" were not corps engineers, and most of the manufacturers were small manufacturers (e.g., owners of small machine shops). None of their fathers were high-ranking officers, administrators, or members of the liberal professions.[24]

The social gulf that separated the Ecoles d'Arts et Métiers from the Ecole Polytechnique and the Ecole Centrale can be seen in details of the functioning of the schools, such as the daily regime and schedule. The Ecoles d'Arts et Métiers, throughout the nineteenth century, were run almost like prisons. Students "were drummed awake at 5:30 A.M." each day, and all their movements were strictly monitored and regulated—often by drumbeat. Although the students put in hard labor every day in the schools' forges and workshops, they were allowed only one shower every three weeks. The regime of strict control extended even to the use of paid "proctors," whose task was to spy on students and report to the authorities.[25] As Day explains:

The proctors served no educational purpose in the schools; rather they were a cross between military policemen and prison guards. . . . They were stationed everywhere—the library, infirmary, workshops, classrooms, court-yard, study hall, bathrooms, and dormitory. Their tiny bedrooms, situated adjacent to the dormitory halls and dubbed "outhouses" [*chiottes*] in the school slang, were equipped with one-way peepholes for spying on their charges. The proctors even spied on the students in the toilets. Each one carried a notebook and reported anonymously to the head proctor, who could pass on the information to the director or censor without the student ever knowing about it.[26]

The students of the Ecoles d'Arts et Métiers worked and studied 12 hours per day, six days a week. Their only time off was Sunday, and even then they were required to study four hours. They were not permitted to celebrate any holiday from October until Easter. They were kept isolated from the outside world during the school year. They were never allowed out of the schools except to go to church on Sunday—under guard, and in military formation. (They did have an eight-week break during the summer, however, and eight days at Easter.) In line with this strict regimen, *gadzarts,* unlike students at the Ecole Centrale and the Ecole des Mines, were never taken to visit any factory or workshop during the three years of their studies.

Polytechniciens also had a highly structured routine, but they were not treated as prisoners. They got up at 6 A.M. and put in $11\frac{1}{2}$ hours per day, but $5\frac{1}{2}$ of those hours were classified as "free study." They got Wednesday afternoon off as well as Sunday, and they could leave the school on their own during those times.[27] Whereas *gadzarts* were expected to clean their own quarters, *polytechniciens* were not. When the American scientist Alexander Dallas Bache visited the Ecole Polytech-nique in 1839, he found that "the pupils do not perform any duties of police; their arms are cleaned and kept by an armourer, and bar-rack-keepers are appointed for the police of the quarters—provisions which I look upon as defects."[28] Since it was considered beneath the dignity of *polytechniciens* to make their own beds or clean their own guns and swords, it is not difficult to imagine how they viewed them-selves in relation to the *gadzarts,* who spent many hours every day engaged in manual work.

The Curricula of the Technical Schools

The social stratification within French technical education was linked to stratification in the kinds of technological knowledge students were exposed to, and also to social and occupational stratification of technical school graduates. The stratification of technological knowledge can be seen by comparing the curricula of French technical schools. As we have already seen, the curriculum of the Ecole Polytechnique concentrated on mathematics, theoretical mechanics, and descriptive geometry. Bache reported in 1837 that over 70 percent of the lessons of the first-year students were devoted to mathematics and theoretical mechanics, while the proportion in the second year was 54 percent. The only course that involved students in any experimental or laboratory work was the chemistry course, which represented only a small part of the curriculum (around 15 percent, with laboratory work occupying only a limited portion of that). Moreover, according to Bache, the method of ranking the students academically placed the most weight on knowledge of mathematics and the least on physics and chemistry; even a student's conduct was weighted more heavily than his knowledge of physics and chemistry.[29] A further problem was a lack of integration between chemistry lectures and lab work, because the chemistry professor did not direct the laboratory work; it was, rather, directed—and even graded—by the laboratory keeper (*conservateur*). Also, laboratory work was infrequent: as of 1851, students did one laboratory assignment every two weeks.[30]

In 1869, the American educator Henry Barnard also visited the Ecole Polytechnique; he reported that the curriculum was still essentially the same as that outlined by Bache in 1837. Barnard further commented on "the chronic dispute which has gone on from the very first year of the school's existence, between the exclusive study of abstract mathematics on the one hand, and their early practical application on the other." He believed that the controversy would produce no real change, that "the traditional teaching of the school will be too strong for legislative interference, and that, in spite of recent enactments, abstract science and analysis will reign in the lecture-rooms and halls of study of the Polytechnic. . . . "[31]

Barnard's intuition proved correct: although there were efforts at reform, particularly in the last two decades of the century, the curriculum kept essentially the same focus throughout the nineteenth century. Little attention was given to experimental work. In the physics course, students did little or no laboratory work before 1880.[32] A book on the Ecole Polytechnique published in 1932 noted that the chemistry laboratories used by the students had undergone no fundamental reorganization or expansion since 1840. By that time, however, a physics laboratory had been established. The book also noted, however, that students spent, on average, 6–7 hours per day in study rooms, going over their course work in small groups. (Group study in small, assigned rooms was a long-standing tradition of the school.) Such a schedule left little time for laboratory work.[33]

The curricula of the affiliated écoles d'application devoted more attention to the immediate needs and details of technological practice than that of the Ecole Polytechnique. At the beginning of the nineteenth century, students at the Ecole des Ponts et Chaussées, which had a three-year program, took lecture courses in mechanics, in stereotomy as applied to construction, in drawing, and in architecture.[34] Bache reported in 1839 that the course work included, in addition to the preceding, construction, mineralogy, geology, administrative law, and foreign languages. The applied mechanics course included hydraulics and hydrodynamics.[35] By 1875, first-year students studied applied mechanics, mineralogy and geology, political economy, agriculture, roads, and construction; second- and third-year students had more specialized courses, covering such topics as bridges, steam engines, railroads, maritime construction, architecture, and administrative law; third-year students had drawing as well.[36] The curriculum in 1888 was very close to that of 1875.[37]

Although the content of some of the courses given at the Ecole des Ponts et Chaussées was mainly descriptive, other courses used and extended the mathematics taught at the Ecole Polytechnique. For example, between 1819 and 1836 the course in applied mechanics at the Ecole des Ponts et Chaussées was taught by Navier, who gave it a highly mathematical orientation. The direction Navier set was continued throughout the nineteenth century.[38] Likewise, the course on construction, with its increasing number of subdivisions dealing with specific topics (such as bridges and railroad construction), although always containing a substantial body of descriptive material, came increasingly to include mathematical theory. This trend, already evident in the first half of the nineteenth century,[39] continued in the second half of the century.[40] Students in 1866–67, for example, covered the basics of the mathematical theory of suspension bridges.[41] By extending the students' mathematical and theoretical training, the schools helped to reinforce this orientation. Barnard saw the effects in the character of the journal published by the Corps des Ponts et Chaussées, which he found remarkable for its "profound theoretical mode of treating all subjects relating to engineering."[42]

The duration of the formal course work at the Ecole des Ponts et Chaussées was initially less than five months per year, but in 1851 it was extended to six months. The other six months were spent in the field, assisting Corps engineers around the country with their work. Much of this summer experience was routine, but it taught the students the daily work practices of Corps engineers.[43]

Laboratory work received much less emphasis than field work at the Ecole des Ponts et Chaussées. A small chemical laboratory was established at the school in 1852, specifically for analyzing cements and other building materials. In 1876 this laboratory was replaced by a larger one in another building, which was used both by students and by a permanent staff who did routine analyses for third parties. As late as 1891, however, this laboratory was used by the students for a total of only 15 days per year.[44] A testing laboratory was established in 1851 and was furnished with machines for testing the strengths of materials. The testing laboratory was moved and enlarged between 1867 and 1871. The larger quarters included chemistry laboratories, a laboratory for testing the strengths of materials, and a small workshop (perhaps 10 feet square) containing a forge and a single set of small-scale machine tools intended for demonstration purposes.[45] These laboratories were still in use in 1891, and had not been significantly expanded. Most of the testing machines in use in 1891 dated from 1871, although some had been replaced. In 1891 the small workshop remained "pretty much in the same state" as in 1871.[46]

The Ecole des Mines in Paris had a two-year curriculum that covered four main subjects: mines and machines, metallurgy, assaying, and geology and mineralogy. Several courses were added to this basic curriculum around 1845: railroads and construction, topography, and industrial economics and legislation. After the Franco-Prussian War of 1870 a course in artillery was added, and in the late 1880s courses in industrial chemistry and in factory organization were established.[47] Other courses were split up; for example, the mines and machines course was split into two courses after 1885.[48] Training at the Ecole des Mines also included laboratory and field work. Bache reported in 1839 that the school possessed chemical laboratories, and that students did laboratory work in the summer of their first year as well as making visits to nearby mining and metallurgical establishments. In the second summer they were sent to establishments in other parts of France.[49] An assay laboratory was constructed in 1845, with the mission of analyzing mineral specimens (at no charge) for industrialists, miners, and ironmasters.[50]

The second tier in the French technical school hierarchy was represented by the Ecole Centrale. Its three-year curriculum was oriented more toward industrial technology than the curricula of the Ecole

Polytechnique and the main *écoles d'applications. Centraliens* devoted significantly less time to mathematics than *polytechniciens,* and they spent at least twice as much time on physics and chemistry. Courses at the Ecole Centrale also emphasized industrial "applications" of science and mathematics. Bache reported in 1839 that students in all years took descriptive geometry, analytical geometry and mechanics, and theory of machines. In addition, the first year covered general physics, chemistry, and "hygiene and natural history applied to the arts." Second-year students took machine construction, industrial physics (with sections on dyeing, distillation, motion of air in pipes, and construction of lightning rods), industrial chemistry, analytical chemistry, architecture and civil engineering, mining, and iron metallurgy. Special courses for the third year concerned steam engines and railroads. No calculus was taught. In 1868 the curriculum was structurally similar to that of 1839, but now calculus was included in the first year and more attention was paid to chemistry and metallurgy.[51]

As these curricula show, the Ecole Centrale devoted much less time and attention to mathematics than the Ecole Polytechnique and the *écoles d'applications.* Yet it devoted significantly more time to laboratory training. In the 1830s, students were required to do weekly laboratory work in both physics and chemistry. In addition, students in the descriptive geometry course had to solve modeling problems using plaster. In the course on industrial physics, they constructed models of chimneys and furnaces. In the courses on machines, public and industrial construction, physics, and chemistry, they had to complete design projects. They also visited construction sites and practiced surveying. Each student's ranking depended not only on his performance in examinations but also on the marks he had received for his class work and laboratory work throughout the year.[52]

The amount of laboratory and practical work appears to have been about the same in 1868 as in 1839. Students were required to do laboratory work in chemistry and physics in the first year. They were also assigned projects in drawing, and during the summer they had to prepare plans of buildings and machines. In the second year they had to do laboratory work on the flow of gases, chimney and oven projects using brick, and "practical exercises in a factory on the construction of machines." They also had to do 27 laboratory exercises in chemical analysis and assaying, and a further project in which they surveyed a watercourse and measured the volume of water in a stream. The second summer had to be spent visiting factories. The practical work for the third year focused on working up a series of design projects, including drawings, cost estimates, and specifications. The problems concerned "subjects connected with machines, buildings, metallurgy, and chemistry," such as the design of girders, locomotives, and boilers.[53]

The third tier of the technical-school hierarchy was represented by the Ecoles d'Arts et Métiers, which had a three-year program and offered primarily manual training. In 1830, and still in 1868, the students spent 7 hours per day in the workshops and foundries, and 5 hours per day in the classroom. The curriculum in 1830 covered arithmetic, algebra, descriptive geometry, basic physics and chemistry, and French composition and grammar. The workshop training included iron and brass founding, wood turning, pattern making, forging, and machine fitting. In 1848, a course in industrial mechanics (which included a dynamometer demonstration) was added, and more emphasis was placed on drafting.[54]

In 1868, the curriculum for the first two years included arithmetic, algebra up to quadratic equations, basic geometry and plane trigonometry, and the elements of descriptive geometry. Drawing was taught in all three years. In the third year, several additional courses were taught: industrial mechanics (with sections of hydraulic motors and steam engines), basic physics, and "a few elements of chemistry." The workshop training was carried out in four shops: a pattern shop, a smithy, a foundry, and a machine shop. In the pattern shop, students learned woodworking and how to make the patterns needed to cast gears and parts for steam engines and other machinery. In the smithy they learned the arts of forging, filing, and fitting, and they learned to use a steam hammer. In the foundry they cast machine parts and statues. In the machine shop they learned first to work carefully with hand tools and then to work with machine tools.[55]

Later in the nineteenth century, the curriculum of the Ecoles d'Arts et Métiers was upgraded. By 1909 it included calculus, electrical physics, and metallurgical chemistry. Students now spent 7 hours every day in the classroom and only 5 in the workshops.[56]

To summarize: The most prestigious French technical schools focused extensively on mathematical theory, while the least prestigious ones focused mainly on workshop training. In particular, students at the Ecole Polytechnique did a lot of mathematics but almost no laboratory work, while students at the *écoles d'application,* in addition to theoretical mechanics and structural theory, did some laboratory work but had no workshop training. The second stratum, represented by the Ecole Centrale, offered an education that combined theoretical and laboratory work, but with less emphasis on mathematics than the first-tier schools and more emphasis on laboratory work and design. Like the first-tier schools, however, the Ecole Centrale provided relatively little workshop training. The third stratum, represented by the Ecoles d'Arts et Métiers, offered mainly workshop training, with only the rudiments of mathematics and theory and no laboratory work. In short, the social stratification that existed within French technical education was paralleled by a stratification of mathematical, laboratory, and workshop training.

Occupational Barriers

The French system of technical education—with its social stratification and its tendency to separate mathematical, laboratory, and workshop training—was also linked to a system of social and occupational barriers among French technologists.[57] The most notable barrier involved the state engineering corps, which recruited only from the Ecole Polytechnique and the *écoles d'application.* In the very beginning of the nineteenth century, when these schools had a virtual monopoly on formal technical education, this rule had at least some justification, although it evoked opposition even from the outset. However, with the evolution of the Ecole Centrale and the Ecoles d'Arts et Métiers, which produced graduates qualified to function as engineers in the private sector, the fact that these graduates were barred from employment in the state corps became more and more incomprehensible, and it engendered political conflict throughout the nineteenth century.

The conflict over recruitment into the state corps provides a good sense of the character and depth of the social and occupational divisions that existed among French technologists. A group of engineers known as *conducteurs* were among those most actively involved in the conflict. The *conducteurs* formed a hierarchy below and legally separate from the official engineering hierarchy of the Corps des Ponts et Chaussées. The number of *conducteurs* grew from 500 in 1818 to almost 5,000 in 1890, while the number of "engineers" remained a little above 500 throughout the century. Thus, by the end of the nineteenth century there were many more *conducteurs* than "engineers."[58]

In the early part of the nineteenth century, *conducteurs* were hired on an *ad hoc* basis, with no significant formal qualifications except the equivalent of a grade-school education (which most people at that time lacked). These *conducteurs* learned on the job. By the 1850s, however, the Corps had established stricter qualifications: prospective *conducteurs* were now selected by competitive examination and had to have mastered logarithms, plane and solid geometry, the use of trigonometric tables, the solution of second-degree equations, computation of square roots, and the calculation of embankments, cuts, and fills, and they had to have knowledge of construction materials and of their strengths, and of surveying.[59]

Many *conducteurs* came out of the Ecoles d'Arts et Métiers. Studies of the employment patterns of graduates of the Ecoles d'Arts et Métiers show that, over the course of the nineteenth century, between 10 percent and 30 percent worked for the Corps des Ponts et Chaussées.[60] In contrast, almost no *conducteurs* were graduates of the Ecole Centrale or the Ecole Polytechnique.[61] The function of the *conducteurs* was to assist the Corps engineers in all their duties and responsibilities: they made surveys, prepared drawings, compiled data and cost estimates, handled accounts, kept logbooks, inspected sites and materials, oversaw foremen and workers, and wrote reports. More generally, they drew up plans for construction projects and then oversaw technical, financial, and administrative details of their execution. In the words of one contemporary observer, *conducteurs* were the "alter egos" of the Corps engineers.[62] Some became more than that: by 1848, 65 *conducteurs* were doing the work of *ingénieurs d'arrondisement*,[63] a position nominally filled only by Corps engineers.[64]

Despite the fact that *conducteurs* clearly functioned as engineers, they were not allowed to advance into the Corps des Ponts et Chaussées. Throughout the nineteenth century, the disparity in status was legally sanctioned and socially reinforced. The inferior status of *conducteurs* was spelled out even in such minute details as the buttons on their coats, as the editor of the *Journal du génie civil* pointed out: ". . . the [state] engineers wear blue coats with red collars, gold embroidery, and gilded buttons; the *conducteurs* wear blue coats with silver embroidery and silvered buttons. Can it not be seen, first of all, that there is something there which conveys a little bit the idea of master and lackey?"[65] On a social level, it was said that *conducteurs* were "treated with haughtiness by the majority of [Corps] engineers, who seem to delight in reminding them at every moment of their inferiority."[66]

The *conducteurs* struggled throughout the nineteenth century to get legislation enacted that would enable them to advance into the engineering corps. The first demands for such legislation were made during the Revolution of 1830, yet no progress was achieved until after the Revolution of 1848. In 1850 and 1851, laws were passed stipulating that up to one-sixth of the Corps engineers might be recruited from the ranks of the *conducteurs*. A *conducteur* could not become eligible for promotion until he had put in 10 years' service, however. In addition, he had to be specifically recommended for promotion by his superiors, and he had to pass a battery of examinations extending over a period of two years.[67]

The law remained a dead letter: not one *conducteur* was admitted into the Corps during the next 20 years. However, by 1868 more than eighty *conducteurs* were actually serving in the capacity of *ingénieur ordinaire*, a low rank in the Corps; but they were not permitted to bear that title or earn the salary that went with it. Further legislation was passed in 1868 and in 1877, and eleven *conducteurs* were admitted into the Corps between 1871 and 1880. However, by 1881 some 112 *conducteurs* were serving in the capacity of *ingénieur ordinaire*, and two were fulfilling, on an interim basis, the duties of *ingénieur-en-chef* (one of the higher ranks in the Corps). But still they were denied the official titles and salaries that went with these positions. It was not until the twentieth century that the treatment of *conducteurs* began to be ameliorated significantly, as a result of legislation passed in 1907 and 1915.[68]

The disparity in status between the *conducteurs* and the Corps' engineers was clearly not rooted in any inherent inability of *conducteurs* to function as engineers. An analysis by Charles R. Day of the occupational attainments of 490 Arts et Métiers alumni who graduated between 1820 and 1880 reveals that, of the 351 who went into industry, 78 percent became owners of, or high-ranking managers and engineers in, medium-size and large firms. In contrast, of the 67 who entered the public services (e.g., as *conducteurs*), 64 percent never went beyond mid-level positions. Of the 63 who became involved in railroads, 76 percent never went beyond mid-level positions.[69] The higher managerial positions in the railway companies were strongly dominated by *polytechniciens:* up to 1870, 66 percent of the directors and chief engineers of the six major French railway companies were *polytechniciens* and another 9 percent were *centraliens*. Since it is implausible that managing large industrial firms demanded less knowledge or ability than managing public-works enterprises or railroads, it can only be concluded that the lower attainment of *gadzarts* in the latter sectors was due to the power of the state engineers to retain control of them.[70]

The social and occupational barriers among French technologists not only kept *gadzarts* out of the state engineering corps; they also tended to keep *polytechniciens* out of machine shops and heavy industry. Machine shops and heavy industry were associated with manual labor and with the working classes (from which most *gadzarts* were recruited), and they accordingly had a certain social stigma. Denis Poulot, a *gadzart* who began his career as a fitter and eventually became the owner of a machine-tool company in Paris, described the humiliation he was made to feel about working with his hands:

To remove the calluses produced by handling the hammer, I took care each Sunday to grind my hands, the pumice stone not having been energetic enough. Placed one day beside a mother and her daughter (a pretty brunette), I responded without malice to the questions that she addressed to me. "What do you do in this machine building firm? she asked me.—Madame, we are setting up a six horse power steam engine, for the exposition at London.—I know, sir, that steam engines are made in your firm; what I ask of you, is the work that you are occupied with there.—But, madame, I had the honor of telling you that I am building a steam engine." (I was happy to be able to affirm the confidence that, while still young, I merited from my boss.) The young woman said to me with a dumbfounded air; "What! You work, you are therefore exposed to all the filth that trade includes?" A bit vexed I responded "But yes, miss, and I dare to believe that none is apparent at this moment." The mother turned her back and the eyes of my beautiful neighbor fell on my well-ground hands, which did not betray me, and she moved away. For her, I was a plague-stricken person. To describe the effect produced by this sign of scorn on a spirit so ardent as mine would be difficult: I felt throbbing in my veins, my face must have changed all the colors of the rainbow.[71]

The association of machine shops and heavy industry with manual labor helped to ensure that *gadzarts* would play an important role in this sector in France. And the status gap between *polytechniciens* and *gadzarts* made it unlikely that the former would go to work in enterprises owned or managed by the latter. Between 80 percent and 85 percent of *polytechniciens* went into the state and military services and remained there. Of those who ended up in the private sector, many went into railroads, mining, and the chemical industry. Data on the class of 1872 show that, among those who moved into the private sector, 27 percent went into the mining industry; 40 percent into the chemical, sugar, glass, paper, and crystal industries; and 18 percent into railroads.[72]

Comparatively few industrial enterprises were founded by *polytechniciens* in the nineteenth century. This was true in the public-works sector,[73] and even truer in the machine-building sector. The relatively small role played by *polytechniciens* and graduates of the *écoles d'application* in the machine-building industry is evident from statistics on the Chambre Syndicale des Mécaniciens, Chaudronniers et Fondeurs, the main trade association of machine builders in France during the second half of the nineteenth century. In his analysis of its membership, James Edmonson found that, of the heads of the firms for which information was available (132 out of 320), only 7 percent were graduates of the Ecole Polytechnique or the *écoles d'application*, whereas 19 percent were graduates of the *écoles d'arts et métiers*. Another 41 percent were graduates of the Ecole Centrale; and 29 percent had no formal technical education, presumably having learned through apprenticeship.[74]

The occupational and social divisions between *gadzarts* and *polytechniciens* were greater than those between *gadzarts* and *centraliens*. As the statistics above suggest, *centraliens* were less inhibited about involvement in machine building, and they entered this sector in significant numbers during the second half of the nineteenth century. *Gadzarts* and *centraliens* were also often united in their opposition to the dominance of the state engineers; both groups wanted the right to be employed within the engineering corps.[75]

At times, social tensions between *gadzarts* and *centraliens* were also evident, however. An example can be seen in events surrounding the Revolution of 1848. In the aftermath of the disturbances, a plan was devised to establish "national workshops" (*ateliers nationaux*) in order to create jobs for the unemployed, and *polytechniciens* and *centraliens* were supposed to cooperate in directing public-works projects. The *polytechniciens* expected the *centraliens* temporarily to adopt the rank of *conducteur,* but the *centraliens* angrily refused, feeling this title to be an insult.[76] Of course, the *centraliens'* stand was predicated mainly on their reluctance to be under the command of the *polytechniciens,* but it also reveals their sense of superiority over the *conducteurs,* many of whom were *gadzarts.* In 1868, when the situation of the *conducteurs* was brought up at a meeting of the Société des Ingénieurs Civils, which was dominated by *centraliens,* the society declined to take an active interest in the issue. A report of the meeting noted that the society, "composed largely of alumni of the Ecole Centrale,"

appears at present to have only a mediocre interest in this question, and the *conducteurs'* position will hardly seem enviable to the alumni of this school as long as they are consigned, by virtue of the conditions for entering and functioning within the Corps des Ponts et Chaussées to an inferiority of position which should only be the consequence of an inferiority of talent.[77]

Engineering Societies and Alumni Organizations

Further evidence for the social barriers that existed among French technologists can be seen in the character of their professional organizations during the nineteenth century. In the United States, technologists created sector-based professional societies: a society for civil engineering, and societies for mechanical engineering, mining engineering, electrical engineering, and chemical engineering. In contrast, French technologists initially created mainly school-based organizations. The Société des Ingénieurs Civils, founded in 1848, was dominated by *centraliens* and was created specifically to counter the power of the engineers in the state corps.[78] The name initially given to the organization, Société Centrale Des Ingénieurs Civils, emphasized the link with the Ecole Centrale. The first meeting of the society, moreover, took place at the Ecole Centrale, and it was attended only by alumni and professors of that school.[79] A year earlier, the *gadzarts* had founded an alumni organization, which came to be known as the Société des Ingénieurs Arts et Métiers; by 1880 it had 2,000 members, and by 1914 it had 9,000. This organization functioned as a mutual-aid society, as a professional society, and as a political pressure group.[80] The Ecole Polytechnique had its own alumni society, established in 1865, and another alumni society was created for graduates of the Ecole des Ponts et Chaussées and the Ecole des Mines. In 1862, *centraliens* founded an alumni society. Though all these organizations functioned as friendly societies and mutual-aid societies, some functioned as professional organizations as well. For example, the alumni organization of the Ecole Centrale was behind the establishment of the technical journal *Le génie civil*.[81]

The first society created by the *centraliens*, the Société des Ingénieurs Civils, eventually became a broader-based professional society for engineers. Yet its membership was composed predominantly of *centraliens* until the 1880s.[82] It did not exclude engineers from other backgrounds, however, and by the 1880s some 20 percent of its members were *gadzarts*. In 1883 a *gadzart* was elected president of the society.[83] Other, sector-based engineering societies began to be established in the 1880s—an association for electrical engineers was created in that decade, and one for chemical engineers after 1900.[84]

Conclusion

This survey has overlooked many aspects of the evolution of technical education and the technological community in nineteenth-century France. Perhaps most significant, it did not take stock of the specialized and regional technical schools that emerged during the last decades of the nineteenth century. Despite these lacunae, the evidence presented does permit several conclusions to be drawn.

First, the Ecole Polytechnique, the *écoles d'application,* the Ecole Centrale, and the Ecoles d'Arts et Métiers together played a significant role in training technologists during the nineteenth century. Each group of schools produced thousands of graduates, and these graduates dominated the state corps, civil engineering (including railways), machine building, and other sectors.

Second, there were enduring occupational barriers among French technologists in the nineteenth century, and these, along with the associated social barriers, influenced the creation of professional associations. Although *polytechniciens, centraliens,* and *gadzarts* were all involved in civil engineering (as *conducteurs,* as civil engineers in the private sector, and as engineers in the Corps des Ponts et Chaussées), they did not easily create broad-based national professional associations; rather, they tended to remain isolated within their separate alumni societies.

Third, France's technical schools were organized into a hierarchy of status, which was manifested both in recruitment patterns and in curricula. Each tier in the hierarchy tended to give preference to a different form of technological learning: mathematical theory for the *polytechniciens,* laboratory and design work for the *centraliens,* shop work for the *gadzarts.*

8

The American Technologists

The American and French technological communities evolved in fundamentally different environments. The overview of the career and work of James Finley in chapters 1–3 pointed to some of these differences, but it is useful to examine them in a more systematic way.

Perhaps most important, the United States at the turn of the nineteenth century was a thinly populated nation expanding rapidly into a wilderness. In 1800 the average population density was a mere four inhabitants per square mile, as opposed to roughly 130 in France. By 1870 the total population of the United States exceeded that of France, but the density was still much less—only a little over 13 inhabitants per square mile, as opposed to 172 in France.[1]

Expansion into a wilderness was a central element in Frederick Jackson Turner's influential interpretation of American history. Turner argued that the frontier was a leveling force that brought Americans back to a simpler and more primitive lifestyle:

Limiting our attention to the Atlantic coast, we have the familiar phenomenon of the evolution of institutions in a limited area, such as the rise of representative government; the differentiation of simple colonial governments into complex organs; the progress from primitive industrial society, without division of labor, up to manufacturing civilization. But we have in addition to this a recurrence of the process of evolution in each western area reached in the process of expansion. Thus American development has exhibited not merely advance along a single line, but a return to primitive conditions on a continually advancing frontier line, and a new development for that area. American social development has been continually beginning over again on the frontier. This perennial rebirth, this fluidity of American life, this expansion westward with its new opportunities, its continuous touch with the simplicity of primitive society, furnish the forces dominating American character.[2]

Later scholars have called into question many dimensions of Turner's analysis, which too readily overlooked the economic links between urban and rural or frontier areas and which too readily discounted the existence of social distinctions in American rural society.[3,4] Nevertheless, life in thinly populated rural regions, although always associated with inequalities in wealth, did not easily support rigid social stratification. The fine social gradations used in France to classify the families of *polytechniciens* could not be applied to regions like Fayette County in the early nineteenth century. Recall Finley's colleague Ephraim Douglass, who had worked as "clerk, scrivener, carpenter, cabinet maker, lumberman, blacksmith, gunsmith, stone mason, and shop keeper."[5] In the French context these were all low-status occupations of the "popular" and lower-middle classes, but Douglass had a high social and economic standing in Fayette County.

We have seen that low population density and lack of rigid social stratification had at least one important effect on Finley's work: in contrast to Navier, Finley did not set his work apart from the intellectual perspectives of artisans. Indeed, he had an incentive to integrate his work in a positive way with craft traditions and knowledge, because often it was only through the voluntary adoption of his plan by craftsmen such as the clockmaker Jacob Blumer or the blacksmith Valentine Shockey that Finley could hope to realize economic gain.

The frontier experience engendered positive attitudes toward manual labor. Americans came to regard manual labor as honorable and as a good means of teaching discipline, self-reliance, and responsibility. Travelers to the United States during the nineteenth century often commented on this trait. Omer Buyse, a Belgian who toured American technical schools shortly after the turn of the twentieth century, asserted that he found "no trace of that prejudice against manual labor, which is ineradicable [in Europe]."[6] And American folklore glorified labor and laborers to a degree that was unparalleled in France. For example, in American popular culture the locomotive driver Casey Jones was portrayed as a heroic, romantic figure. In France there was no popular mythology celebrating locomotive drivers; in fact, they were referred to as *gueules noires* (black mugs)—a term that emphasized the griminess of their occupation, which was considered distasteful.[7]

In the 1820s and the 1830s, Americans' positive attitudes toward manual labor were manifested in a widespread interest in combining manual labor with formal education. In 1831 a "Society for Promoting Manual Labor in Literary Institutions" was formed, and it sent a representative around the country to lecture on the benefits of "uniting labor with study" in academies and colleges. The lecturer,

Theodore Weld, who was involved with the Oneida Institute of Science and Industry (a school that followed this model), traveled many thousands of miles and gave hundreds of lectures on the benefits of combining manual labor with a general education. He outlined for his listeners the benefits that manual labor would bring to the student:

(1) The manual labor system furnishes exercise natural to man. (2) It furnishes exercise adapted to interest the mind. (3) Its moral effect would be peculiarly happy. (4) It would furnish the student with important practical acquisitions. (5) It would promote habits of industry. (6) It would promote independence of character. (7) It would promote originality. (8) It is adapted to render permanent all the manlier features of character. . . . (12) It would tend to do away with those absurd distinctions in society which make the occupation of an individual the standard of his worth. (13) It would have a tendency to render permanent our republican institutions.[8]

In 1851, a supporter of Weld's ideas—a cabinetmaker by the name of Harrison Howard—drew up a proposal for a People's College that would combine intellectual and manual labor. Howard was able to interest some reformers in his plan, and a People's College Association was formed in New York to collect funding and lobby for the creation of the college. The project got as far as a cornerstone laying in 1858 in Havana, New York. One of the speakers at the event envisioned that, with the creation of the college, the study of "the Differential and Integral Calculus" would "commingle with the ring of the anvil and the whir of the machine shop."[9]

The American Technological Community

The absence of a rigid social hierarchy and the relatively positive attitude toward manual labor made the American technological community less stratified than the French. Omer Buyse observed that "America knows neither social hierarchy nor diploma fetishism; a man is judged there according to what he is capable of doing; intelligent, practical work is classed above purely intellectual abilities. The engineer is above all a practitioner and a worker."[10] At least one of Buyse's impressions—the absence of "diploma fetishism"—was correct: status and occupational attainments were not strongly tied to academic credentials within the American technological community. This characteristic extended even to the U.S. Army Corps of Engineers and the Naval Engineer Corps. Unlike their French counterparts, they did not restrict recruitment to graduates of particular schools. Admission was based on a competency examination and on prior experience. Practical experience in the private sector, including workshop experience, was considered an advantage. Moreover, there were no rigid social or occupational distinctions between government and private-sector engineers in the United States; individuals moved freely between the two sectors.[11]

In France, the concern with credentials led to restrictions on who could use the title of *ingénieur* and to low status for those who were denied it. For most of the nineteenth century graduates of the Ecoles d'Arts et Métiers were not allowed to use it; they had to be content with the title of *mécanicien*. Moreover, *gadzarts*—and machine builders who learned through apprenticeship—had low social status, at least during the first half of the nineteenth century.[12] In contrast, American machine builders and mechanics—including those who worked or did apprenticeships in machine shops—did not have separate titles and low status; they were accepted as the equals of other kinds of engineers. Monte Calvert's study *The Mechanical Engineer in America* makes clear that machine builders and machine-shop owners were often among the social elite of their communities.[13] Examples can be seen in the careers of William Sellers, Coleman Sellers, and Alexander Lyman Holley.

William Sellers (1824–1905) was born into a family whose landholdings in Upper Darby, Pennsylvania, dated back to 1682. As a boy Sellers was educated at a private school established by relatives. At the age of 14 he began an apprenticeship with his uncle John Morton Poole, a machinist in Wilmington, Delaware. After seven years, he established his own machine-tool manufactory in Philadelphia. Sellers became a highly reputed designer and innovator, and he took out some ninety patents over the course of his career, either alone or with co-inventors. Socially, he was among the elite of Philadelphia. He became a trustee of the University of Pennsylvania and a member of the American Philosophical Society, the National Academy of Sciences, and the Union Club of Philadelphia (an elite businessmen's club).[14]

Coleman Sellers (1827–1907), a cousin of William, was born in Philadelphia. His father and other relatives on his father's side were mechanics; his grandfather on his mother's side was Charles Willson Peale, a bridge designer (mentioned in chapters 1 and 2), a portrait painter, and a member of the American Philosophical Society. Like his cousin William, Coleman Sellers attended private schools. At the age of 17 he went to live and work for two years on a relative's farm, after which he took a job in the Globe Rolling Mill in Cincinnati, which was run by his two older brothers. There, "his mechanical ingenuity quickly asserted itself."[15] One of his projects was to redesign and rebuild the wire mill. In 1856 he accepted the position of chief engineer in William Sellers's firm, and in that capacity took out a number of patents. Later in his career he became a consultant for the International Niagara Commission, which was involved in the development of hydroelectric power at Niagara Falls.

Like William Sellers, Coleman Sellers became a member of the American Philosophical Society, serving for a period as vice-president. He also became a professor of "engineering practice" at the Stevens Institute of Technology, a respected engineering school in Hoboken, New Jersey. At the same time, however, Sellers continued to advocate the apprenticeship system, and he favored bringing manual training into formal technical education. He contended that "any system of education that to mental culture adds manual skill and impresses the student with the fact that there is dignity in manual labor, and that implants in him habits of industry and frugality, is to be encouraged."[16]

Alexander Lyman Holley (1832–1882) came from a well-known family of very comfortable means. His father was a businessman, a banker, and a governor of the state of Connecticut. The father had originally wanted Holley to receive a classical education at Yale, but the son was more interested in science and technology. He was allowed to attend Brown University, where a program in civil engineering had recently been established. The young Holley worked in a machine shop throughout the duration of his studies, and after receiving a bachelor's degree in 1853 he began work in the Corliss steam engine works. Later in his career, Holley became a prominent engineer in the steel industry. Career patterns like this were much less common in France than in America. Whereas most *polytechniciens* would never have dreamed of going to work as a machinist, even temporarily, in the United States it was not unusual for young men from distinguished families to be apprenticed to machine builders.[17]

The relative lack of credentialism and of social and occupational barriers within the American technological community of the nineteenth century can be seen in the character of the professional societies that were formed. American technologists organized nationally, according to field, rather than (as their French counterparts did) on the basis of where they had been schooled. Among the societies formed were the American Society of Civil Engineers (founded in 1852 and revived in 1867 after having become moribund), the American Institute of Mining Engineers (1871), the American Society of Mechanical Engineers (1880), and the American Institute of Electrical Engineers (1884). In later years, more specialized societies were established for heating and ventilating engineers (1894), electrochemical engineers (1902), and so on.[18] Some societies allowed anyone interested in a particular domain of practice to become a member. The more professionally oriented societies created two or more ranks of membership, limiting full membership to those with direct involvement in engineering work as evidenced by their ability to design and to direct engineering projects. (Such criteria were meant to exclude those who were simply managers or businessman.) Yet, whatever their standards for membership, the American engineering societies tended to be heterogeneous in terms of education, geographical origin, and occupation. In both the ASCE and the ASME, engineers with college degrees participated alongside others who lacked formal degrees but who had risen through the ranks in construction projects, machine shops, or who had become prominent inventors or designers.[19]

The American Society of Civil Engineers, which initially sought to be a comprehensive society for all engineers, was in some respects the most professionally oriented and elitist of the engineering societies. The main criteria for full membership were to be in "responsible charge" of engineering work for a specified period and to be qualified to design and direct engineering work.[20]

A look at the educational backgrounds of ASCE engineers between 1870 and 1892 shows that they were drawn from many schools and many geographical areas, and that a substantial number of them had no formal engineering degree. Of those who became members in the period 1870–1874, 40 percent had no formal degree. The remainder were graduates of more than nine schools, the Rensselaer Polytechnic Institute having the highest contribution (20 percent of the total). Of those who became members in the later period (1885–1892), about 27 percent had no formal degree. Of those who did have degrees, membership was widely distributed among more than 42 schools, with the Rensselaer Polytechnic Institute again claiming the highest portion (14 percent).[21]

The occupations of ASCE members also varied widely. The largest portion (around 18 percent) were civil, mechanical, and mining engineers in consulting or private practice. Around 15.5 percent were chief engineers of railways, canals, cities, waterworks, and "superintendents

of motive power." Assistant engineers (about 12 percent) and government and military engineers (about 6 percent) were also represented.
And 21 percent were either professors or persons in occupations
related to engineering (e.g., manufacturing).[22]

The frontier experience and the American economic environment
not only discouraged social barriers and credentialism, they also
shaped people's intellectual perspectives. They provided a fertile environment for the empirical, experimental, problem-solving approach
that was associated with Common Sense philosophy. Experimentation
and testing were physical, material approaches to the creation of
knowledge, and often they did not presuppose a great deal of formal
education. The tenets of Common Sense philosophy, moreover, rested
on the assumption that nature's laws were knowable to whoever was
willing to do the physical work of making the observations or carrying
out the experiments; in short, this was a pragmatic, non-elitist research
philosophy.

Common Sense philosophy came to have a pervasive influence on
American thought and education. According to the intellectual histo-
rian Henry F. May, it became the standard philosophy propagated in
American schools throughout much of the nineteenth century:

> . . . before the Revolution, increasingly after it, and with growing volume
> through at least the first half of the nineteenth century, a specific kind of
> Scottish thought acquired a massive influence in America. This was the phi-
> losophy of common sense, whose central thinker was Thomas Reid. . . . At
> least until the Civil War, and in some places long after, the Common Sense
> philosophy reigned supreme in American colleges, driving out skepticism and
> Berkeleyan idealism and delaying the advent of Kant. At first it was inculcated
> directly from Scottish books. Then, beginning in the twenties, these were
> replaced by American textbooks of moral philosophy that simplified the Scot-
> tish principles and used them explicitly to validate republican and Protestant
> institutions.[23]

Travelers to America in the nineteenth century also noticed the growing dominance of Common Sense philosophy. One individual who
immigrated to the United States in 1868 "was struck by the Yankee
aptitude for practical observation and invention and concluded that
realism based on common sense might very well become the distinctive
American philosophy."[24]

The comparatively undeveloped material and institutional infrastructure of the frontier necessitated on-the-job training and career
mobility of the kind shown by Ephraim Douglass and James Finley.

People from many backgrounds turned their hands to technological development in early-nineteenth-century America, and the majority did so with little or no academic technological training. Apprenticeship and self-study flourished. Under the apprenticeship system in civil engineering that existed in New England well into the second half of the nineteenth century, a "pupil" would become the assistant of an engineer, generally for three years. One individual trained under this system recalled paying "tuition" of $100 per year. In return, he was paid $12\frac{1}{2}$¢ per hour for his work in the field, and nothing for office work. The engineer, besides collecting the tuition fee, made a profit from a student's labor, for which clients were billed between $2 and $6 per day. The education acquired was not systematic; it came hit-or-miss through the work process. The former student cited above explained that the instruction "consisted chiefly in the student's privilege to ask a question if he chose and get an answer if he could; in addition to which he had the further privilege of learning what he could from the work assigned to him in field or office." He continued:

Of formal instruction, of recognized or regular supervision there was none. If the pupil wanted to learn he could, and if he didn't, he needn't. In any case he could hardly fail to learn something from the field and office work assigned to him. The system had some advantages of its own, along with many disadvantages. . . . [Knowledge] was acquired simultaneously with practical experience, and was largely in the direct line of that experience. It had cost the student some exertion, instead of being stuffed down his throat like a sausage, and was generally learned "for keeps." Above all, although there was little formal instruction, there were daily hints and bits of information or advice from (presumably) a well qualified and experienced engineer.[25]

In his book *The American Civil Engineer,* Daniel Calhoun shows that large civil engineering projects—canals and railroads—also became an important training ground for American civil engineers. According to Calhoun's statistics, 65 of the 87 civil engineers in America in 1837 (that is, 75 percent) were trained "on the job" by rising through the hierarchy in large civil engineering projects.[26] Calhoun's definition of civil engineer does not include local road and bridge builders, but most of them were trained on the job as well. The same pattern holds for mechanical inventors and machine builders.[27] Among the prominent American inventors and engineers who had no academic engineering degree were Thomas Edison, Thomas Blanchard, Elmer Sperry, James Francis, Uriah Boyden, James Eads, George Corliss, Elihu Thomson, James Westinghouse, Henry Ford, and Wilbur and Orville Wright.

American Technical Education, 1800–1860

Although by some estimates technologists who were self-taught or trained through apprenticeship remained in the majority in America throughout the nineteenth century, formal engineering training did become increasingly important. Yet its evolution followed a different pattern from that in France. American education, especially in the first half of the century, was characterized by localism and decentralization; in line with this characteristic, American technical education was notable above all for its complexity and variety.

Three principal models for formal engineering training emerged in this period. The first, which might be termed the "military" model, was represented earliest and most importantly by the establishment of the U.S. Military Academy at West Point in 1802. In terms of the numbers of engineers it produced, West Point was the most important engineering school in America during the first half of the nineteenth century. Nevertheless, most West Point graduates did not became engineers; most went into farming, law, politics, or banking. Accordingly, West Point did not produce nearly as many engineers as the Ecole Polytechnique. From 1802 to 1814 West Point produced only five civil engineers while the Ecole Polytechnique produced at least 250.[28] From 1815 to 1830 West Point produced fewer than 70 civil engineers; the Ecole Polytechnique produced more than four times as many, although the population of France was only about three times that of the United States. Between 1831 and 1848, when the population of France was only about twice that of the United States, West Point produced around 100 civil engineers while the Ecole Polytechnique produced about 600.[29]

West Point is often referred to as "America's Ecole Polytechnique," because in some ways it was deliberately modeled after the French institution. Yet the schools differed in two fundamental respects. First, unlike the Ecole Polytechnique, with its long and difficult entrance examination, West Point had no entrance requirements beyond basic literacy and a knowledge of arithmetic.[30] It was felt that "to raise the standard of admission would be to exclude many young men of worth, whose early education has been neglected."[31] This rule remained in force throughout the nineteenth century. Second, the curriculum at West Point, although considered by Americans to be quite mathematics-oriented, was not comparable to the curricula of the Ecole Polytechnique and the *écoles d'application*.[32] During West Point's first decade, when the total enrollment averaged only about twenty, graduates were supposed to have learned basic algebra, geometry, and trigonometry, logarithms, conic sections, surveying, "planometry," and "stereometry." They also had to know basic French, and some military basics.[33] Following an Act of Congress in 1812, the school was reorganized and expanded, particularly under the superintendence of Sylvanus Thayer, who took up his appointment in 1817 and remained until 1833. The upgraded, four-year curriculum established in 1816 did not include calculus as a required subject, but Thayer later made it mandatory.

A guide to West Point written in 1840 indicates how the curriculum had evolved by that time. First-year students took algebra, geometry, trigonometry, descriptive geometry, mensuration, and French. Second-year courses included theory of perspective, analytic geometry, calculus, surveying, French, geography, English grammar, and drawing. In the third year, students took chemistry, "natural and experimental philosophy" (including mechanics, astronomy, optics, and acoustics), and drawing. Fourth-year students took artillery, fortifications, pyrotechny, civil and military engineering, mineralogy and geology, rhetoric, moral philosophy, and political science.[34] This curriculum remained in effect until 1854.

Before 1854 no time was set aside for laboratory work. Students in the chemistry and experimental philosophy courses did not perform experiments. Yet in the latter course they did learn to use various instruments, including a sextant and a zenith telescope. Beginning in 1851, there was a course in "practical military engineering" taught by a member of the Army Corps of Engineers. In this course, which was not graded, the students built military bridges and scale models of field fortifications and did a great deal of fieldwork in surveying and construction.[35]

Between 1833 and 1854, West Pointers had only one semester of calculus and one of mechanics; *polytechniciens* had two years of each. And those *polytechniciens* who went on to the Ecole des Ponts et Chaussées also studied applied mechanics, comprising the theory of machines (two years), the strength of materials (1 year), and hydraulics (one year), none of which was covered at West Point except perhaps in the mechanics course.

Like the *polytechniciens*, however, West Pointers had a highly structured, military-style routine. They arose at dawn and studied from sunrise until 7 A.M. Classes ran from 8 A.M. until 4 P.M., with an hour for dinner at 1 P.M.; then from sunset until 9:30 P.M. there was more studying. One difference with the Ecole Polytechnique, however, was that West Pointers, in line with the American idea that labor builds character and self-reliance, had to clean their own rooms and guns each morning.[36] These tasks were specified in the official charts of 1823 and 1832, which outlined the "distribution of studies and employment of time during the day."[37] Hugh Reed, the author of an account of cadet life at West Point published in 1896, likewise recalled cadets' taking "bundles, buckets, and brooms to barracks" and cleaning their guns upon being awakened at 5:45.[38]

By the 1830s, West Point had gained a reputation as mainly benefitting wealthy families, because student selection was a matter of political patronage: one vacancy was reserved for each congressional district. The idea that ordinary taxpayers were subsidizing the education of the rich aroused much popular opposition. In 1833 and 1834 the Tennessee and Ohio legislatures passed resolutions demanding that the school be abolished. The Tennessee resolution complained that "a few young men, sons of distinguished and wealthy families, through the intervention of members of Congress, are educated at this institution at the expense of the great body of the American people, which entitles them to privileges, and elevates them above their fellow citizens, who have not been so fortunate as to be educated under the patronage of this aristocratical institution."[39] In 1837 an angry report by a nine-member committee of Congress established to investigate West Point similarly demanded the abolition of the military academy, recommending the establishment of a "school of practice" in its place.[40]

In response to such criticisms, data were compiled between 1842 and 1879 on the "circumstances of parents of cadets." These data showed that 12.6 percent of the students came from families in "indigent" or "reduced" circumstances, 86.1 percent from families in "moderate" circumstances, and 4.2 percent from families in "affluent" circumstances. Those in "moderate" circumstances included artisans, small businessmen, bank clerks, doctors, lawyers, schoolteachers, and manufacturers. Overall, the statistics show that the social origins of West Point students were more varied than those of *polytechniciens*, and that a much larger percentage of West Pointers were drawn from the "popular" classes. Nevertheless, the fact that admission to West Point was a matter of political patronage meant that occupations with political connections or influence—especially lawyers and army officers—were overrepresented.[41]

Despite the criticisms, West Point persisted, and the "military" model it had established was adopted by a number of schools founded in the antebellum period, most of which turned out a few engineers. Among these was the Literary, Philosophical, and Military Academy in Norwich, Vermont, founded in 1820 by the West Point graduate Alden Partridge. By 1826 that school, which had a looser organization than West Point and a less rigorous curriculum in mathematics and science, was offering a formal program in civil engineering. Six military colleges founded in the South also offered engineering courses.[42]

The second model for American technical education in the first half of the nineteenth century was the lyceum. The lyceum model was closer to the kind of "school of practice" which the committee investigating West Point had wanted. It drew inspiration from a long tradition of popular interest in "self- improvement" and in the application of the sciences to the "useful arts." In the eighteenth century this tradition led to the founding of a number of scientific societies in America, most notably the American Philosophical Society.[43] In the 1820s and the 1830s the tradition was given a more popular orientation through the establishment of schools, institutes, and societies, among them the Franklin Institute,[44] the American Lyceum,[45] the Gardiner Lyceum, and the Rensselaer School.

The American Lyceum, the mother organization for a network of local associations, was the brainchild of Josiah Holbrook. Holbrook studied at Yale College from 1806 to 1810, where he took chemistry under Benjamin Silliman. In 1819 Holbrook took charge of his father's farm in Derby, Connecticut. In 1824, influenced by the manual-labor movement, he opened a school of agriculture and manual labor on his farm, with the goal of linking chemistry and mechanics directly to farming practice.[46] In 1826 Holbrook began a more ambitious endeavor. He developed and published a master plan for establishing a network of "associations for mutual instruction in the sciences, and in useful knowledge generally."[47] His idea quickly took root, and by the mid 1830s some 3,000 local societies were offering public lectures and discussions. (In principle these organizations were all branches of the American Lyceum, which was the name given to the national organization.)

One of Holbrook's principal goals was "to apply the sciences and the various branches of education to the domestic and useful arts, and to all the common purposes of life."[48] Behind this vision was a hands-on approach to learning. Holbrook expected that the "mutual instruction" associations would make extensive use of tools and equipment to teach mathematics, mechanics, chemistry, and astronomy. By 1830 Holbrook had convinced Timothy Claxton, a mechanic and the owner of a small machine shop, to manufacture apparatus for schools and lyceums. Later Holbrook went into this business on his own, devoting much energy to designing, producing, and selling educational tools and equipment: "eolopiles," orreries, arithmometers, sets of "mechanical powers" (e.g., lever, wheel, screw), hydrostatic bellows, geometrical solids, globes, pyrometers, "pneumatic cisterns," "conductometers," concave reflectors, glass tubing, retorts, flasks, and so on. Holbrook was so successful in this business that his surname became a household word and a legal generic name for such apparatus.[49]

Neither the Franklin Institute nor the lyceums that grew out of Holbrook's plan were involved to any great degree in formal technical education, but the guiding ideas behind these institutions did foster two early technical schools, the Gardiner Lyceum and the Rensselaer School, the latter of which became quite prominent. The Gardiner Lyceum, established in 1823 in Gardiner, Maine, was a cross between a college and a manual-labor school. It had a full three-year scientific curriculum, including calculus and special courses in engineering. It also emphasized hands-on laboratory and field work: "Surveying and Levelling are taught not only in recitation room but in the field; the pupil in chemistry is carried into the laboratory, and allowed to perform experiments."[50] The influence of the manual-labor movement could be seen in the school's "large and commodious work shop," in which "the ingenious and industrious" were said to be able to "earn sufficient to pay their board."[51]

The Gardiner Lyceum relied on funding from the Maine legislature. When this funding was withdrawn after 1832, the school had to close. The Rensselaer School survived, at least in part because it had a more secure source of funds: the wealthy New York State landowner Stephen Van Rensselaer. The guiding force behind the school was its principal teacher, Amos Eaton, a man of Finley's generation with a passion for popularizing science and "bringing it to the people." One of Eaton's students recalled that he wore old, unfashionable clothes, old boots, and a squirrel-skin cap, and looked like the kind of person who would be "hooted out of town."[52] Eaton described himself as a "popular howler" who was most at home "among the rabble,"[53] but he had begun his professional career as a lawyer. Born in Chatham, New York, on an 80-acre farm, Eaton had ties, through relatives, to the artisan community. One relative was a blacksmith, and Amos was sent to live with him at the age of 14. Amos's brother became a coachmaker. Amos, however, attended Williams College, and then studied and practiced law. He became interested in science, and while still working as a lawyer he studied botany, chemistry, and natural philosophy with two professors in the area of New York City. Having decided to leave the law profession, Eaton opened a surveying and land-agency office in Catskill. There, in 1810, he was charged and found guilty of forgery, and sentenced to life imprisonment, but he was released five years later through the intercession of a wealthy merchant, William Torrey. After his release, Eaton attended Yale for two years, where, like Josiah Holbrook, he studied chemistry and mineralogy with Benjamin Silliman.

Like Holbrook, Eaton was caught up in the movement to make science useful and bring it to the people. Eaton's thinking was also shaped by the principles of Common Sense philosophy and by its emphasis on discovering nature's laws through observation and experiment.[54] He attempted to apply these principles in a new way at Rensselaer. At Yale, Benjamin Silliman had not included laboratory work by students as a structured part of the curriculum, but the germ of the idea was there in several ways. First, Silliman had what was, for that era, a large and excellently equipped chemical laboratory. Second, he made experiments the centerpiece of his teaching. One student recalled:

During the lecture hour there was no lull or intermission; all was . . . a constant appeal to the delighted senses. Here were broad irradiations of emerald phosphorescence, there the vivid spangles of burning iron, or the blinding effulgence of the compound blow-pipe, or the galvanic deflagrator. Strange sounds saluted the ear, from the singing hydrogen tube, the crackling decrepitation up to the loud explosions of mingled gases and detonating fulminates. As forms of matter once regarded simple were torn in to their elements, or these again compounded in manifold ways, the very kaleidoscope of changes came into view, of which the greatest was the transformation of the whole seeming phantasy into science, through the lucid rationale of the gifted lecturer.[55]

Third, Silliman used many student assistants to set up and carry out his experimental demonstrations and to help him in his own research. Amos Eaton was one of those assistants.[56]

At Rensselaer, Eaton took the Silliman approach a step further. The entrance requirements of the school were minimal: reading, writing, and arithmetic, and an age of at least 13. With a curriculum focused almost exclusively on science and technology, the school's stated aim was to give "an opportunity to the farmer, the mechanic, the clergyman, the lawyer, the physician, the merchant, and in short to the man of business or leisure, of any calling whatever, to become *practically scientific*."[57] One of the earliest announcements of the school, dating from 1826, specified that students would spend the entire afternoon every day doing projects, field work, shop work, or laboratory work. The students were expected to do about 1,600 "experiments." Some of the activities listed included collecting specimens and analyzing them under the microscope, surveying, performing hydraulic experiments, "making and using a set of mechanical powers," "making and using galvanic batteries and piles," making and using a camera obscura, "constructing and using air thermometers and hygrometers," and "taking specific gravities." In the school's workshop, the students made their own laboratory apparatus.[58] Students were taken on excursions as well. One student's diary from 1826 shows that in the course of a semester he had visited a brick kiln, a plaster mill, a fulling mill, a bleaching works, a tannery, a smithy, a printing office, a paint shop, a millstone factory, and a stoneware factory. Each student was required to describe in detail all the operations observed at each site. The aforementioned student also reported that Eaton had hired a barge and had traveled with students on the Erie Canal.[59]

Approximately 70 percent of the Rensselaer students were drawn from New York State, most of the others from nearby states. Most were from middle-class or farm backgrounds. The fact that a substantial proportion of the students (sometimes as many as half) had already completed college has led some to call Rensselaer the first American graduate school.[60]

In 1828 Eaton organized an evening course for mechanics. With permission from Stephen Van Rensselaer, he was to instruct "40 members of the Troy Mechanics' Society upon the Rensselaerean plan." Each evening ten of them were required to present their own lecture-demonstrations, performing all the experiments with their own hands.[61]

In 1835 the Rensselaer Institute, as it was now called, began to offer a degree course in engineering, yet even before that time the school had turned out graduates who became engineers. Of the 149 graduates between 1826 and 1840, 31 became civil engineers, five became surveyors, one became a mechanical engineer, and 32 became science teachers; others became doctors, lawyers, farmers, and businessmen.[62]

Between 1840 and 1860 Rensselaer turned out another 37 engineering graduates. From 1861 to 1880, the school produced a small but steady stream of engineering graduates, averaging 20 per year.[63]

The engineering curriculum established in 1835 followed the general spirit of the Rensselaer School's earlier program. During the 40-week course, afternoons were to be devoted to surveying, use of instruments, chemistry experiments, and the like. The first eight weeks of the 12-week winter term were devoted to "practical Mathematics, Arithmetical and Geometrical," the remaining four weeks to "Logic, Rhetoric, Geology, Geography, and History." The civil engineering program took up the rest of the course. Eight weeks were devoted to learning the use of various instruments, then eight weeks to mechanics and construction. Students were to learn such things as "calculating the height of the Atmosphere by twilight, and its whole weight on any given portion of the Earth, its pressure on Hills and in Valleys as affecting the height for fixing the lower valve of a Pump." The next four weeks were to be spent on hydraulics, and the final four weeks on steam, electricity, and geology (with visits to shops, mills, and factories).[64]

Apart from West Point, the Rensselaer Institute was the most important technical school in the United States during the first half of the nineteenth century. However, by the 1840s there was a third model for formal engineering training: the establishment of engineering courses and programs at traditional colleges. Among the American colleges and universities that instituted courses or programs in engineering between 1840 and 1860 were Brown, Harvard, Yale, Wesleyan, the University of Michigan, New York University, Dartmouth, Rutgers, Indiana University, Cincinnati College, the University of Pennsylvania, the University of Virginia, the University of Georgia, and the University of Maryland.[65]

Up to 1860 the entrance requirements at American schools and colleges generally remained low, but by the end of the nineteenth century many institutions had established much stiffer requirements. Yet, because of the complexity, variety, and regionalism of American technical education, entrance requirements of schools awarding engineering degrees varied widely. A survey of 89 engineering colleges in 1896 found that 31 colleges (35 percent) required "algebra through Quadratics, plane geometry, solid geometry *or* plane trigonometry, one year of a foreign language, and moderately high requirements in English." Another 33 colleges (37 percent) required plane geometry and algebra through quadratic equations, and 25 colleges (28 percent) had still lower requirements. The better colleges usually had higher requirements, but this was not universally the case. Rensselaer and Brown, which had good reputations, were in the second group, while the Worcester Polytechnic Institute, often designated a trade school because of its heavy emphasis on manual shop training, was in the first category. And the third category included a number of respectable engineering schools, such as Purdue University and the Colorado School of Mines.[66]

American Technical Education, 1860–1900

In general, American technical education in the second half of the nineteenth century still manifested enough variety and regionalism to preclude the emergence of a strict hierarchy among the schools, such as existed in France. Yet the landscape of technical education had changed in three ways since the earlier period: the "lyceum" model had evolved into a more professionally oriented "polytechnic" model, the "college adjunct" model had become increasingly prominent (especially since 1862, when the Morrill Act was passed), and the "military" model had became more marginalized as a training ground for engineers.

The Rensselaer Institute played a leading role in the emergence of the "polytechnic" model. The transition came about first through Rensselaer's evolution into a degree-granting institution in the 1830s, and then through the complete reorganization of the curriculum that was carried out in 1849–50, several years after the death of Amos Eaton. Drawing inspiration from European technical schools (notably the Ecole Centrale in Paris), Rensselaer's new academic director, B. Franklin Greene—a graduate of the school—revamped the curriculum and added "Polytechnic" to the name. Whereas the old engineering degree had required only a year to complete, the new program comprised 3 years of mathematics, science, and engineering courses, with systematic examinations in every course.[67]

The polytechnic model, characterized by degree-granting institutions that specialized in science and engineering, gained momentum in the 1850s and after. Among the schools that were established were the Polytechnic College of Pennsylvania (1853), the Brooklyn Polytechnic Institute (1855), the Cooper Union (1857), the Massachusetts Institute of Technology (1861), the Worcester Polytechnic Institute

(1868), the Stevens Institute of Technology (1871), Alabama Polytechnic (1872), Rose Polytechnic (1874), and the Georgia School of Technology (1888).[68]

Besides the polytechnics, engineering education also expanded in general colleges and universities—particularly after 1862, when the Morrill Act stipulated that each state would receive 30,000 acres of public land for each senator and congressman. The proceeds from the sale of the lands had to be used for "the endowment, support and maintenance of at least one college where the leading object shall be . . . to teach such branches of learning as are related to agriculture and other mechanic arts." Each state was supposed to have at least one college meeting this description within five years.[69] The conception underlying the Morrill Act was for democratic schools (having low tuition and low or moderate entrance requirements) to fulfill the educational and vocational needs of the people of each state. The Morrill Act, and the popular movement that led to it, drew significantly on ideas stemming from the earlier manual-labor and lyceum movements.[70]

The Morrill Act promoted a massive growth in engineering education. New colleges and universities were founded, and existing ones added or expanded their engineering programs to meet the requirements of the law. Among the existing institutions that received land-grant aid were MIT, Rutgers, Yale, and the University of Michigan. Among the many new institutions created were Cornell, Purdue, the University of Nebraska, Ohio State University, and Iowa State University. Between 1862 and 1872 dozens of engineering schools and programs were created, and the number of engineering graduates rose rapidly. According to one listing, the total number of engineering students receiving degrees was less than 30 in 1862; by 1876 the number was above 300; by 1892 it was over 900.[71] According to another estimate, there were between 9,000 and 10,000 students enrolled in engineering programs in the United States in 1886–87.[72] Although the 1862 figure (which does not include West Point and other military schools) is an underestimate, these statistics give a sense of the rapid growth in formal engineering education that occurred in the second half of the nineteenth century.

Besides promoting a quantitative growth of engineering programs, the Morrill Act helped to shape the character of American technical education. Together with the growth of the polytechnics, it speeded the decline of the military schools, especially West Point, as important training grounds for engineers.[73] And, by requiring the provision of education in the "mechanic arts," the Morrill Act reinforced a process by which all branches of engineering education were brought into the university. Frederick Jackson Turner saw this process as a further manifestation of the frontier experience. From "pioneer ideals," according to Turner, "came the fuller recognition of scientific studies, and especially those of applied science devoted to the conquest of nature; the breaking down of the traditional required curriculum; the union of vocational and college work in the same institution; the development of agricultural and engineering colleges and business courses."[74]

The American social and cultural environment was significant not only in that it brought engineering into the university but also in that it created a context in which (mathematical) theoretical training, laboratory training, and workshop training could all be integrated into engineering curricula.[75]

According to Charles Bennett's *History of Manual and Industrial Education up to 1870,* formal workshop training was first brought into academic engineering curricula in the United States through the founding of the Worcester (Massachusetts) Polytechnic Institute in 1868. The initial endowment for this school was a bequest of $100,000 by one John Boynton. To this endowment was added $75,000 by a Worcester manufacturer, Ichabod Washburn, which was to be used to equip a set of full-scale industrial-style workshops to train mechanics. Among the stated premises underlying the program at the school were that "all mechanical engineers" would benefit by "going through a work-shop training," and that "work-shop instruction is best given in a genuine manufacturing machine shop where work is done that is to be sold in open market and in unprotected competition with the products of other shops."[76]

Although the establishment of the Worcester Polytechnic Institute was an important precedent in America for bringing the machine shop into the university, it was not the only one. The idea had been clearly expressed in the first half of the nineteenth century by proponents of the manual-labor movement, and had even been materialized in the Gardiner Lyceum and the Rensselaer School. In addition, other engineering schools had workshop facilities either earlier than or at about the same time as WPI, although their curricula were not centered around extensive workshop training. MIT, when its main building in Boston was completed in 1866, apparently did not include extensive workshop training in the curriculum but nevertheless had a large workshop among the teaching laboratories. The new main building had two chemistry laboratories, two physics laboratories, a photography laboratory, a mineralogical and blowpipe laboratory, a metallurgical laboratory, and a workshop. The latter, more than 1,000 square feet in size, was larger than the quantitative chemistry laboratory. In the fall of 1876, an additional large building (5,000 square feet) was built to house a large machine shop and two more chemistry laboratories. This machine shop was at first used for a special "School of Mechanic Arts," but it was also used for the regular engineering students, especially after the school of Mechanic Arts was discontinued in 1889.[77] Similarly, when Cornell University's Sibley College of Engineering was established in 1871, the facilities established included machine shops.[78]

As a general rule, workshops—including forges, foundries, and machine shops for wood and metal—came into technical schools with the growth of mechanical engineering programs in the 1870s and the 1880s. In America workshops were at the heart of mechanical engineering. In view of the hands-on American approach to technical education and the whole climate of opinion that led to the Morrill Act, it was natural to bring the shops into the university. By the end of the nineteenth century, many engineering schools, including some of the most highly respected, offered and required shop training.[79]

Omer Buyse, the Belgian engineer who toured American technical schools shortly after the turn of the twentieth century, reported that shop work was a universal feature of higher technical education in the United States.[80] The amount of shop work required of American engineering students varied considerably, however. A survey of eleven major engineering schools carried out in 1899 showed that the amount of shop work required in the mechanical engineering programs ranged from 252 hours at Lehigh University to 1,170 hours at Cornell and 1,629 hours at WPI, for an average of 765 hours.[81]

Shop training, once established in the universities, spread beyond mechanical engineering. By the early 1890s, electrical, mining, and sometimes even civil engineering students had to take shop courses. In an "ideal" civil engineering curriculum sketched out by a civil engineering professor from Cornell in 1895, one shop course was required.[82] Another educator felt that "the principles and methods involved in forging, molding, pattern-making, casting, rolling and ordinary machine work should be understood by the civil as well as the mechanical engineer," but he believed that less shop work was necessary for the former than the latter.[83] A survey of civil engineering curricula in 1892 showed that many required significant amounts of shop work.[84]

In electrical engineering, substantial shop work was the norm. A study carried out in 1899 of 18 electrical engineering curricula showed that required shop work averaged about 19 credit hours, distributed among forge, foundry, woodworking, and metal-turning shops.[85] Another study of ten engineering schools found that the amount of shop work in electrical engineering courses varied from none (at Lehigh) to 1,170 hours at Cornell and 1,479 hours at Worcester Polytechnic, for an average of 584 hours.[86]

Although engineering educators were largely united in recognizing a need for shop training for engineering students (in mechanical and electrical engineering, at least), they did not always agree as to the function it was to serve. Some felt that shop work should be, as much as possible, a real apprenticeship-like training in a large, production-oriented machine shop. The Worcester Polytechnic Institute followed this model, and other schools did as well. A professor of mechanical engineering at the Case Institute of Technology argued that

the discipline, methods, and routine, in fact the whole atmosphere of the college shop, should be that of a well-ordered modern manufacturing establishment. . . . This argument for a practical shop, in contrast with a toy shop, is not based on any consideration of dollars and cents. The writer never has known a piece of machinery built by students that did not cost the college more than its market value. But he believes that the only way to give young engineers the necessary familiarity with shop practices and methods is to make the college shop a real shop where work is done as in outside establishments.[87]

Many educators, however, came increasingly to see shop work as a kind of field work or laboratory work—"laboratory work with shop appliances," as one educator expressed it—intended to acquaint the students with the tools and practices of their trade in a systematic way.

The importance of shop work in educators' eyes had several dimensions, two of which are particularly worth noting. First, shop training was seen to be necessary in order to train the prospective engineer-manager to "direct the labor of others intelligently"—that is, to "direct the labor of other men, and the machines to do it, in the best and cheapest way possible."[88] Second and more important, shop training was seen as a necessary foundation for good design. Designers needed to consider, in the design process, how the machine or part could be most efficiently and economically produced in the forges, foundries, and machine shops, and to be able to do this they had to be acquainted with machine-shop practice: " . . . the great demand is for young men who are capable of designing and constructing improved machinery in accordance not only with correct mechanical principles, but with modern shop practice."[89] A professor of mechanical engineering at Cornell explained that students had to be made strongly aware of the commercial dimensions of machine design, and that linking design and shop-work courses in engineering curricula was a way to do this:

A properly arranged course in the school shops . . . is a most valuable adjunct to a course in design. I believe that . . . the departments of design and shop work should be in close touch. . . . The designer must study processes of construction; he must be familiar with the methods and limitations of the shops, especially with the relative efficiencies of the equipments and the workmen in the various departments of his own works. These so-called "practical" matters (and they certainly are practical) play the star parts in actual designing of the most successful works.[90]

Another educator, in order to achieve this same end, required each student not only to design machines but also to detail specific production processes—e.g., to indicate "how he would chuck them on the machine, [and] how mould them." To do this, students had to have had some shop training.[91]

Viewed most broadly, shop work was a kind of field work for American mechanical engineering students. Field work had long been required of civil engineering students. Since the acknowledged terrain of the mechanical engineer was the shops, it made sense to require work and experience in this domain. Specialized field work also came to be expected of students in other domains, such as mining engineering. But mines could not be brought into the universities, of course, so the students went to them. At the Columbia School of Mines it was decided that the students needed to devote at least a month each summer to working full time in a mine, with ordinary workmen. Summer schools were organized for this purpose in 1876, and were still conducted in the 1890s. The president of the Columbia School of Mines explained that mining students

must be as familiar as possible with the practical details of mine work . . . For this purpose summer schools of practical mining are necessary—not mere excursions to mines, spending a few hours under ground and on the surface at a new mine each day, and the bulk of the time in hotel parlors and railroad cars. The student will learn but little practical mining in this way. A regular and systematic plan of instruction should be organized, the class spending a month or more at one mine, going underground each morning with the miners, spending the day with them at their work, taking up one subject at a time, shaft sinking one week, timbering another week, etc., working and eating with the men, and absorbing as much as possible of the mine dirt and the mine atmosphere that he may become quite at home in his new and unfamiliar surroundings. Started in this way in the school, the graduate will feel the importance and necessity of completing his practical education when he leaves, and will know how to set about it.[92]

Another important feature of American technical education as it evolved in the second half of the nineteenth century was the attention given to laboratory training. Omer Buyse wrote:

The most striking characteristic of the physical organization of American higher technical schools is the richness, the perfect order, the surprising number of laboratories in which students carry out themselves a series of experiments which constitute the basis for a course. . . . The influence of laboratories is such that no higher technical school can preserve its position without possessing a complete set of apparatus for research into and demonstrations of fundamental scientific and technological facts.[93]

Two aspects of the laboratories in American technical education were especially noteworthy. The first was their extent and variety. By the 1870s and the 1880s, their development had gone beyond the traditional chemistry and materials-testing laboratories to include innovative laboratories of hydraulics, metallurgy, applied mechanics, steam engineering, and so on. Robert Thurston established a research-oriented mechanical engineering lab at the Stevens Institute of Technology in 1871, equipping it first with machines for testing materials and then with other kinds of machinery and apparatus. Between 1875 and 1878 a "series of investigations of the copper-tin-zinc alloys and a number of other extensive lines of experimental work were carried on" in this lab.[94] Among the latter was a study of the properties of lubricating oils. At first this laboratory was used only for research, but it later became a teaching and research laboratory within the Department of Experimental Engineering. Other schools began to establish laboratories as well, equipping them with steam engines, materials-testing machines, electrical equipment, and a host of other apparatus. At least one university, Purdue, had an experimental locomotive.[95]

The Massachusetts Institute of Technology was among the most innovative schools in laboratory development. The aforementioned building in Boston, completed in 1866, had two floors of laboratory and workshop space (about 12,500 square feet in all), and a large workshop and laboratory building was soon added. More buildings and laboratories were added in 1876, in 1883 (two new buildings with laboratories and mechanical shops), in 1888 (a large six-story building with two floors for mechanical engineering laboratories), in 1892, and in 1898.[96] By the time Omer Buyse visited MIT, it had laboratories of applied mechanics, steam technology, hydraulics, mining and metallurgy, materials testing, general chemistry, industrial chemistry, physical chemistry, general electricity, industrial electricity, biology, sanitation, sewage testing, general physics, heat, physical measurement, physico-chemical measurement, electro-chemical measurement, geology, and mineralogy, and a geodesic observatory. In addition to these laboratories (and the private laboratories of the professors) there were a number of workshops.[97] Many of the laboratories and workshops were very large, and to judge from the descriptions, inventories, and photographs presented by Buyse they were extremely well equipped. No French technical school had comparable research facilities at the time.[98]

The second notable aspect of American technical schools' laboratories was their importance in the educational process.[99] Buyse's impression was that the labs and workshops had become the principal foundation on which American technical education was built, reflecting an American philosophy of education through "empirical re-discovery":

[American] professors hold that education in general, and especially scientific education, cannot be fruitful if the students are not trained to discover truths and resolve scientific questions themselves. The teaching of pure and applied sciences is imbued with the principles of the method of "rediscovery," practised in the laboratories and workshops. Lectures, of a very reduced importance, prepare, accompany, or confirm the practical studies of the laboratory and workshop, which are the center of interest of the institutions. The laboratory and workshop notebooks, in which the events and the phenomena observed by the students are recorded . . . constitute the touchstone of value of the studies. No value is placed on lecture notes, which play such a large role in our schools. The student must wrest the secret of the phenomena and the laws which govern them directly from the experimental apparatus and equipment.[100]

To judge by their comments, many American engineering educators would have agreed with Buyse. In the 1890s several of them commented that American technical schools' laboratories were more developed and varied than those of the European schools.[101] De Volson

Wood, in his presidential address to the Society for the Promotion of Engineering Education in 1894, expressed a philosophy very much like that noted by Buyse: "The student should be led to discover truth as if it had not been discovered a thousand times before. The value of this method has been recognized and resulted in the establishment of many laboratories and shops."[102]

As with shop work, the amount of time spent on laboratory work in American technical schools varied significantly, but in a number of schools students spent more time working in the laboratories and shops than attending lectures. At Cornell University's Sibley College of Engineering around 1892, the four-year program in mechanical engineering included almost 1,700 hours of laboratory and shop work[103]; around 1900, according to Buyse, Cornell's mechanical engineering program included some 2,160 hours of laboratory and workshop experience and only about 1,700 hours of lectures.[104] An 1892 survey of mechanical engineering curricula indicates that the proportion of time devoted to lectures varied between 32 percent and 81 percent, with most schools requiring students to spend between 40 and 60 percent of their time in lecture courses. Table 8.1 gives some specific examples of percentages of student time devoted to lecture courses in the mechanical engineering curricula of various schools in 1892.

Like shop work, engineering laboratory work in American technical schools came to have multiple purposes. In many schools, laboratory work was closely coordinated with lectures and provided a means of reinforcing students' understanding of material presented in lectures. Laboratories were also seen as an important means of recreating, in the school environment, the outside world of engineering practice, so that students could learn how standard tests were conducted, how particular equipment was used, how records were kept, and so on. Laboratories were likewise seen as a means of instilling systematic habits of observation and of leading students in the direction of creative engineering research. Finally, laboratory experience, in which students worked with, took apart, and tested machines, was seen as an important aid to design: "The laboratory is one of the foremost (perhaps the first) of the essential elements of an ideal training in design; for nothing so impresses a man with the good and bad features of a machine (especially the latter) as the actual manipulation of that machine."[105]

Table 8.1
Percentage of time devoted to lectures in mechanical engineering curricula, 1892.

Yale University	60
University of Michigan	54
MIT	51
Lehigh University	74
Tufts	47
Washington University	54
Worcester Polytechnic	37
University of Pennsylvania	68
Stevens	47
University of Wisconsin	40
Cornell University	41
University of California	47
University of Kansas	39
University of Minnesota	33
University of Illinois	63
Texas Agricultural College	54
University of Georgia	42
Haverford	58

Source: "The engineering schools of the United States, XXIII," *Engineering News and American Railway Journal* 28 (September 22, 1892), p. 268.

Conclusion

Relative to its French counterpart, the American technological community was fluid, with comparatively few social and occupational barriers. There was no strong hierarchy among American technical schools, whereas in France the prestigious schools were accessible only to the upper classes because of entrance requirements that presupposed years of expensive preparatory studies. American technical schools did vary in prestige, but many good ones did not have demanding entrance requirements. American technical education was characterized by diversity and regionalism, although after the Morrill Act this regionalism occurred within the framework of a national system that specifically encouraged education in the "mechanic arts." Schools in different regions competed with one another for reputation, and this friendly competition was clearly an invigorating force in American technical education.

Among the most noteworthy features of American technical education was the integration of theoretical, laboratory, and workshop training. The facility with which Americans instituted this pattern was rooted in several precedents and traditions: the manual-labor movement and the schemes it engendered for combining intellectual and manual training; the lyceum movement and the mechanics' movement, and the hands-on approach to learning which they fostered; the tradition of Common Sense philosophy, with its focus on experiment; and the fluidity of the American technological community, which accepted machine builders as the equals of other kinds of engineers and which had long regarded apprenticeship and on-the-job training as an acceptable way of becoming an engineer. All these precedents were linked in some way to America's frontier experience, and all contributed to the expansion and integration of workshop and laboratory training in American engineering schools.

The roles of workshop and laboratory training changed over time, however. The manual-labor movement incorporated the idea of bringing the workshop into the college, but the initial reasons for doing so were not only educational. A different aim was to allow people who needed to work for a living to combine work and study. Manual labor was also viewed as a means of helping students build character. At the Rensselaer School, the workshop was intended to allow students to build apparatus that might not otherwise be readily available at reasonable cost. No doubt the students learned a lot by building their own apparatus, but doing so was partly a means to another end.

By the time the workshop came into American engineering schools, in the second half of the nineteenth century, the aims were quite different, and they continued to evolve thereafter. The basic aim now was educational: workshop training was to help students learn what they needed to know to become good engineers. Beyond that, the workshop began to be put to a variety of educational uses. Some used it primarily as a kind of "fieldwork" training, to give engineering students systematic exposure to industrial machine-shop practice; others treated is as a variety of engineering laboratory; still others used it as a means of sensitizing students to the commercial and manufacturing dimensions of design.

The evolution of laboratory training in American technical schools was similar in its multiplicity of uses and roles. For some, the laboratory, like the shop, was to provide "fieldwork" training by help-ing students learn how testing and research were done in industrial and government laboratories. For others, it was to provide training for exploratory research; or it was to be a means to reinforce book learn-ing; or it was to serve as an aid to good design. But in any case, the vitality and creativity of the workshop and laboratory traditions in American technical education reflected a long-standing taste for hands-on education.

9

Ideology and Technological Practice

The preceding two chapters have surveyed the distinct social structures of the French and American technological communities and their distinct systems of technical training. The aim of this chapter is to consider how these differences shaped technological practice. In particular, how did these differences shape the ideologies and processes of technological research and design in each country? How did they affect the way engineering was done and the way technological problems were formulated and solved?

The case of France is complex because of the divisions within the engineering community. As a first approximation, it is possible to identify two distinct approaches to (or philosophies of) engineering in France, one associated with the corps engineers and the other associated with the *centraliens, gadzarts,* and others who worked in the private sector.

Before discussing the corps engineers' approach, it is important to know more about their role and their activities. It has been suggested (e.g., by Terry Shinn) that the corps engineers were really scientist-administrators who had no direct involvement in technological research and development.[1] Such a view is incorrect. Corps engineers were by no means only administrators. Those who served in the Corps des Ponts et Chaussées were directly involved in designing, planning, and directing a wide range of engineering works throughout the nineteenth century, including ports, dams, lighthouses, canals, roads, bridges, railroads, and water-supply, irrigation, and sewage systems. In addition, they often performed functions commonly associated with the position of consulting engineer. Among other things, they had a legal mandate to review and oversee every civil engineering project that was proposed or undertaken anywhere in the country. This involved analyzing engineering plans and designs for feasibility and safety, and proposing changes if necessary. These were not simply administrative tasks, for they required extensive and up-to-date technical knowledge. Corps engineers also had to set design and safety standards and specifications, and they periodically had to inspect existing structures to assess their safety and general condition, all of which required professional expertise.[2]

It is nevertheless true that corps engineers were part of a highly structured bureaucracy and accordingly *were* deeply involved in administrative tasks, such as collecting industrial statistics, carrying out technical and economic surveys, managing projects, and handling bureaucratic paperwork.[3] But, as numerous studies have emphasized, bureaucracy has become an integral feature of engineering since the industrial revolution, because large, complex technical undertakings involve not only research and development efforts but also the organization of work on a large scale; political, financial, and legal negotiations; and so on.[4] Accordingly, engineers in both the public and the private sector increasingly had to function as both technicians and administrators. Perhaps the corps engineers were more deeply involved in administrative affairs than their counterparts in the private sector, but if this was the case the difference was one of degree rather than kind.

The further notion that corps engineers were not involved in technological research is incorrect; there is abundant evidence throughout the nineteenth century that corps engineers were involved in research on technological problems. Some of this work centered on empirical tests and experiments, such as towing-tank experiments in ship design; materials testing; experiments and tests on steam engines, water wheels, and turbines; artillery tests; empirical research in hydraulics; research on methods of sewage treatment; and experiments on the use of hydraulic cement.[5] The main focus of corps engineers, though, was on theoretical rather than empirical research. Yet this work cannot be labeled "science" rather than "engineering" simply because it involved principles of mechanics or sophisticated mathematical analysis— particularly since much of it concerned technological issues. Rather, what must be acknowledged is that the corps engineers evolved a particular conception of technological research that placed the greatest emphasis on theoretical, mathematical analysis. In their view, mathematical theory constituted an indispensable foundation for the development of useful and reliable technological knowledge.

The reward system within the government corps explicitly fostered this kind of approach: theoretical accomplishments were a means of rising above a banal technical or administrative career. Jules Jacquart, who was a member of the Corps des Ponts et Chaussées during the First World War, characterized two types of Ponts-et-Chaussées engineers: the type who was content to be a "cog" and who ended his career as *ingénieur-en-chef* in some obscure province, and the more successful *"ingénieur technicien,"* who managed to escape the fate of the cogs by directing important engineering works, publishing in the *Annales des Ponts et Chaussées,* and ultimately winning a professorial position. Within the context of the state corps, such achievements called for mathematical expertise.[6]

One of the pinnacles of achievement for a corps engineer, as we saw in the case of Navier, was to be elected to the Académie des Sciences. An analysis of the academy's membership over the course of the nineteenth century reveals that corps engineers held a significant percentage of the positions in the sections most relevant to technology. The mechanics section had 29 members between 1816 and 1900 (not including corresponding members). Of this total, 23 were drawn from the engineering community, and 16 of the latter were members of the state engineering corps.[7]

The lure of election to the Académie des Sciences meant that ambitious corps engineers were pulled in two directions. In their day-to-day work they had to deal with technological issues, yet theoretical accomplishments were necessary to achieve recognition from the academy and its members. Ultimately they achieved a synthesis of these two demands by finding ways to apply mathematical analysis to engineering and, more broadly, by attempting to develop the (mathematical) theoretical foundations of the subjects relevant to their work.

In the Corps des Ponts et Chaussées, engineers studied the mathematical theory of elasticity and strength of materials, the theory of structures, soil mechanics, hydraulics, the theory of traction (related to railways), theoretical issues related to telegraphy and electric light and power, and so forth. Navier was an important exemplar of the mathematical-theoretical approach in the early decades of the nineteenth century, and many other corps engineers followed the same path in subsequent decades. Maurice Lévy (1838–1910), who was involved particularly with navigation on the Seine, developed, among other things, a special winch for raising and lowering weirs and equipment for plowing snow. He was also a consultant on many bridge-construction projects and on international projects to lay undersea telegraph cables between France and North Africa. At the same time, Lévy did extensive mathematical research in the engineering sciences, including work on the theory of elasticity and strength of materials, the theory of retaining walls, the theory of dynamos, and the theory of undersea telegraph cables. He was elected to the Académie des Sciences in 1883. In 1894 he attained the rank of *inspecteur général* in the Corps.[8]

Jean Résal (1854–1919), who also reached the rank of *inspecteur général* in the Corps des Ponts et Chaussées, designed and directed the construction of several bridges, including the famous Pont Alexandre III in Paris. At the same time, Résal published mathematical theories, particularly concerning the stability of stone and metal bridges. Still another member of this corps, André Blondel, did extensive theoretical work on electric lighting and machinery: arc lights, synchronous and asynchronous motors, high-voltage, alternating current transmission lines, and so on. His research on the last topic grew out of a plan proposed by him and two other engineers to establish a hydroelectric power plant at Genissiat, the power from which was to be transmitted to Paris over high-voltage lines.[9]

Engineers in the Corps du Génie Maritime studied, most notably, hydrodynamics, the theory of hulls and ships, the theory of tides, the theory of floating bodies, the theory of marine engines, and thermodynamics. Charles Dupin (1784–1873) directed a number of important engineering works, such as the building of a port on the island of Corfu; he was also involved in designing a new kind of frigate and in other engineering projects. He became director of naval construction and ultimately attained the rank of *inspecteur général* in the Corps. At the same time, throughout his career he did research in the engineering sciences: he developed a new mathematical theory of floating bodies, did research on strength of materials, and published a work on road construction (in particular, cuts and fills, in which he "found the chance to prove some ingenious theorems in Pure Geometry").[10]

Louis de Bussy (1822–1903), also a member of the Corps du Génie Maritime, carried out experimental research that led to the development of a new generation of steel-hulled warships. At the same time, he published research on the theory of ships, including a work in which he analyzed the angular motion of a ship as a function of its speed and its hull size. De Bussy was elected to the Académie des Sciences in 1888 and ended his career in the Corps du Génie Maritime with the rank of *inspecteur général*.

Emile Bertin (1840–1924), who also combined achievements in naval construction with contributions to the theory of hulls and ships, was characterized by contemporaries as "the very model of the great French engineer, both profound in theoretical research and skillful in practical interpretations and applications." Among his achievements was a new design system for battleships comprising a network of watertight compartments beneath an armored deck. The system was said to have been adopted in many countries; it was used in the first Japanese war fleet, for example. Bertin was elected to the Académie des Sciences in 1903 and achieved the rank of *directeur du Génie Maritime*.[11]

Engineers in the Corps d'Artillerie and in the Administration des Poudres et Saltpêtres did extensive theoretical work in ballistics, strength of materials, and various branches of applied mechanics, as well as research on the theory of firearms and gunpowder. Emile Sarrau (1837–1904) began his career in Poudres et Saltpêtres working in one of the government gunpowder factories. By day he carried out tests and experiments at the factory and by night he read the mathematical works of Cauchy. Eventually he published articles in the *Journal de mathématiques pures et appliquées*. Sarrau's work in the gunpowder factory convinced him of the importance of experimentation, but for him its primary role was merely to test the correctness of mathematical theories. As a result of his participation in the Franco-Prussian War of 1870, Sarrau became interested in ballistics, and thereafter he did mathematical research in that area and in other branches of applied mechanics. One of his important contributions was to establish "complete formulas" for constructing the various components of guns. He also taught a course on the "theory of explosives" at the Ecole d'Application des Poudres et Saltpêtres. He was elected to the Académie des Sciences in 1857.[12]

In the Corps d'Artillerie, Guillaume Piobert (1793–1871) invented a new type of cannon and helped in other ways to modernize the French artillery system. At the same time, Piobert carried out theoretical research on artillery problems. A particular phenomenon that plagued gunners was the rapid deterioration of large-bore cannon when they were fired at full charge. Piobert "submitted the principle aspects of the phenomenon to mathematical analysis" and eventually wrote a memoir to the Minister of War outlining his conclusions and recommendations. Later he generalized this work to achieve "a complete solution of the problem." He also did theoretical work on the effect of air resistance on projectile motion, and he published a "theoretical and practical treatise" on artillery. He eventually became a member of the artillery board and attained the rank of general. He was also elected to the Académie des Sciences. Other members of the artillery corps, including Georges Halphen (1844–1889), Hippolyte Sebert (1839–1930), and Pierre Henri Hugoniot (1851–1887), also carried out theoretical research on engineering questions.[13]

Engineers in the Corps des Mines, apart from research in chemistry, mineralogy and geology, did research in hydraulics, thermodynamics, applied mechanics, theory of machines, and theory of railways. Edouard Phillips (1821–1889) began his career in the Corps des Mines and subsequently took a position as *ingénieur du matériel* with a railway company, where he remained for 20 years. (He did not officially leave the Corps des Mines, however.) At the same time, Phillips did theoretical work in applied mechanics, and he was eventually offered a professorship at the Ecole Centrale. As an aid to locomotive design, Phillips worked out a mathematical theory of springs, from which he derived simplified formulas that could be easily applied. He also analyzed the problem of vibration in metal bridges and came up with a fourth-degree partial differential equation which he solved by methods of approximation. He was elected to the Académie des Sciences in 1868, and in the Corps des Mines he reached the rank of *inspecteur général*.[14]

These capsule biographies illustrate the point that the ideal type of engineer in the state corps was one who could analyze technological systems mathematically and, conversely, who could successfully use mathematical theory to guide engineering design. There were at least three reasons why this particular model became dominant. First, as was suggested in preceding chapters, the corps engineers' approach was partly an outgrowth of prevailing social attitudes concerning intellectual and manual labor. Second, their approach was fostered by their education and institutional position. Their mathematical training prepared them most effectively for theoretical research; in addition, the fact that they were salaried employees meant that they could afford to carry out research that would perhaps bring no immediate economic benefit. Third, the corps engineers' approach was the product of a historical tradition whose roots, as we have seen, can be traced back to the eighteenth century. Traditions have to be protected and nurtured in order to endure, however, and among the forces that helped to keep the "analytical ideal" intact was the ongoing rivalry between the corps engineers and the *conducteurs* and *ingénieurs civils*. Since the only formal distinction between the corps engineers and their rivals was the nature of their education, it followed that the only definite criterion the corps engineers could point to in defense of their position was their advanced training in theoretical science and mathematics. When *conducteurs* in 1848 demanded the right to be promoted into the Corps des Ponts et Chaussées, the Corps' director attempted to counter their demands by arguing that only those who knew calculus could do innovative research in engineering:

> . . . let no one believe that occasions to apply science are as rare as certain individuals like to suggest. On the contrary, these occasions present themselves very frequently, whether it be in construction projects, in the building or use of machinery, or yet again in the numerous issues that relate to water supply and to factories. I speak above all of mathematical science, for it is the study of this which receives greatest development at the Ecole Polytechnique and the utility of which is the most contested. When it sometimes happens to engineers, even the most scientific, to fail in the solution of this or that problem, it is not the inutility of science that is to blame, but rather its insufficiency and the fact that it is not advanced enough in certain branches, notably hydrodynamics. . . . Here theory cannot be separated from practice . . . [and] the progress that [hydrodynamics] has been able to make in the last years is due to the research of a number of distinguished engineers. Yet this research . . . can only be undertaken by men familiar with all the difficulties and all the possibilities of integral calculus.[15]

A common extension of this line of reasoning was that only a solid training in mathematics and theory could produce engineers capable of functioning effectively in all areas of technology. A training that focused more on the details of technological practice, it was argued, would produce narrow technicians unable to move beyond the confines of their particular specialties.[16]

In short: the corps engineers nourished their mathematical approach as a hallmark of their special position and as a defense against continuous pressure and opposition from technologists outside the corps who wanted to curtail their power and influence.

French Engineers in the Private Sector

In contrast to the analytical ideal of the corps engineers, the *ingénieurs civils* tended to favor a more empirical and experimental approach to engineering. Indeed, in their view, the corps engineers' reliance on mathematical theory was misguided. The *conducteurs* consistently supported their demand for the right to enter the Corps des Ponts et Chaussées by arguing that theoretical knowledge was not a sufficient guide to engineering practice and that theoretical training alone did not make an engineer. They saw the training at the Ecole Polytechnique as too theoretical to foster engineering ability:

Pure theory has taken up too large a part of the studies. . . . It has the disadvantage of fruitlessly tiring the students' minds, of giving them too much confidence in theories, of turning them away from the study of facts, and often of inspiring in them disdain for manual and practical things. There, it must be admitted, is the origin of the reproaches that have been addressed, not without some reason, to the engineers of the Ponts-et-Chaussées. The admission of some *conducteurs* to the rank of engineer would have the effect of making it understood that the state does not only demand science in the public services, but principally practice enlightened by science.[17]

The view of the *conducteurs* was echoed by others in the private sector, who pointed to the successes of self-taught British and American engineers as proof that technological achievements were not built on theoretical analysis. One French engineer who visited the United States toward the close of the nineteenth century reported:

It is not engineers backed by fifteen years of specialized studies who created the establishments I visited. Neither Baldwin, nor Carnegie, nor Westinghouse was destined by his family to become a *savant*; they did not waste their youth in front of blackboards scratching out scientific problems; but, exposed at the age of fifteen or sixteen to all the practical problems that arise in business, they invented simpler methods to attain a particular goal, and mastered the necessary knowledge as they needed it.[18]

The position of the *ingénieurs civils* was born as much out of necessity as choice; as independent entrepreneurs and inventors, or as engineers engaged directly in technological work, their success (like that of their American counterparts) had to be won in the marketplace. When an entrepreneur undertook to build a bridge in France, his profit came from the difference between the value of the tolls collected and the cost of building the bridge: the higher the costs, the less the profit. (In contrast, when an engineer in the Corps des Ponts et Chaussées undertook to build a bridge or some other public work, the construction costs had no effect on his salary, which was fixed by the state.) The economic pressures faced by the *ingénieurs civils* directed their attention toward finding the most economical ways of doing things, and they often viewed empirical, experimental research as a more efficient means of achieving their goals than mathematical theory.

   The career of Marc Seguin, one of the most noted French inventors of the nineteenth century, provides a case in point. Seguin was the son of a cloth manufacturer and the nephew of the famous balloonist Joseph Montgolfier. He had no formal technical education, but he did receive some informal scientific training from his uncle. Seguin became noted both as a railroad builder and a bridge builder. Like Navier, he became especially interested in suspension bridges, and he built a suspension bridge over the Rhône, between Tain and Tournon, at the very time that Navier undertook to build the Pont des Invalides (figures 9.1 and 9.2). The way that Seguin approached the problem of suspension bridges showed a marked contrast with Navier, however.[19]

   Seguin felt he had little use for analyses such as Navier's, because, in his words, they rested "only upon theoretical notions, or on mathematical solutions which [are supposed] *a priori* to be admissible."[20] He further complained that engineers like Navier "produced books filled with abstractions, and neglected to descend to the matter-of-fact study of facts, the knowledge of which is alone able to illuminate practice."[21] Seguin eventually wrote his own book on suspension bridges; it was published a year after Navier's. A comparison of the two works highlights the differences in the approaches of the two men.

   Whereas the focal point of Navier's work was mathematical analysis, Seguin relegated mathematics to footnotes and appendixes and concentrated on experimental evidence. He had done experiments comparing the strengths of bar iron and iron wire, in order to determine which would make the best cables. Seguin found that iron wire had a mean strength about 50 percent greater than that of bar iron, and also that it had much greater resistance to fatigue. A major drawback of iron wire, however, was its variable quality and strength due either to flaws in the wire or to differences in the quality of iron coming from different manufacturers. Seguin could not prevent such variability, but he was able to gain a quantitative assessment of its limits by testing the tensile strengths of hundreds of samples of commercial iron wire from many different manufacturers. The specimens were loaded gradually, in stages, and the amount of elongation that occurred at each stage was recorded. When a specimen finally ruptured, its crystal structure was examined through a magnifying glass or under a microscope. These tests and experiments showed that wire cables had a number of economic and technical advantages over chain cables.[22]

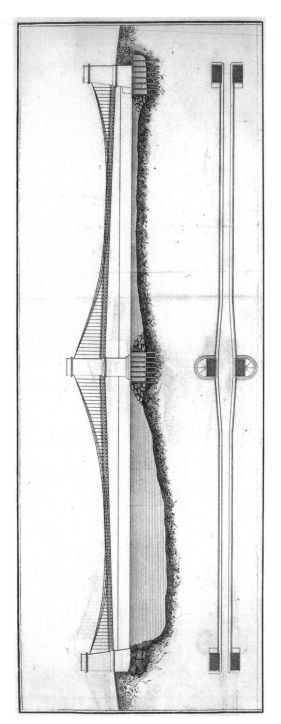

Figure 9.1
Marc Seguin's Tain-Tournon suspension bridge, constructed in 1824–25. Source: Marc Seguin, *Des ponts en fil de fer*, second edition (Paris: Bachelier, 1826).

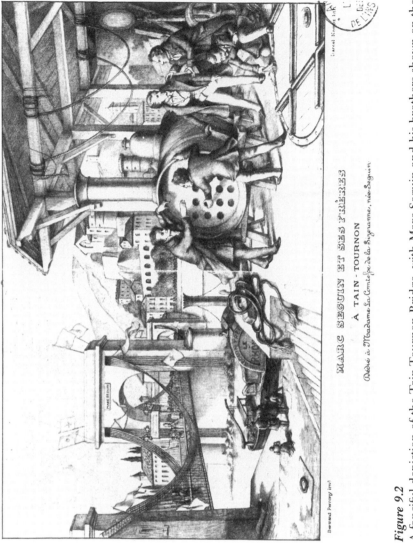

Figure 9.2
A fanciful depiction of the Tain-Tournon Bridge, with Marc Seguin and his brothers shown on the right. Source: Archives de l'Académie des Sciences de Paris.

Seguin's orientation not only led him to investigate questions that had been overlooked by Navier, it also led him to ignore certain problems that Navier had treated in detail. Referring to Navier's mathematical analysis of the precise curve of a suspension bridge cable, Seguin commented that "the object to which this research could be applied would be the determination of the vertical cords [i.e. the determination of the hanger lengths], which it seems to me would be simpler and surer to determine experimentally. . . . "[23] The method Seguin used to determine hanger lengths was similar to Finley's and Telford's: he loaded a cable so as to simulate the way it would be loaded in the finished bridge; then he simply measured the correct length for each hanger.

Another question to which Navier had devoted considerable attention was how much the roadway of a suspension bridge would deflect as a result of traffic loads. His aim was to understand how the amount of deflection varied with the length of the span, assuming the roadway to be unstiffened. Seguin, on the other hand, was interested to know how much to stiffen a roadway in order to prevent it from deflecting—a question Navier had ignored. When Seguin eventually tested one of his bridges, he found that, whereas Navier's theory predicted a deflection of 11 centimeters (for a load of 500 kilograms placed at the middle of the span), the actual deflection of his stiffened roadway was negligible. On the basis of his empirical research, Seguin was able to work out a more economical system for designing suspension bridges (elements of which he patented). With the help of his brothers, he oversaw the construction of dozens of bridges based on this plan.[24]

Interaction of the French Engineering Traditions

Numerically, the corps engineers were a minority within the French engineering community. And while the number of engineers working in the state corps remained fairly stable over the nineteenth century, the number of engineers working in the private sector grew substantially.[25] It might be thought, therefore, that the mathematical approach of the corps engineers would have had a small and diminishing influence; yet in fact its influence was, and remained, quite strong. Not only were the corps engineers an extremely powerful group who exercised control over broad areas of technological development, but they also stood at the top of a *social* hierarchy, which gave their values and approach authority. As a result, the values, the ideology, and the research achievements of the corps engineers became prominent elements of the engineering culture—elements that those outside the corps could hardly ignore.

Engineers in the private sector often had to pay attention to the work of corps engineers in order to enable their own projects to win approval. Part of the reason was that mathematical-theoretical considerations became an important part of the debate over the viability of engineering designs, and private-sector engineers often needed to have some understanding of theory in order to function effectively. When Seguin first submitted a proposal to the Corps des Ponts et Chaussées in 1822 to build a suspension bridge, the plan was transmitted for review to Navier, who immediately rejected it, partly on *theoretical* grounds. Seguin submitted a revised plan later the same year, but this time he "took care to draw on formulas derived by Navier."[26] Similarly, although Seguin took a rather dim view of Navier's theory of suspension bridges, he nevertheless took it carefully into account in preparing his own book on suspension bridges.

A focus on mathematical theory also made itself felt in other contexts that affected engineers in the private sector, such as prize committees and committees of the Académie des Sciences to evaluate inventions and technical research. When the Société d'Encouragement pour l'Industrie Nationale decided in 1829 to sponsor a competition for hydraulic turbine designs, the problem was formulated in such a way as virtually to require potential competitors to have a good understanding of turbine theory. The announcement of the competition began by explaining that theory indicated that hydraulic turbines had important advantages over other hydraulic prime movers and that it was "therefore essential to realize completely and economically in practice the valuable results [indicated by] theory."[27] Entries (which could refer only to turbines actually constructed and put into operation) had to be accompanied by an explanatory memoir. Of course, an inventor could in principle have entered a purely empirical design, but eventually the prize was awarded to one who combined theoretical and design achievements: Benôit Fourneyron.

There was also, however, an undercurrent of opposition to the corps engineers' analytical ideal. Many in the private sector lost no opportunity to ridicule the corps engineers and their mathematical calculations. This occurred, as we have seen, when a problem emerged with the anchorages of the Pont des Invalides. A similar situation occurred decades later when a dam built by the Corps des Ponts et Chaussées collapsed. Immediately the issue of the theoretical versus the practical ability of corps engineers was brought to the fore. The headlines of one newspaper read: "The Bouzey catastrophe; . . . The Ecole Polytechnique in the dock." The article reported that fissures preceding the collapse had been explained away by the engineer-in-chief on the grounds that they were both predicted and necessary, which brought the suggestion that the dam had split open "polytechnically." The article proposed that the corps engineers had predicted the entire catastrophe mathematically: "The savant engineers of the sacrosanct Ecole, knowing the danger for having ascertained it in memoirs filled with numbers, knowing nearly to the penny how much the damages for the destruction of entire villages would amount to, and what the loss of human life would cost to the state, still filled the menacing reservoir up to the brim, until the definitive crack came to confirm the mathematical exactitude of their previsions."[28] The *conducteur* assigned to the project was portrayed, in contrast, as a hero who had tried in vain to alert the administration to the impending disaster: "He never ceased to raise the cry of alarm. . . . The pessimistic *conducteur* was transferred . . . his successor was drowned. . . . " Another newspaper, which gave a more reserved account of the disaster, nevertheless bemoaned the fact that *polytechniciens* were "only theoreticians" upon their graduation, and that many "remained theoreticians all their lives."[29]

Despite this culture of opposition, many engineers in the private sector chose, consciously or not, to emulate the values, the ideology, and ultimately the approach of their rivals, because these constituted an accepted means of gaining recognition and status in France. Many became excessively concerned with achieving scientific recognition from the Académie des Sciences. Marc Seguin again provides a good example. In 1823, when he was involved in trying to get permission from the Corps des Ponts et Chaussées to build his first suspension bridge, he wrote to his father that he had been "toying with the idea of pushing ahead into a scientific career and taking a place among scientists, since I think it's indispensable in our position to have a name that is known."[30] On more than one occasion he sent project descriptions or accounts of his engineering research to the Académie des Sciences to be read and judged.

Seguin further revealed his desire (one might say obsession) to win scientific recognition in a series of letters written to a lifelong friend, Henry Desgrand, between 1820 and 1860. He confessed to Desgrand that he had worked to become rich in order to gain a foothold within the scientific establishment, for, as he put it, "wealth is a sign which makes the buyers resolve to go to you in preference to those who cannot put up such a display."[31] He told Desgrand more specifically of his hopes and efforts to be elected to the Académie des Sciences, efforts which spanned a decade and ultimately brought him the position of corresponding member.[32] Once elected, Seguin further related to Desgrand his attempts to gain the esteem of the scientists. But his comments and confessions leave no doubt that he remained an outsider in their eyes. At one point he wrote:

. . . Several times the thought has come to me to approach [Biot] but I always recoiled before the certainty that the bent of his character was making him refuse not only to read but even to notice my writings, so often has he manifested the opinion that science was not my affair, that it was already more than enough that he, Biot, recognized me as the world's premiere industrialist, but that to want to suggest the desire to place myself beside him as a savant was an idea that he would never accept and which it was necessary to give up without hope of ever bringing it up again.[33]

In a more hopeful moment, Seguin told Desgrand of a plan to woo the *savants:*

As time costs me nothing, I am following the course that I adopted. I am setting up my observatory, making contacts with astronomers. I bought a pendulum for 1500 francs which I lacked. I built a beautiful greenhouse where there will by lovely parlors and musical instruments and when the railroad is finished I will bring the savants here and make good dinners; I will show them my instruments; they will not be able to suppose that a host who treats them so well is utterly unworthy of their attention and that may make me bolder and them softer in my regard. Small measures as you know sometimes make big things happen.[34]

Seguin was not the only industrialist eager for recognition from the Académie des Sciences. In 1843 Benôit Fourneyron presented himself as a candidate for a vacancy in the mechanics section of the Académie des Sciences created by the death of Gaspard Coriolis. In the first round of voting Fourneyron came out ahead; however, by the third and final round of voting, Arthur Morin, a *polytechnicien* and a member of the Corps of d'Artillerie, received the most votes. Jules Guillemin, in an obituary notice written after Fourneyron's death in 1867, stated: "We may simply say that he would have been elected [to the Académie des Sciences] if he had been a *polytechnicien.*"[35]

Another way in which engineers in the private sector emulated the corps engineers was by assimilating the latter's view that mathematical theory was the preeminent force in the development of engineering.

This position was explicitly adopted by the president of the Société des Ingénieurs Civils in a speech at that society's annual banquet in 1862:

The most substantial assurance for the future, the indispensable token of authority, of success, of respect in the engineering profession, is knowledge of the mathematical sciences. He who lacks it will be unable, in his life, to employ metal, in whatever manner, without calling upon the cooperation of those who possess [such knowledge]. However ingenious he may be, when it is a question of mechanical applications, or of the physical properties of matter, he will crawl along imitating or throw himself into the contingencies of cut-and-try. The majority of the books of the masters will be closed to him.[36]

The analytical ideal acquired such prestige in France that sometimes even those who consciously sought to encourage a more empirical style of engineering accepted a model of technological development that pictured innovation as flowing principally from mathematical analysis. This attitude can be seen in comments by Jacques Eugène Armengaud, who taught machine drawing at the Conservatoire d'Arts et Métiers. Armengaud had close links with the machine-building community, since he was a partner in the machine-building firm of his father-in-law.[37] He also edited the journal *Publication Industrielle,* which, in his words, was intended to provide "all industrial classes in general, consequently all eminently working classes" with up-to-date technical information of a practical nature: "simple, practical rules," because "we know how precious their time is, how few moments they can devote to theoretical studies or research."[38] In short, Armengaud was one of the major promoters of the interests of engineers and mechanics in the private sector.

Yet when this same Armengaud wrote an obituary of Eugène Pihet, one of the leading machine builders in Paris during the first half of the nineteenth century, his words left the impression that the development of machine tools owed more to mathematicians such as Lagrange than to practitioners such as Pihet. After introducing the obituary with a review of progress made in the design of machine tools up to the time of the 1844 Exhibition, Armengaud continued:

Since then, all this complicated machinery has been further augmented and completed, above all with the aid of the mathematical sciences which better determine the fundamental principles of dynamics; but it must not be inferred that these well-ordered developments have all been the work of Archimedeses, Newtons, or Lagranges; if these latter have set down the immutable bases for the laws of mechanics, others less scientific, more limited in their conceptions, have also known how to establish the relationships between different rotating parts and lever forces; artisans with hands hardened by the handle of the gouge or the hammer, but with a mind quick to comprehend the best way . . . have often in their turn perfected or invented a portion of these machines which are so perfectly contrived. These men of a simple nature, too often unknown by their contemporaries, sometimes even sadly compensated for their efforts, nevertheless have the right, if not to a keen admiration, at least to all our gratitude and all our esteem. . . . One of these deserving and modest workers [was] M. Eugène Pihet. . . .[39]

In fact, Pihet's shop was one of the largest and best-equipped in Paris in its day. It had about 500 employees, and according to Armengaud it became a kind of school for training mechanics. It produced a full range of machines for preparing, spinning, and weaving cotton, many of which were either invented or improved by Pihet; it also produced steam engines, turbines, and firearms. One of Pihet's important commissions was to produce 90,000 iron beds of a unique design for the French army. Pihet invented an entire range of new machine tools to carry out this project: mechanical presses, automatic lathes, drilling machines, screw cutters, stamping machines, and so forth. At the Paris Exhibitions of 1834 and 1844 Pihet was awarded gold medals for his contributions to machine building.[40]

Perhaps Armengaud was only mouthing the ideology of the corps engineers; however, others actually followed its precepts and became, themselves, adept theoreticians. Benôit Fourneyron provides a case in point. Fourneyron received his education from the Ecole des Mineurs in Saint-Etienne, a two-year school with a vocational orientation. The entrance requirements (arithmetic, drawing, elementary algebra and geometry) were low by French standards, and the curriculum was not much more sophisticated than those of the Ecoles d'Arts et Métiers.[41] After graduating, Fourneyron held several positions. He worked first as a mine manager and then as an engineer for an ironworks. But his spare time was devoted to mathematics and theoretical mechanics, and he eventually mastered the works of Euler, Borda, Navier, and others. When Fourneyron later (in 1834) submitted an entry in the hydraulic turbine competition of the Société d'Encouragement pour l'Industrie Nationale, the centerpiece of his accompanying memoir was a sophisticated mathematical theory.[42]

Another leading French engineer from the private sector who favored a theoretical approach was Gustave Eiffel (1832–1923). Eiffel graduated from the Ecole Centrale in 1855, after which he worked for a railway equipment manufacturer. Later he founded his own civil engineering firm and built "hundreds of major iron structures including bridges, railway stations, exhibition halls, gas works, reservoirs, cranes, factories, and department stores."[43] Eiffel and his engineering staff always calculated their designs mathematically, and in some cases they structured the designs so as to simplify the calculations. When discussing the design of his monumental tower, Eiffel explained in broad terms how a mathematical analysis of the combined forces of the weight of the structure and the wind was used to determine the curve of the legs.[44] In Eiffel's view, this kind of mathematical approach was superior to the more empirical design traditions of the British: "English engineers have almost entirely by-passed calculations and they fix dimensions of their members by trial and error and by experiments . . . and by small-scale models. . . . [They] went far ahead of us in their practice, but we have had the honor, in France, to surpass them by far in the theory and to create methods which opened up a sure path to progress, disengaged from all empiricism." The value of theory for Eiffel was that it permitted "exact calculations [from which come] structures which are much lighter and at the same time are stronger than those built earlier."[45] Eiffel's comments, published in 1888, closely paralleled comments made earlier in the nineteenth century by Navier, who held that "the true engineer calculates, and endeavors to proportion the strength of each member to the forces that it must resist."[46]

One further example of a leading inventor and engineer in the private sector who assimilated the approach of the corps engineers was Henri Giffard (1825–1882). Giffard was noted for his work in aeronautics and for his invention of the steam injector. (See pp. 287–290 for an explanation and illustrations of this device.) Giffard had no formal technical education, although he indirectly followed courses at the Ecole Centrale by studying the notes of friends who were students there. His knowledge was mostly acquired through self-study and on-the-job training, however: he started work in a machine shop and was also a locomotive driver for a time. By inclination Giffard was an inventor and a tinkerer, yet within the French environment he also became a theoretician. When Giffard built a steam-powered dirigible in 1850, he used theoretical analysis to work out design features to an extent and in a way that would have been quite unusual for an American self-taught inventor of his generation. For example, he developed a mathematical theory of propellers, which he then used to proportion and angle the blades of the dirigible's screw propeller. Later, when he developed and patented the steam injector, he published a comprehensive theoretical analysis of that device as well.[47,48]

Correspondence from Giffard to the *secrétaire perpétuel* of the Académie des Sciences, Jean-Baptiste Dumas, reveals that, like Fourneyron and Seguin, Giffard was eager for scientific recognition and approval. When Giffard constructed a gigantic captive balloon for the Paris Exhibition of 1878, he received a letter of praise from Dumas. Giffard responded deferentially, and was careful to give primacy to science as the basis for his achievement: "I am excessively flattered by the only too benevolent letter which I had the honor to receive from you. In constructing this captive balloon with the aid of my fellow workers, I have only, so to speak, put to use the principles of Science, of which you are for so long, Sir, the most illustrious representative; and your encouragements, as well as those of the eminent savants who have with interest examined this construction, are for me the highest reward which I could venture to claim."[49]

The tendency of French engineers in the private sector to emulate the corps engineers became manifest not only at an individual level but also at an institutional level. The Ecole Centrale, although created in opposition to the model of the Ecole Polytechnique, nevertheless was shaped by that model. *Centraliens* chose to wear uniforms resembling those of *polytechniciens*.[50] And although the Ecole Centrale's curriculum supposedly favored practical studies, it was carefully pointed out in contemporary literature that "practical" here did not imply manual studies but merely the application of theory to specific problems: "The practical attainments [of *centraliens*] do not consist so much in the manual practice of different mechanical arts, like at the Ecoles d'Arts et Métiers, but rather in the application of theoretical principles to the preparation of projects which embrace the entire range of the industrial arts."[51] As this passage suggests, an effort was made to ally the Ecole Centrale with the traditions of the Ecole Polytechnique and correspondingly to distance it from the Ecoles d'Arts et Métiers.

Other French engineering schools founded later in the nineteenth century or early in the twentieth century—schools for electricity and for aeronautics—were also influenced by the model of the Ecole Polytechnique. A cornerstone of the early curriculum of the Ecole Supérieure d'Electricité, founded in 1894, was a course in "general electrotechnics" taught by Paul Janet, who had previously taught industrial electricity at Grenoble. This course drew extensively on the mathematical research of corps engineers, including André Blondel; in fact, the textbook Janet wrote for the course was entirely within that tradition. The extensive section on alternating-current motors (ca. 150

pages) was all abstract mathematical theory; it made no mention of any actual motor design.[52]

The influence of the mathematical approach among French engineers was not without limits; there were factors that militated against acceptance of this tradition. Robert Fox has provided evidence that independent, regional technical traditions acted as a counterweight to the traditions of the state corps.[53] Yet France was a very centralized country, and this worked against independent, provincial engineering traditions in a number of ways. First, nationwide recruiting was a structural mechanism by which some of the brightest young men from the provinces were socialized into the scientific, technical, and administrative cultures with which the *grandes écoles* were associated. Equally significant, the state engineering corps functioned throughout the country. Every province had its contingent of Ponts et Chaussées engineers, and many were members of local scientific societies.[54] Moreover, it is clear that many educators and private-sector engineers who worked in the provinces were deeply influenced by the traditions and attitudes fostered by the corps and by the Ecole Polytechnique. Fourneyron, Seguin, and Janet provide three notable examples.

My own research suggests that the influence of the corps engineers' theoretical tradition was delimited more by economic sector than by geography. Specifically, the analytical ideal, with its emphasis on mathematical theory, tended to predominate in sectors where the corps engineers kept a high profile (civil engineering and "high" technology), whereas more empirical traditions predominated in sectors where the influence of the corps engineers was more limited (the machine-tool, metalworking, and textile industries, and some areas of "low" technology).

Social position was also an important factor in defining the boundaries of particular research traditions. The state corps and the Ecole Polytechnique were the center of this tradition, and those who most closely approached members of these groups in status were the most influenced by it. Thus, *centraliens* embraced the analytical ideal more fully than *gadzarts*. Proponents of the analytical ideal also published in prestigious journals, which gave their mathematical approach a high level of national and international visibility.

The American Engineering Tradition

The American tradition (and ideology) of engineering research differed in several ways from what we have seen in the case of France.

To begin with, among American engineers there was no powerful elite with a vested interest in mathematical theory; such a group simply did not exist there. Certainly the American engineering corps played no such role. Besides lacking anything like the power and influence of their French counterparts, they were not as a group particularly oriented toward mathematical research in engineering. In fact, one of the most noted government engineers of the nineteenth century— Benjamin Isherwood, Engineer-in-Chief of the U.S. Navy's Bureau of Steam Engineering during the Civil War, was aggressively hostile to theory. In the words of Edwin Layton Jr., Isherwood "denounced the use of all deductive methods and mathematical theory in engineering."[55]

The relations between the scientific and technological communities were also different in France than in the United States. In France, as we have seen, the scientific and technological communities evolved together in a unified hierarchy, within which mathematical research tended to be accorded higher status than empirical research or design. In contrast, the scientific and technological communities in the United States developed more independently of one another. Each community evolved its own institutional structure and reward system, and among engineers a higher value was placed on designing and building than on theorizing. In addition, design-oriented, empirical research was often accorded more weight among American engineers than mathematical theory.[56] For the most part, American engineers did not look toward scientific institutions for recognition and approval. Some were elected to the National Academy of Sciences; yet as a group they did not try to shape their work to fit the priorities of that institution, which in any event did not especially privilege mathematical research.

In short, most of the factors which led French engineers toward an interest in mathematical theory were not operative in the United States in the nineteenth century. Theory did not become part of the "landscape" in the United States, as it did in France. It had no independent status value, as it did in France—that is, theoretical engineering research was not valued for its own sake, as a contribution to knowledge. In France, when the engineer Louis Le Chatelier (1815–1873) worked out a mathematical analysis of a problem relating to locomotive stability, one of the first things he did was prepare a memoir to submit to the Académie des Sciences.[57] This kind of reaction, common among French engineers, would have seemed unusual if not out of place in the United States.

The principal concern of American engineers was to achieve material success—that is, to create "successful" artifacts and to achieve commercial success in the market. Mathematical theories were considered important only insofar there were essential in achieving these aims. In 1881 the steam engineer Alfred Wolff suggested in an address before the American Society of Mechanical Engineers that American engineers had neglected the "mechanical theory of heat" because they felt it had no direct utility for their work: "From personal experience, as well as from the experience of others, I am led to believe that engineers, as a class, look upon the knowledge of the mechanical theory of heat not much unlike that of the department of 'belles-lettres,' as possibly adding to refinement, to a broader view of things in general, to a fair drilling of the mind, but of no practical or only slight practical value in the ordinary, or even extraordinary, exercise of their profession."[58] In a more humorous depiction of some inventors' attitudes toward thermodynamics, Coleman Sellers told of a mechanic's complaint that the "scientific experts" who had made a trial of one of his inventions "had put some thermo-dynamics or some other scientific stuff into the boiler, on purpose to prevent [the] device from operating."[59]

The attitude of many nineteenth-century American engineers toward mathematical theory was not just one of neglect but one of mistrust bordering on hostility. The French engineer Emile Malézieux commented after a visit to the United States in 1870 that American engineers "distrust an analytical formula as a tool that does not feel right in the hand; [for them] it is a cane that often hampers their advance rather than assuring it."[60]

Edwin Layton contends that this anti-mathematical bias was a structural feature of the American philosophy of engineering. He cites as typical the attitude of John Trautwine, author of the influential *Civil Engineers Pocket-Book* (some 150,000 copies of which were sold over the nineteenth century). In his preface, Trautwine proclaims that the efforts of engineering theorists represent "little more than striking instances of how completely the most simple facts may be buried out of sight under heaps of mathematical rubbish."[61] More extreme was the view of Frederick Taylor, who "privately endorsed a sweeping denunciation . . . of all theoretical physics from Newton to thermodynamics."[62] According to Layton, the American antipathy toward theory was due partly to the fact that mathematical theory was potentially subversive in the American environment: the rise of a theoretical tradition would have militated against American traditions of on-the-job training, a "close alliance with employers," and "upward mobility from craftsman to engineer."[63] In France just the opposite was true: an empirical, experimental tradition was potentially subversive, because it promised to undermine the traditions, power, and privileges of the state engineering corps.

To the extent that Americans conceptualized a necessary relationship between theory and practice, their perception of its nature was different from what prevailed in France. This can be seen by contrasting Armengaud's ideas on the relation between mathematical theory and practice in the machine-building industry with the view of the American engineer Coleman Sellers. Armengaud's conception (as expressed in his obituary of Pihet, discussed above) might be termed a trickle-down view: mathematical theory was developed on high by the Lagranges, and the results trickled down to the Pihets, enabling them to devise better machine tools. Sellers's conception was, in contrast, a "grass-roots" view: theory arose from the shop floor. Moreover, when Sellers spoke of "fundamental principles that underlie the theory of construction of machine tools," he did not mean mathematical theories at all; he meant general concepts arising from the experience of using and designing machine tools.

Sellers spoke of "the theory and practice of self-adjusting journal bearings with extended surface for wear, as applied to transmission of power by line shafting." Or he spoke of the "straight-line principle of steam-engine building" developed by Edson Sweet, according to which the lines of strain in an engine were first determined and then the parts and material placed in such a way as to accommodate these strains most efficiently. Sellers also discussed "the theory of broad-finishing cut," which referred to a concept of designing machine tools so that rates of feed, cutting tools, and so forth could be switched rapidly in order to minimize the time needed to prepare for a cut.[64] French machine builders such as Pihet probably also made discoveries of this type, yet in France they would have been considered to be of such low stature as to be hardly worth mentioning.

When it came to formal theories in physics, mechanics, and the like, there was also a conceptual and ideological difference between French and American engineers. To the extent that American engineers acknowledged the importance of such theories, they tended—in line with the tenets of Common Sense philosophy—to regard them as mere "condensations" of experience: " . . . after all what is theory but practice reduced to a connected system of laws or principles? Without facts, without phenomena, in short, without practice as a foundation, how could theory ever be conceived or established? . . . Theory is based on facts, hypothesis on speculation; hypothesis changes to theory when speculation changes to facts."[65] Such a principle was often embodied in laboratory training in American technical schools.[66] When taken to its limit, however, this line of reasoning led to the idea that one needed only to compile masses of data in a coherent way in order to uncover new laws or develop new theories.

In France, in contrast, theories were considered to be conceptual and mathematical formulations that did not simply "condense" experience but could go beyond it in the sense of providing better, more sophisticated foundations for interpreting experience. Of significance for French engineers was the idea that theories could reveal phenomena or functional characteristics of technologies that could not be intuited or fully understood from practice. In this sense, French engineers attempted to use mathematical analysis as a mechanism for discovery—that is, as a means of revealing previously unknown laws and relationships. Navier, as we have seen, discussed how this could be done and applied his ideas in his theory of suspension bridges.

The French also commonly used mathematical theory as a substitute for experience in the process of invention and design—a role the Americans were more hesitant to accord it. Giffard confidently designed a steam engine for his dirigible without a centrifugal governor, because he had determined theoretically that the screw propeller would regulate the engine's speed. In many other cases too, French engineers worked out essential features of new designs on the basis of theory. In fact, their ideal was to plan a new design as far as possible on the basis of mathematical theory, construct an exemplar, and then see how its behavior accorded with theoretical predictions. In contrast, American engineers characteristically worked out new designs by building and testing a series of models and prototypes. Henry Ford, who built numerous trial internal-combustion engines, and the Wright brothers, who built experimental gliders, are two notable examples.[67] They did not ignore theory, but for them it was no substitute for experience. Henry Ford said that, although he read everything he could find on gas engines, "the greatest knowledge came from the work": "There is an immense amount to be learned simply by tinkering with things. It is not possible to learn from books how everything is made—and a real mechanic ought to know how nearly everything is made. Machines are to a mechanic what books are to a writer. He gets ideas from them, and if he has any brains he will apply those ideas."[68] Ford built his first four-stroke gas engine in 1887 just to see if he "understood the principles," and he continued to build many trial models in the following years until he "fairly knew [his] way about." At first he built single-cylinder engines, but he gradually worked his way up to two-cylinder and then to four-cylinder engines. Ford's ultimate aim, of course, was to build automobiles, and accordingly he was at the same time designing and testing his engines as components of larger systems.[69]

In general, American engineers gave less attention to mathematical theory as a means of generating new designs and new knowledge and more attention to direct tests and experiments than their French counterparts. During the first half of the nineteenth century, independent inventors as well as societies such as the Franklin Institute in Philadelphia (a society for "the Promotion of the Mechanic Arts") undertook sophisticated experimental research programs. During the early 1830s the Franklin Institute sponsored an extensive series of water-wheel tests and a series of experiments and tests related to the strength and safety of steam boilers. In each case, the work gained international recognition.[70] A somewhat later example was a series of hydraulic experiments and turbine tests undertaken by James Francis on behalf of the Proprietors of Locks and Canals of Lowell, Massachusetts. The results were published in 1855 in a major treatise, *Lowell Hydraulic Experiments,* which became a classic in its field. Francis's research was renowned for its thoroughness and precision, and nothing fully comparable had been published in France although that country had an internationally recognized tradition of hydraulic research.[71] Francis's work went through several editions and was still being cited by European engineers early in the twentieth century.[72]

Examples of major experimental or empirical research and testing programs in the United States from the late nineteenth century are legion (and certainly much more common than French examples). The work of Thomas Edison is a well-known example. Another example is a series of experiments on metal cutting carried out by Frederick Taylor over a period of 26 years, beginning in 1880. According to Taylor, "800,000 pounds of steel and iron were cut up into chips" and "between 30,000 and 50,000 experiments were carefully recorded." Through this research, Taylor isolated twelve independent variables in metal cutting and carefully studied the influence of each. Among the variables were the chemical composition of the cutting tool, the thickness of the shaving, the contour of the tool's cutting edge, and the amount of water used to cool the work.[73] Some of these experiments provided the basis for Taylor's development of high-speed tool steel, a fundamental innovation. To my knowledge, no comparable research program was undertaken in France on such a scale.

In France, the high status and impressive development of mathematical theory often led to experimental research's being treated merely as an adjunct to theory—a means of testing theoretical results and predictions. In the United States, experimental research was more exploratory: it was directed toward identifying and studying the relevant variables of a given problem, and was generally not confined by a narrowly specified theoretical framework. The French tradition of mathematical engineering science also promoted a certain acceptance of idealization and abstraction. These were necessary prerequisites for theoretical analysis, because many phenomena (e.g., turbulence) could not be handled mathematically in their full complexity. In contrast, the American emphasis on exploratory research tended to foster attempts to come to grips with the complex behavior of technologies as they functioned in the real world. It also led to a concern to know in detail the limits of validity of any theory. The American engineering educator Dugald Jackson emphasized the need to make students aware of the limits of theory: "While higher mathematics is a useful aid in each of the divisions [of the electrical engineering curriculum], its limitations as an agent must be carefully shown in the class room and laboratory. For the purpose of educating the judgment of a student, and fully defining the limitations of theories and mathematical deductions, the laboratory is indispensable, and the equipment should be selected with this object continually in view."[74]

French engineers, guided by mathematical theory, tended to treat the hydraulic turbine as a "black box," paying attention only to conditions at the points where water entered or exited the machine (which were the significant parameters of the theory). In contrast, American engineers and millwrights attempted to learn empirically what happened to the water during its trajectory through the machine. Toward this end they devised experiments using glass-walled testing flumes and small floating pellets, with the aim of discovering empirically how to alleviate turbulence (which impaired turbine performance).[75] American turbine research and testing led to fundamental design changes of international importance. French turbine development stagnated during the second half of the nineteenth century, at least partly (it has been suggested) because French research was hampered by reliance on overidealized (two-dimensional) theories.[76]

On a different level, there is evidence that the high status of theory in France also diverted attention away from patenting and from the kind of "mundane" empirical research and testing that provides the basis for many patents. In any event, many if not most of the corps engineers were more interested in publishing than in patenting. (Theories cannot be patented, of course.) And some of the most noted engineers and inventors in the private sector devoted at least as much time to publishing theories as to carrying out design-oriented, empirical research. The most noted American inventors invariably took out more patents than the most noted French inventors, and published less. And, whereas there is abundant evidence that American inventors carefully followed the patent literature, studied every new patent in their area of interest, and directed their research according to trends in patenting, there is less evidence of such a degree of concern and interest among French inventors.

An indication of the difference between the patenting cultures of France and the United States in the nineteenth century can be seen in the fact that already during the first half of the century American patents were both written and printed up in a standardized fashion, and copies could be purchased by any interested person. Inventors generally obtained copies of all patents issued in their area of interest. In contrast, French patents were unique, handwritten manuscripts, not standardized at all. Some were accompanied by long, discursive essays; others were short and descriptive. Even today, one who wishes to examine a nineteenth-century French patent must consult the original manuscript.

Case Study: The Steam Injector

The distinct ideologies and research traditions of French and American technologists ultimately led to differences in the form and content of the knowledge and artifacts they produced. A comparison of the research and design activities of French and American technologists involving a particular innovation—the steam injector—serves to illustrate this point. (As noted earlier, the steam injector was a device invented in France in 1858 to feed water into steam boilers.)

In order to understand the relevance of the steam injector as an example of French and American research and design traditions, it is helpful to have a basic idea of how an injector functioned. It consisted

Injecteur pour les chaudières à vapeur.

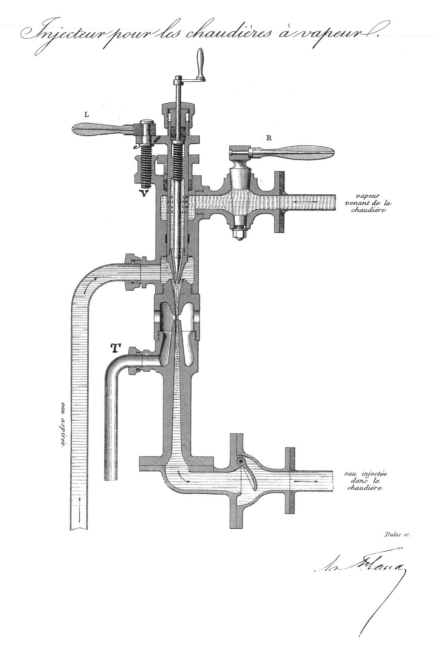

Figure 9.3
The steam injector patented by Henri Giffard in 1858. This illustration is signed by the manufacturer, Henry Flaud. Source: author's collection.

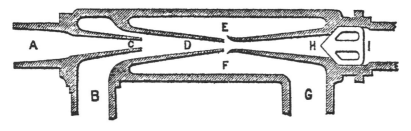

Figure 9.4
A schematic of a steam injector. Source: *Journal of the Franklin Institute*, third series, 56 (1868), p. 54.

of three nozzles: a steam nozzle (A–C), a combining tube (D), and a delivery tube (H), the latter two being separated by a small gap (E–F) which was open to the atmosphere. The device was fitted to the side of a boiler, and a water-supply pipe (B) led from the combining tube to a feedwater reservoir. Steam from the boiler entered the device through the steam nozzle and then passed into the combining tube, causing water to rise from the reservoir into the combining tube. As more steam entered this tube, it collided with the cold feedwater there and condensed. The mixture then passed into the delivery tube, where it forced open a valve (I) and made its way into the boiler. The pipe (G) carried any overflow from the gap (E–F) back down to the feed-water reservoir.

In functional terms, the steam injector was simply a pump. Yet its operation simultaneously involved principles of mechanics (collision between steam and feedwater); principles of fluid mechanics (e.g., the Bernoulli Theorem), and principles of thermodynamics (steam flow through nozzles, conversion of heat into work). The device was there-fore complex from a theoretical point of view. It was also complex from an operational point of view, because small changes in the design of the device significantly affected its operation. If the water was not cold enough to make all the steam condense, the efficiency of the device was affected. If the shapes and dimensions of the nozzles were slightly altered, the device performed differently. And to work at very high steam pressures, the steam nozzles had to be redesigned to look like diffusers, with their diameters increasing in the direction of flow rather than decreasing. There were many other tricky aspects of injector design and performance, too.

The steam injector became a standard piece of equipment on almost all locomotive, naval, and stationary steam engines during the last half of the nineteenth century. The device was therefore economically important[77] as well as challenging from the perspectives of mathematical theory, design, and performance. For these reasons, it provides a good test case for comparing the research traditions of French and American technologists. Technologists in both countries had both incentive and opportunity to do research on this device, in view of its economic importance, its theoretical complexity, and its complex design and performance characteristics. Moreover, this comparison, by focusing on the overall bodies of knowledge that evolved around the technology, captures research practices at a group level rather than only at an individual level.

In order to make a meaningful comparison, it is evidently necessary to have a good sample of the research that was done in the two countries from 1858 to the end of the century. Since there is no comprehensive subject index or citation index for technical literature of the nineteenth century, it is difficult to be certain that one has found all of the relevant literature on a particular topic. However, I used several methods to ensure finding the most significant literature on

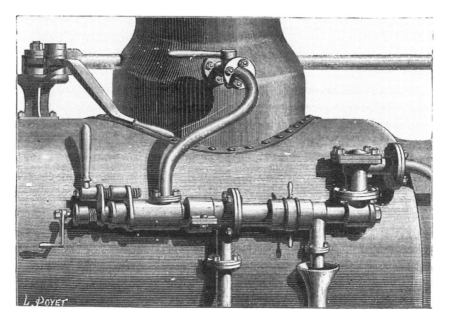

Figure 9.5
A steam injector mounted on a locomotive boiler. Source: *Scientific American* 47 (1882), September 9, p. 163.

injectors. These included tracing injector patents in France and the United States, going through the major French and American engineering journals year by year for the period 1858–1900, searching for injector trade catalogs,[78] and visiting libraries and archives in France and America that were likely to have important literature on injectors. For the United States these included, most notably, the Library of Congress, the National Museum of American History, the Franklin Institute, the University of Pennsylvania, the Hagley Museum's Eleutherian Mills Historical Library, Purdue University's Goss Library, the John Crerar library, and the New York Public library. For France these included the library and archival collections of the Académie des Sciences, the Ecole des Ponts et Chaussées, the Ecole des Mines, the Ecole Centrale, the Archives Nationales, the Conservatoire des Arts et Métiers, the Bibliothèque Nationale, the Bibliothèque Forney, and the library of the Centre de Recherches sur la Civilisation Industrielle at Le Creusot. In addition, I traced the citations in contemporary books and articles about injectors. The bulk of the literature turned up for both countries was simply descriptive, but there also emerged for each country a set of books and papers that concerned original research on injectors. The evidence presented in this section is based on this body of research literature, which includes more than thirty items.[79]

What does this literature show? To begin with, there was a noticeable difference in character in the literature produced in the two countries. The French articles were filled with pages of complex equations, whereas most of the American articles had no equations at all or, at most, a few simple algebraic manipulations or numerical examples. In fact, during these four decades not one American published original mathematical research on the injector, but at least a dozen individuals in France carried out such research. (This number may not seem very significant today, but for nineteenth-century France it represents considerable manpower.)

When it came to experimental work, the roles were reversed: the American literature was replete with sophisticated graphs and lengthy tables of experimental data, whereas the French literature contained almost none. The American literature, comprising articles, books, and unpublished studies and including the most important work on steam injectors by American researchers (15 items in all), contained 93 tables and 30 graphs. In contrast, the 17 French items, including the most important work on injectors by French researchers, contained only 18 tables of experimental data and no graphs at all. The French items did contain eight tables of theoretical data, and one article included a short table of injector capacities (how much water could be pumped per hour) deduced from theory.[80] Americans also published tables of injector capacities, but in every instance these were based directly on tests. Not one of the 93 American tables was deduced from theory.

Another way to compare the amounts of experimental work done by French and American researchers is to look at the number and the range of groups who were involved in such research. For the United States, there is evidence that extensive experiments and tests on injectors were being done at universities (including the University of Wisconsin, Cornell University, the University of Pennsylvania, the University of Michigan, and the Stevens Institute of Technology),[81] by professional societies (such as the Franklin Institute, which carried out tests of particular injectors as well as comparative tests), by trade organizations (such as the American Railway Master Mechanics' Association, which carried out carefully monitored road tests comparing the efficiency and behavior of injectors and steam pumps), by the U.S. Navy, by numerous railroad companies, by private consulting firms, and by the manufacturers of injectors.[82]

Some injector manufacturers set up veritable industrial research laboratories. William Sellers, who obtained the American patent rights to Giffard's injector, set up a department to conduct tests and experiments. He hired Strickland Kneass, a graduate of the Rensselaer Polytechnic Institute, to run the department, and Kneass carried out experimental research on injectors for decades. The department eventually came to be known as the Sellers Research Laboratory.[83]

There is no evidence that such a wide range of groups were involved in French injector research. Some research was done by railroad and mining companies, by injector manufacturers, and by one engineer in the Corps des Mines. But there is no evidence that engineering schools (Ecole Centrale, Ecole des Mines, Ecole des Ponts et Chaussées, Ecoles d'Arts et Métiers) were involved in any experimental research on injectors, nor have I found evidence that professional societies or trade associations undertook such research, nor does it seem that anything approaching an industrial research laboratory was established by any French injector company.[84]

It appears that American engineers were more likely to publish the results of experiments and tests than French engineers, presumably because such research was more highly esteemed in the United States. The extent of the difference can be seen by comparing the leading

French and American injector manufacturers: H. Flaud & Cie. of Paris
and the Sellers Company of Philadelphia. The former was the original
French manufacturer of the Giffard injector. Its owner, Henry Flaud,
a *gadzart,* worked closely with Henri Giffard from the beginning and
continued to make improvements in later years. In the 1870s, Giffard
and Flaud developed a special injector for the French navy.[85] It is
evident that the two men must have done experimental work and
testing, but this was never discussed in publications. Giffard's "theoretical and practical" treatise on the injector, although nearly 90 pages
long, contained only two short tables of data, each consisting of
only one row, and there was no explanation of how the data were
obtained.[86]

In contrast, experimental work and tests done by the Sellers Company were discussed extensively in the publications of Strickland
Kneass. Kneass's articles and his treatise on the "practice and theory"
of the injector contained 19 tables of data, 13 graphs, and extensive
discussions of a wide range of experiments and tests. For example, in
order to design more efficient steam nozzles, Kneass built a variety of
nozzle shapes and tested them under various conditions to see which
shape produced the least turbulence. He also carried out many sophisticated experiments on steam flow. And he carried out experiments to
determine the best techniques for producing the special bronzes
needed for injectors.[87] Other tests and experiments were also discussed
in trade catalogs published by the Sellers Company. One early trade
brochure (ca. 1860) included a lengthy table of capacities (showing the
amount of water an injector could pump into a boiler per hour relative
to its size and the steam pressure within the boiler).[88] A Flaud brochure
of approximately the same date contained no such table—yet it was
deemed important in the Flaud brochure to cite "a remarkable article
containing the theory of the injector" that had been published by a
member of the Académie des Sciences.[89]

If the amount of experimental work on injectors varied between
France and America, so too did its focus. American injector research
was, for the most part, aimed at empirical questions of design and
performance: how design changes in particular components affected
an injector's behavior, how a specific injector performed under various
conditions, how different injectors performed relative to one another
under a specific range of conditions, and so on. French injector research usually aimed either at verifying general theoretical predictions
about all injectors or at exploring fundamental thermodynamic phe-

nomena. In fact none of the published French experimental work that I was able to locate included comparative performance data, and I found only two tables that gave performance data for particular injector designs.[90] In the United States such tables were published by the dozens.

Even French engineers who worked in practice-oriented environments (e.g., for railways or mining companies) did theoretically oriented experimental research. An example can be seen in a paper by Deloy, an engineer with the Paris-Mediterranean Railway: nearly 10 of the 17 pages were devoted to a mathematical theory of the injector. A significant proportion of the experiments were, moreover, aimed at testing theoretical predictions. One example: The injector's action involved a collision between hot steam and cold feedwater. In principle, it was possible to deduce the post-collision temperature theoretically, given the initial temperatures and the flow rates of steam and water. One of Deloy's experiments was to see if the final temperatures actually measured agreed with the theoretical predictions. American researchers paid almost no attention to such questions.[91]

The evolution of injector design also followed distinct paths in France and America. My intention here is not to detail this history, but merely to point out that French injectors evolved less rapidly and less extensively than American ones. The evidence can be seen in many comments by technical writers of the period. The author of a French treatise on locomotives published in 1898 noted that a number of important French railroads had "for many years" used injectors that were "remarkably little different from the primitive Giffard type."[92] Strickland Kneass, in an overview of injector development, characterized the design changes introduced by French engineers as "comparatively unimportant."[93] It is evident that this view was correct: aside from the original invention, no French injector design attained international importance or reached the American market.[94] By the end of the nineteenth century, French injectors could not even compete in the French market. By that time, most of the French railways had begun to employ Sellers injectors.[95]

Most of the design changes introduced in France and in the United States were embodied in patents. Therefore, some further indication of the attention given to injector design in each country can be gained by comparing the number of patents taken out by the leading French and American injector manufacturers. Looking once again at the engineers associated with the leading injector company of each country (Henri Giffard and Henri Flaud in France, William Sellers and

Strickland Kneass in the United States), we find a notable difference in patenting activity. Giffard took out three injector patents, including his patent on the original invention; Flaud took out one.[96] Sellers took out nine injector patents; Kneass took out 25.[97] Thus, Sellers and Kneass together took out about eight times as many patents as Giffard and Flaud. Some of these were of minor significance, but a number were fundamental, and taken as a group they help to account for the eventual success of Sellers injectors in the French market.[98]

Several conclusions emerge from this overview of French and American injector research. Broadly, the French research produced mathematical-theoretical knowledge of a universal nature—general theoretical truths about the thermodynamic behavior of all injectors. The American research generally produced design-oriented, empirical knowledge, such as knowledge about how certain design changes would affect the performance of a certain class of injectors. French knowledge characteristically took the form of mathematical equations; American knowledge generally took the form of tables and graphs. French research was preferentially embodied in scientific publications; American research tended to be oriented more toward patents and was often described in trade catalogs.[99] Finally, French injector designs evolved less—and less rapidly—than American ones.

Thus, the steam injector provides a striking example of how differences in traditions and ideologies of research led French and American engineers down different paths of technological practice and development. Of course this example cannot be taken as representative of all technological development in the two countries. In some fields (such as chemical engineering) experimentation played a more prominent role in French research, and in other fields (such as electrical engineering) the elaboration of mathematical theory played a more prominent role in American research. Nevertheless, the case of the steam injector exemplifies a broad trend that involved many other technologies, including hydraulic motors, steam engines, bridges, and electrical machinery.

French and American Engineering Traditions: Continuity and Change

In attempting to analyze some of the underlying characteristics of the French and American technological traditions, my principal focus has been on continuity rather than change. Nevertheless, changes occurred, and they should be taken into account at least briefly. It is clear that today the French and American traditions of engineering are both more complex and more similar than they were in the nineteenth century. American engineers today routinely employ mathematical techniques and theories that the vast majority of their nineteenth-century counterparts would have considered utterly abstruse and without practical value, and French engineers have fully embraced the methods of industrial research.

Several forces have gradually helped to broaden the scope and loosen the hold of earlier traditions. Most important, the increases in the scale and in the complexity of technology have come to require a closer intermingling of mathematical and experimental research and empirical testing.

It has been shown time and again that technologies cannot be developed merely on the basis of mathematical theory: every theory of necessity simplifies and idealizes the phenomena or systems it is supposed to model, and the degree of idealization is generally determined as much by the possibilities of calculation as by the needs of design. In other words, accurate theoretical models are often too complex to handle mathematically, while models that can be handled are often not accurate enough to be an adequate guide to practice. Thus, the French corps engineers' analytical ideal, which envisioned mathematical theory as the mainspring of technological development was a chimera. This became evident already in the nineteenth century when, as was frequently the case, empirical research proved to be an indispensable prerequisite for improving the performance and reliability of technologies.

However, the scientification of technology also brought the limits of empiricism to the fore. Complex, large-scale systems and structures cannot always be fully tested in advance, yet at the same time they cannot simply be built on the basis of trial and error. Engineers increasingly needed mathematical-theoretical guidelines to structure and proportion their designs. Thus, the belief of many American engineers that empirical research was a sufficient guide for practice was also an illusion. Edison publicly railed against theorists, but he found it necessary to hire a mathematician. James Buchanan Eads (1820–1887), designer of the St. Louis Bridge (1867), also accepted a need for mathematical theory. Eads was not an adept mathematician, and he worked out the basic structural character of his design for the St. Louis Bridge through physical insights derived from a series of conceptual, non-mathematical models. Yet he commissioned German-trained engineers to carry out extensive mathematical calculations to bring the design to the blueprint stage.[100]

The broadening of the French and American engineering traditions was also fostered by structural and institutional changes. In the United States, the growth and proliferation of engineering schools provided an environment in which a theoretical tradition of research could be created and diffused (although, as we have seen, American engineering educators were just as interested in laboratory research and workshop training). In France, the growing percentage of engineers working in the private sector and (more generally) the expansion and influence of private enterprise created an environment that helped to foster and legitimize empirical, design-oriented research. In addition, both countries were influenced by the internationalization of industry and trade. International competition revealed the strengths and weaknesses of each country's research traditions and thereby provided an incentive for change. Thus, in the latter decades of the nineteenth century there were more and more warnings by American engineers that too much faith in empiricism threatened to place American technology at a disadvantage in international competition, and there were warnings by French engineers that too little attention to empirical research threatened to place French technology at a disadvantage.[101]

Despite these forces of change, however, the differences between the French and American technological traditions have never been fully eliminated. France still has a notably more stratified system of technical education than the United States, and French engineers are still noted for a particular emphasis on formalistic, mathematical theories. American engineers are still noted for their attention to empirical, design-oriented research. In short, both countries still bear traces of their history embedded in the character and structure of their institutions and, less tangibly, in their ideologies and values.

Conclusion to Part II

Chapters 7–9 surveyed several features of French and American engineering cultures and technological practices as they evolved in the nineteenth century. The comparisons highlighted fundamental differences between these cultures with respect to the social structures of the technological communities, the evolution and role of technical education, the ideologies of technological research, and the patterns of technological practice. The comparisons also focused on interactions among elements of engineering culture: how the social structure of each country's technological community shaped the evolution of technical education and how research ideologies shaped the practices of research and design.

The analysis revealed that the American engineering community was characterized by a relatively high degree of social and occupational mobility, while the French community was more stratified and hierarchical. The American frontier experience created a context in which a premium was placed on mobility and adaptability. Particularly in the first half of the nineteenth century, Americans often took up multiple occupations or learned several trades, and they often moved back and forth between technological pursuits and agricultural or commercial pursuits. France had a more finely graded social structure, which was partly a heritage of the social system of the *ancien régime* but which was also due to the fact that France was a more populous and economically developed society than the United States. The more rigid and finely differentiated social structure in France led to a more hierarchical technological community. It also led to more concern for formal credentials—and for status through credentials—than its American counterpart.

The hierarchical structure of the French technological community was manifested particularly in the existence and traditions of the powerful state engineering corps, which exercised a tremendous influence within the French technological community. The effect of this hierarchical structure was twofold. On the one hand it meant that each stratum of the French engineering community tended to evolve its own distinct traditions of practice and its own pattern of technical education. Yet on the other hand it created a context in which the traditions of the most elite sector—the state corps—had higher status, greater visibility, and a disproportionate influence on the character and direction of technological research. Among other things, the corps' extensive legal authority over engineering activities in France provided a context in which its criteria and its values (as opposed to, e.g., market forces) could materially shape the work of engineers in the private sector. In the United States there was no system of centralized administrative control over engineering activity, nor was there any powerful, salaried elite that could set priorities or shape research activities in the way that the corps engineers in France were able to do. As a consequence, much of American engineering research was more oriented toward economic and design issues than French research.

The evolution of technical training also followed different paths in France and the United States. In France, formal technical education already played a major role in the first half of the nineteenth century: every corps engineer had a formal technical education, and companies in the public-works sector, were generally dominated by academically trained engineers. In the machine-building sector, apprenticeship did remain dominant in France during the first half of the nineteenth century, but *gadzarts* (and, later, *centraliens*) also played a significant role in this sector.

In the United States, formal technical education played a minor role before the latter half of the nineteenth century. Most American technologists, even in the public-works sector, learned through apprenticeship, self-study, and on-the-job training, although West Point played some role in training engineers. In this context, it is not surprising that American technologists did not do much research in mathematical engineering science in the first half of the nineteenth century. This kind of research, to be of much value in engineering design, required years of preparatory study. This was made evident in the case of Navier and his theory of suspension bridges. Navier's long formal education and his salaried position in the state corps made it possible for him to undertake his mathematical research on suspension bridges. Most American technologists in the first half of the nineteenth century had neither the time, nor the inclination, nor the institutional support to be able to do this, however.

In the second half of the nineteenth century formal technical education became much more prominent in the United States than it had been, but it developed along different lines than its counterpart in France. Apart from the structural differences, the contents and philosophies of the curricula differed profoundly. The French system of technical education was hierarchical, with schools at different levels recruiting from distinct social strata. The elite schools had such demanding entrance requirements that usually only individuals who could afford special preparatory education could hope to gain admission. French technologists consolidated this system of social stratification by establishing school-based alumni associations, which appear to have inhibited the development of sector-based professional societies such as were created in the United States.

American technical schools were much less hierarchically organized. They were organized more along regional lines, but they increasingly competed with one another for reputation. The entrance requirements of the American technical schools tended to be more moderate than those of the elite French schools. Nevertheless, the Worcester Polytechnic Institute, which more than any other American technical school resembled the Ecoles d'Arts et Métiers, had comparatively high entrance requirements, while the Rensselaer Polytechnic Institute, a respected school for civil engineering, had relatively low admission requirements (even in the last decade of the nineteenth century, when there was a movement among American schools to raise admission standards).

In France, mathematical, experimental, and manual studies were separated hierarchically, whereas in the United States they were united. The American practice of bringing the workshop into the university reflected the fluid character of the American social system and the higher status of manual labor in the United States relative to France. It also reflected the strong influence of apprenticeship training in the field of mechanical engineering: the workshop was seen by many not only as the economic center of that field but also as its intellectual and educational heart. More broadly, the strong traditions of apprenticeship and on-the-job training in the United States tended to promote the idea (although there was debate on this point) that both workshop and laboratory training should be undertaken in contexts that replicated as closely as possible the world of industrial practice.

On another level, the degree of attention given to laboratory and workshop training in American technical schools reflected the strength of the tradition of Common Sense philosophy in American society and education. This intellectual orientation was linked with the principle of "empirical rediscovery," which the Belgian observer Omer Buyse saw as figuring so prominently in American technical education. In any event, the large amount of time that students devoted to laboratory and workshop training meant that mathematics and mathematical engineering science were usually covered in less depth in American technical schools than in the high-ranking technical schools of France.

Another difference between the content of French and American technical education concerned design. The curricula of the Ecoles d'Arts et Métiers do not appear to have given much attention to this subject *per se,* although some experience in design must have been gained through the workshop training and the study of technical drawing. The Ecole Centrale and the *écoles d'application* did give attention to design; however, to the extent that they did, the study of design was linked primarily to mathematical theory and technical drawing.

In the United States, with the integration of workshop training and laboratory training into the curricula of technical schools, design came to be taught not only in relation to theory and technical drawing but also in relation to workshop practice and laboratory testing. Students studied the relationship between machine design and performance through extensive comparative testing in the laboratory, and they studied the relationship between machine design and production through workshop training. They were encouraged to bring all of the knowledge they acquired to bear in the design process. I have no evidence that French engineering students were taught design in this way, and some evidence that they were not. Coleman Sellers suggested that technical schools in France and Germany were not teaching prospective engineers to pay attention to the production implications of their designs, and that this affected the kinds of designs they produced:

I have seen a locomotive . . . that required for its construction a specification covering the leaves of a large folio book of many pages, telling the maker how each minute part is to be constructed, regardless of the facilities that the maker might have to construct economically on his own system of shop sizes. The cylinders seem to be the work of one scientific expert on cylinders; the valve motion the work of another, the boiler was the brain work of a third, perhaps; and the whole brought together into one machine—well, I confess I know not how, but looking like too much science and not enough practice.[1]

French and American technologists also differed in their ideologies of research. French engineers tended to take the view that mathematical theory should *precede* and *guide* practice, whereas American engineers tended to hold that theory should *emerge from* practice. And whereas French engineers tended to treat experimentation as an adjunct to theory, American engineers were inclined toward the opposite view. In general, a higher status was accorded to theoretical than to experimental or empirical research in France; the opposite was the case in the United States. Moreover, whereas "theory" in the context of French engineering ideology almost always meant mathematical theory, in the context of American engineering ideology it often meant conceptual or physical insights that were non-mathematical in character.

All these differences serve to highlight the fact that the scientification of engineering that occurred during the nineteenth century took different forms in different social and institutional contexts. And this shaped engineering practices, for it makes a great deal of difference whether engineering is "scientized" through a greater emphasis on mathematical analysis or through a greater emphasis on experimental research and testing. The conclusion to be drawn, I think, is that we need to begin to examine more specifically how engineers in different contexts have incorporated mathematical-theoretical and experimental methods in their research and design work, and what bodies of knowledge they have created. For example, did French engineers utilize thermodynamics in a different way than American engineers, and did they extend this body of knowledge in different directions? It may be that answers to such questions will reveal a situation similar to what we saw in the case of the steam injector—namely, that bodies of knowledge and traditions of practice vary considerably from nation to nation, in part because distinct ideologies lead engineers in different environments to orient their research in consistently different ways.

Of course, when it is a question of examining how the process of scientification varied between nations, it is important to consider the problem not only in relation to ideology but also in relation to institutional structures. As we have seen, the institutional relations between science and engineering were in some respects closer in France than in the United States. In particular, whereas in France the Académie des Sciences was a central institution in both the scientific community and the engineering community, in the United States the National Academy of Sciences was not. One effect was that leading engineers in France, unlike their counterparts in the United States, tended to orient their work toward the priorities of science as much as toward design and innovation.

The influence of engineering cultures on rates of technological and economic development needs to be addressed, at least briefly. Certainly the question of what determines rates of development is complex and involves a range of economic and environmental variables. The aim here is not to attempt to analyze the problem but merely to point out that, in principle, the social and institutional factors that define engineering cultures can affect these processes. There are two interrelated reasons why this should be the case. The first reason is simply that different engineering cultures are not necessarily equally conducive to technological and economic growth. The second reason is that there is no objective basis for determining *a priori* precisely what kind of engineering culture constitutes the optimal environment for technological or economic development (although research and the lessons of experience have begun to provide a deeper understanding of the problem). It is clear, however, that in the nineteenth century decisions about such things as how to organize technical education, how to structure curricula, and whether and how to manage engineering activities at a national level were never made solely on the basis of objective economic criteria. Such decisions were inevitably shaped by ideology, by political and social realities, and by history. Thus, the overall system (for an engineering culture can indeed be regarded as a system) was not really the product of a conscious, rational choice at all, even if an objectively optimal choice could have been made.

It is generally agreed that French technological development over the second half of the nineteenth century was in certain respects less dynamic than its American counterpart.[2] I would argue that the difference between the engineering cultures of the two countries was a factor in this. At any rate, aspects of the French and American engineering traditions did have different technological and economic implications. Perhaps most notable, whereas the patents and design innovations that resulted from American testing and experimental research had immediate economic and proprietary value, the scientific publications that resulted from French theoretical research had no proprietary value. In addition, whereas the theoretical results in French publications were available to all to make use of, the empirical knowledge and data underlying American patents often were not. Americans did make use of French theoretical work to their own economic advantage, moreover. Yet, when the French used American patents or bought American machines (as in the case of the steam injector), much of the economic advantage went not to the French but to the Americans. Of course, French engineers developed and patented some important fundamental inventions on the basis of theoretical and experimental research. But these inventions generally became reliable only after years of further testing and experimental research, which was more likely to be done by American than by French researchers. Thus, I would argue that, in the short run, American empirical research led to more rapid technological and economic development than French mathematical theory. In the long run, however, the greatest advantage was bound to go to the nations that could encourage both approaches simultaneously and foster their interaction. French and American engineers increasingly came to recognize this by the end of the nineteenth century, and both attempted to reshape their engineering traditions accordingly.

General Conclusion

This book emerged from a perception that the landscapes of engineering in nineteenth-century France and America differed in significant ways. To a degree, the origin of this perception fits the American Common Sense model, although in this case the theories arose not from the shop floor or the laboratory but from the library stacks.

My research began not as a comparative analysis but as a series of more "universal" studies of specific technological developments (hydraulic turbines, injectors, and suspension bridges) based on the assumption that, at least in the industrial era, technology and knowledge were universal. The information that came from the library stacks sent a different message, however.

The French literature on these technologies was generally more mathematical than the American, and the American literature was generally more experimental and oriented toward physical testing than the French. Explaining why this was the case became one of the goals of my research. It soon became evident that the difference could not have been rooted entirely in a specific type of technology, because this difference appeared in all the technologies I studied. I concluded that it had to be rooted in conditions within the technological communities of each country, or in conditions that immediately affected those communities. Accordingly, the object of my research shifted to discovering, through systematic comparative analysis, what those conditions were, and, more generally, how the environments in which French and American technologists worked shaped their research and design practices.

The comparative approach proved to be an excellent tool for learning "what society is made of." It encouraged a systematic accounting and ordering of elements that differentiated French society and American society in the nineteenth century, and it provided a context for exploring how those elements shaped the social and institutional environments of technological practice. The analysis showed that the French and American environments differed in significant ways.

In the United States, the frontier experience—associated throughout the first half of the nineteenth century with low population density, rapid expansion into a wilderness, and concomitant lack of material and institutional infrastructure—helped to create a context in which people from many backgrounds were encouraged to take up technological pursuits and could easily shift occupations; a context in which craftsmen had a high social status and manual labor was not stigmatized; a context in which self-study, on-the-job training, and apprenticeship predominated among engineers, at least until 1850; a context in which research practices rooted in craft traditions (such as the building of models and the establishment of design rules) were considered to have intellectual value; and a context in which Common Sense philosophy, which encouraged an accessible, hands-on approach to learning and research, found broad acceptance.

The French environment was characterized by higher population density, a more stratified social structure, a greater availability of specially skilled labor, and (throughout the first half of the nineteenth century) a better-developed material and institutional infrastructure. Within this institutional infrastructure, the state engineering corps played several important roles. The corps reflected a tradition of extensive, centralized control over technological activity; they helped to build and preserve a tradition of monumental architecture in public works; and they helped to formulate and maintain an ideology which sought to make technology deducible from mathematical theory and which opposed research traditions associated with craft practices (such as model building). More generally, social stratification in France created a context in which craftsmen and their intellectual perspectives had low social status and manual labor was stigmatized. It also created a context in which credentialism and formal technical education were already important within the technological community during the first half of the nineteenth century.

The study of Finley and Navier revealed how some of these environmental characteristics shaped their respective efforts at bridge design. Finley lived in a rural, frontier-like area; he served as a county commissioner and knew the level of funding that counties like his own could commit to bridge building; he was aware of the lack of specially skilled labor in rural regions, and he knew that in order for his design to be widely applicable it had to be easy to build. Finley also was aware of, and drew on, the tenets of Common Sense philosophy; he knew that justifying his design in relation to those tenets would help to gain acceptance for it. Navier, educated in the *grandes écoles,* was strongly committed to the Corps des Ponts et Chaussées and its traditions; he wanted to further its tradition of monumental architecture and to extend the analytical ideal; he was aware of social stratification within his society, and his design catered explicitly to the upper strata of Parisian society; he was committed to the analytical ideal, and he knew that a design that could be justified in relation to that ideal would be respected by the hierarchy of the Corps.

Finley and Navier responded to characteristics of their environments through their design goals, through their decisions about how to prioritize those goals, and through their decisions about where and how to seek knowledge that might help in their realization. Finley sought knowledge through encyclopedias; probably through discussions with blacksmiths and other bridge builders; through the building of a trial bridge in his home county; and, in line with the tenets of Common Sense philosophy, through experiments with cables and pulleys and tests of the rupture strength of iron. Navier sought knowledge through two voyages to Britain, where he spoke with suspension bridge designers, read existing literature on suspension bridges, collected experimental data on the strength of iron, and visited several completed suspension bridges. In line with the analytical ideal, Navier also sought knowledge through extensive reading in mathematics and theoretical mechanics and through mastery of the new technique of Fourier analysis.

Armed with their respective bodies of knowledge, Finley and Navier established social, economic, and aesthetic design goals, and prioritized those goals. Their goals and their knowledge, in turn, guided their efforts to steer among conflicting technical possibilities. Finley's environment and its organizational systems for bridge building led him to establish low cost, ease of construction, and structural calculability from experimentally derived laws as predominant design goals. Navier's environment and its organizational system for bridge building led him to establish monumentality and structural calculability from mathematical theory as predominant design goals. Low cost was also among Navier's stated goals; however, in his case this meant low cost relative to other types of monumental bridges, such as stone arches, not low cost relative to other suspension bridges or to other inexpensive bridges.

The technical features of Finley's and Navier's designs—their choices of hardware, and the technical means of interconnecting the various elements of the structures—simultaneously embodied their knowledge, their goals, and their ideas about how the structures would be built. The technical details of the designs embodied tradeoffs among these elements, and specific choices often undercut other goals or influenced other elements of the design. Finley's decision to adopt a sag/span ratio of 1/7 simplified the calculations that builders needed to carry out, and ensured that no iron would be "wasted" to accommodate additional stresses due to greater tautness; yet that ratio sacrificed stability. Finley's simple tower design helped to make the bridge easy to build; but it effectively limited the length of the spans that could be built, and it led to greater reliance on multiple spans. Reliance on multiple spans, in turn, sacrificed stability. Finley's cable design, with only one link between hangers, saved iron and made the structure cheaper, but it necessitated the design of two types of "keys" to connect the hangers to the cables.

Navier's decision to adopt a sag/span ratio of 1/10 for the Pont des Invalides helped to ensure greater stability. His compound cable enhanced the monumentality of the design but sacrificed economy and ease of construction. His method for determining hanger lengths enhanced reliance on mathematical theory, but it too sacrificed economy and ease of construction (in the sense of requiring more highly skilled labor in the design-to-construction process in order to carry out the necessary mathematical calculations). Navier's deck design attempted to achieve stability primarily through weight rather than stiffness. (Recall that the dead weight per linear foot of his bridge was greater than that of any previous suspension bridge.) Using weight to achieve stability tied the Invalides design—at least in Navier's view—more closely to the precepts of his theory. Yet it sacrificed economy, because the added weight increased the cost.

Whereas in part I of the book comparative analysis was used to reveal how the French and American environments shaped technological design, the role of comparative analysis in part II was to show how these environments were linked to engineering research ideologies and practices and, more broadly, to the structures of the technological communities and their systems of technical education. This analysis showed that the comparatively high level of social stratification in French society was reproduced within the technological community and within its system of technical education. Technical schools preserved and sometimes enhanced social, occupational, and cognitive barriers among technologists. Particular schools were explicitly organized to draw students from given social strata and to train them for specific, predetermined occupations, which were often hierarchically separated. The Ecoles d'Arts et Métiers were designed to recruit from the lower and lower-middle classes, and to produce foremen and *conducteurs* rather than corps engineers and high-level managers; the Ecole Polytechnique was designed to recruit mainly from the upper strata of French society. The divisions fostered by this hierarchical system of technical education were further preserved through the establishment of alumni societies, which helped to create a distinctive *esprit de corps* for the graduates of each school or set of schools.

The American technological community and the American system of technical education were less hierarchically organized. During the first half of the century, technical schools played a small role in technical training, and they were not organized so explicitly to recruit from specific social backgrounds or to supply technical personnel at a predetermined social level. Most graduates of West Point, although trained for service in the military engineering corps, went into other occupations. More generally, the American technological community was characterized by social and occupational mobility, and this was reflected in the geographical, educational, and occupational heterogeneity of members of professional technical societies such as the ASCE and the ASME.

Comparative analysis also called attention to the significance of the intellectual content and philosophy of technical schools' curricula. So far historians have devoted relatively little attention to exactly what was taught and how it was taught, to how curricula developed over time, or to how curricula varied regionally and internationally.[1] In the absence of such information, there is a tendency to assume that engineers everywhere were taught roughly the same thing in roughly the same way—for example, that basic courses in thermodynamics, or programs in mechanical or electrical engineering, or the attention given to laboratory and workshop training followed a kind of universal standard. Yet when careful international comparisons are made, such assumptions immediately fall by the wayside. It becomes evident that curricula too were socially shaped.

The characteristics of the American environment created a context in which there were early efforts to bring the machine shop into the university. Such efforts not only shaped the focus of the Morrill Act of 1862; they also resulted in the establishment of mechanical engineering with workshop training as a comprehensive program alongside civil engineering in many leading technical schools. The American social and intellectual environment created a context in which a particular emphasis was placed on experimentation and hands-on education, a trend that was already evident in the curriculum of the Rensselaer School. By the end of the nineteenth century, this orientation had fostered the establishment of large, diverse, well-equipped laboratories and workshops in American technical schools. The American environment also created a context in which there was an effort to reproduce the world of industrial practice within the university, and to link the study of design to industrial workshop practice and to the tradition of comparative testing (which was also brought into the university from the outside world of technological practice).

The French environment created a context in which mathematical-theoretical, experimental, and workshop training tended to be separated hierarchically among different schools rather than integrated. The elite schools offered little or no workshop training, while the lower-status schools offered little more than workshop training. Within the French social and intellectual environment, the principle of "empirical rediscovery," which figured so prominently in American technical education and which led to extensive laboratory training, carried much less weight. The elite schools devoted the most attention to mathematical theory and much less time to laboratory training. The lower-status Ecoles d'Arts et Métiers also largely ignored laboratory training. Only the Ecole Centrale gave laboratory training a central place in the curriculum during most of the nineteenth century.[2] Although French technical schools pioneered the introduction of laboratory training for students, this trend stagnated during the second half of the nineteenth century, for both social and cognitive reasons. By the end of the nineteenth century, French technical schools had fewer and less extensive laboratory and workshop facilities than American ones. Finally, the French environment created a context in which design was taught in the more elite schools primarily in relation to mathematical theory and drawing technique, although at the Ecole Centrale it was also taught in relation to laboratory work. But because students in these schools had little or no workshop training, they had less opportunity to study design in relation to workshop practice.

Besides the content of technical education, comparative analysis showed that the distinct environments of France and the United States shaped the two countries' ideologies of engineering research. As with technical curricula, however, there has been no major attempt by historians and sociologists of technology to define or compare engineering research ideologies in precise terms—specifically, the ways of conceptualizing the relationships among theory, experiment, testing, design, and other forms of practice. In the absence of such comparisons, it is easy to assume that engineers everywhere held roughly similar views, or that, if their views differed, they had little effect on the way engineering was done. The comparison between France and the United States showed that this was by no means the case, however.

In France, the research ideology promoted within the state engineering corps held mathematical theory to be a necessary foundation for good design. Not all engineers within or outside the corps accepted this ideology, but the imperialistic position of the state corps helped to give that ideology stature within the broader technological community.

The predominant ideology of technological research in the United States followed the tenets of Common Sense philosophy and put primary emphasis on experiments and tests to gain knowledge relevant to design. This ideology did not privilege mathematical theory; rather, it tended to give experimental evidence independent or even dominant status in relation to mathematical theory. American ideology also gave more prominence to non-mathematical concepts and principles; these were more liable to be accorded the status of theory in the American than in the French environment.

Neither technical training nor research ideologies fully determine patterns of technological practice, but they certainly bear an important relation to technological practice. Technical training reflects existing traditions of technological practice, but it may also enhance those traditions. As was shown in chapter 4, the mathematical education given at the Ecole Polytechnique reflected a trend in which mathematics became an important element in the tool kit of French corps engineers. This trend was already evident in the careers of Gauthey and Prony. Yet, through the establishment of the Ecole Polytechnique and the requirement that all prospective corps members undergo an extensive mathematical training in this school, this trend was further enhanced and reinforced. In the United States, an extensive focus on laboratory training in technical schools reflected a tradition—already evident in the first quarter of the nineteenth century—in which technologists sought knowledge relevant to design through systematic experimentation. Yet the expansion of formal technical education requiring significant laboratory training further reinforced and enhanced this tradition.

Similarly, while there is always some gap between ideology and practice, French and American ideologies did shape research practices. American ideology served to justify a situation in which the majority of technologists (before the 1850s) had little practical possibility or incentive for undertaking years of study in order to do original mathematical research on technological problems. Yet this ideology also encouraged American engineers to devote time and energy to do creative experiments and design-oriented, comparative tests. In France, the ideology of the state engineering corps served to justify a strong educational focus on mathematics in the *grandes écoles,* and to justify the practice of recruiting corps engineers only from graduates of the Ecole Polytechnique. Yet this ideology also encouraged French engineers to devote time and energy to develop mathematical theories relevant to their domain.

Because of these differences in social organization, training, and ideology, French and American technologists tended to apportion their research time differently, and to link mathematical research, experimental research, and testing in different ways. American engineers spent a great deal of time doing experiments and tests, often independent of any mathematical-theoretical framework. Leading French engineers in the state corps spent a great deal of time developing mathematical theories of the technologies within their domains, as well as mathematical theories of natural phenomena that affected technological design, such as friction, fluid flow, and elasticity. Many corps engineers did experiments and tests too, but often these were structured in relation to mathematical theories rather than in relation to design issues. Within the private sector, some technologists, including Giffard and Fourneyron, also spent a significant amount of time developing mathematical theories (after spending time gaining the necessary background knowledge). Others, however, did place greater reliance on experiments to gain knowledge for the purposes of design.

Yet French engineers on the whole do not appear to have devoted as much time and energy to systematic, comparative testing and experimentation as American engineers.

The differences in the research practices of French and American engineers were evidenced by the bodies of knowledge they created around the steam injector. French research was highly mathematical, and most of the French injector experiments were carried out in relation to mathematical theory. American injector research, in contrast, was mainly experimental. American injector researchers focused extensively on comparative testing, publishing dozens of tables and graphs. Hardly any comparative test data on injector performance were published in France, however. Finally, the rapid evolution of injector design in the United States provides evidence that much design-oriented research was done there, whereas the slow evolution of injector design in France indicates that comparatively little design-oriented research was done there.

In addition to the importance of comparative analysis for discerning the social shaping of technological research and design, the evidence presented in this book shows that this approach has implications for another historiographic issue: technological diffusion. The diffusion of suspension bridge technology from China to the United States and from Britain to France was considered in part I, and the diffusion of French models of technical education to the United States was considered in part II. In each case, comparative analysis made it possible to go beyond the initial outward evidence that diffusion had occurred, and to show clearly that the diffusion process involved a sifting and adapting of information so as to integrate it with goals, values, and perspectives stemming from the receiving environment.

Finley learned about Chinese chain bridges, but he also learned about stone arch bridges. He rejected the latter because he did not perceive that they fit in any meaningful way into his own environment. (Economically, his perception was on the mark: stone arch bridges in France cost millions of francs and often took years to build.) Finley did see potential in the idea of a chain bridge, but he adapted the Chinese form to the vehicular traffic it had to be able to accommodate in his environment. He further adapted the idea of the suspension bridge to a particular social system for construction involving the patent system; an agent (John Templeman); carpenters, blacksmiths, and

other craftsmen; local investors; and county commissioners or municipal officials.

Navier too sifted the knowledge and precedents he observed or read
about in Britain and adapted them to his environment. Fortunately
he left a clear record of what he saw, what he learned about, and what
he subsequently did, and by using this record within the context of a
comparative approach it is possible to see how Navier selectively borrowed the knowledge to which he had access. For example, Navier
carefully observed Samuel Brown's Union Bridge, but much of what
he saw—including elements of cable design, deck design, saddle design, and anchorage design—he did not borrow. Navier preferred to
use and combine other technological elements in such as way as to
accord with aspects of his mathematical theory, and to adapt the
suspension bridge to the French tradition of monumental architecture.

More generally, we saw that the whole *problématique* around suspension bridge technology was transformed in Navier's hands. Whereas
British engineers had focused extensively on cable design, and on
developing new kinds of hardware, Navier gave comparatively little
attention to these matters. His mathematical deduction that wooden
cables would be seven or eight times as economical as iron cables never
led him to build and test such a cable, or even to describe how such a
cable might be built. Whereas British designers built models and did
tests, Navier did not. Whereas both Finley and British builders determined hanger lengths empirically, Navier chose to develop a method
based on mathematical analysis. Whereas British builders and engineers did extensive experiments on the strength of iron, Navier did
not, but instead incorporated their observations into his mathematical
theory. Whereas Navier's reading and observations in Britain provided
clear evidence of the possibility of vertical deck oscillations due to
wind, and evidence that British builders were using technological
measures to prevent such vertical oscillations, Navier largely ignored
this evidence and instead carried out a 30-page mathematical analysis
of cable vibrations due to moving traffic loads—an analysis that was
geared to current research concerns within the Paris mathematical
community.[3] Navier's sifting of the knowledge and evidence he came
across in Britain was fundamentally shaped by his commitment to the
"analytical ideal," which was an important perspective within his own
environment but which had not motivated the bridge designers he met
in Britain.

Evidence of this kind of sifting can also be seen in the realm of technical education—in particular, in the adapting of the model of the Ecole Polytechnique to West Point. Certainly West Point drew on the model of the Ecole Polytechnique, but again comparative analysis reveals that knowledge of this model was sifted and adapted in line with goals, priorities, and conditions that prevailed within the American environment, and as much of the original model was rejected as was accepted. Most important, the whole idea of high entrance standards, which was such an important foundation for the role and meaning of the Ecole Polytechnique in France, was explicitly rejected in the United States, and moreover throughout the entire nineteenth century. The idea that students at such a school should have armorers and barracks keepers to keep their guns and rooms clean was also explicitly rejected. In the American environment, doing these tasks was seen as a way of learning discipline, responsibility, and self-reliance, and they were written into the school's formal timetable. The curriculum of the Ecole Polytechnique also was by no means taken up intact at West Point. It too was sifted and adapted. Finally, the role of the Ecole Polytechnique as a feeder to other schools and to a host of engineering corps was not taken up in its entirety at West Point.

Historians generally have direct access to the outward evidence of technological diffusion, but the processes of sifting and adaptation that occur with diffusion can often be discerned only through comparative analysis. Most borrowers do not fully explain what information they have rejected or ignored, and why. Such knowledge must be gained through a careful comparison of what information the borrower had access to, what information was actually used, how it was used, and toward what ends. Only in this way is it possible to retrace the selection process that has been carried out, and to show that it was not random but was structured by goals, priorities, and values rooted in a specific environment.

Notes

Introduction

1. Alexis de Tocqueville, *Democracy in America* (Knopf, 1945), volume II, pp. 42, 45–47.

2. John A. Kouwenhoven, *The Arts in Modern American Civilization* (1948; Norton, 1967); Hugo A. Meier, "Technology and democracy, 1800–1860," *Mississippi Valley Historical Review* 43 (1959), March, pp. 618–640; Eugene S. Ferguson, "The American-ness of American technology," *Technology and Culture* 20 (1979), no. 1, pp. 3–24; John F. Kasson, *Civilizing the Machine: Technology and Republican Values in America, 1776–1900* (Penguin, 1976).

3. Of the studies by Tocqueville, Kouwenhoven, Meier, Kasson, and Ferguson, only Tocqueville's attempts to be comparative in any systematic way. (The others refer to European traditions, or in some cases to specific European countries, but their references serve only as a foil by which to highlight the distinctiveness of American society.) Tocqueville's comparisons, however, are actually between two ideal types—democracy and aristocracy—although it is evident that America and France provided the models for his ideal types. Robert Nisbet examines Tocqueville's use of ideal types in his article "Tocqueville's ideal types," in *Reconsidering Tocqueville's Democracy in America*, ed. A. Eisenstadt (Rutgers University Press, 1988). The other essays in that book also provide useful insights into Tocqueville's work.

4. Other contributions to the history of technology have also begun to fill this gap, a notable example being Thomas P. Hughes's *Networks of Power: Electrification in Western Society, 1880–1930* (Johns Hopkins University Press, 1983). Neither Hughes's nor any of the other studies, however, deals with national (or regional) research traditions and how they shape specific technologies or patterns of technological development.

5. Tocqueville simply asserted the existence of these differing traditions but gave no systematic evidence to justify his observations. Another characteristic of Tocqueville's analysis is that it was framed within the "applied science" model of technology. He assumed that the knowledge used by technologists

was scientific knowledge. Thus, Tocqueville would categorize the engineering sciences as being part of scientific knowledge, whereas I treat them as being part of technological knowledge. As a consequence, I view the production of even theoretical knowledge in engineering as a dimension of technological practice; Tocqueville does not. What this means, more broadly, is that Tocqueville's analysis is framed not in terms of different traditions of technological knowledge and research, but in terms of different traditions of scientific knowledge.

6. Only Tocqueville makes any comparative analysis of knowledge and research traditions. Kouwenhoven and Meier touch on the question of American empiricism, but neither goes beyond the ideas or the approach of Tocqueville. Ferguson and Kasson do not deal with the question at all. With regard to France, Terry Shinn has attempted to explain the French orientation toward theory in terms of institutional structures, and his work has been an important resource for my own analysis. However, he has not examined the way in which French research traditions have shaped technological design. See Terry Shinn, "Des corps de l'état au secteur industriel: genèse de la profession d'ingénieur, 1750–1920," *Revue française de sociologie* 19 (1978), pp. 39–71; see also the revised English version of this paper in *The Organization of Science and Technology in France, 1808–1914*, ed. R. Fox and G. Weisz (Cambridge University Press, 1980); see also "Raisonnement déductif: formation et déformation," *Bulletin de la Société française de physique*, n.s., 42 (1981), pp. 32–34; "Reactionary technologists: The struggle over the Ecole Polytechnique, 1880–1914," *Minerva* 22 (1984), no. 3–4, pp. 329–345; Terry W. Shinn and Harry W. Paul, "The structure and state of science in France," *Contemporary French Civilization* 6 (1981–82), no. 1–2, pp. 153–193; Shinn, *Savoir scientifique et pouvoir sociale: L'Ecole Polytechnique, 1794–1914* (Presses de la Fondation Nationale des Sciences Politiques, 1980).

7. Ulrich Wengenroth, "The comparative study of the history of technology," presented to Conference on Critical Problems and Research Frontiers in History of Technology and History of Science, Madison, Wisconsin, 1991.

8. Michel Callon and Bruno Latour are among those who have particularly emphasized the network-building and society-rearranging dimensions of socio-technical development. See Callon, "Society in the making: The study of technology as a tool for sociological analysis," in *The Social Construction of Technological Systems: New Directions in the Sociology and History of Technology*, ed. W. Bijker et al. (MIT Press, 1987); Latour, *Science in Action: How to Follow Scientists and Engineers through Society* (Open University Press, 1987); Latour, "Mixing humans and nonhumans together: The sociology of a door closer," *Social Problems* 35 (1988), no. 3, pp. 298–310. See also, however, a critique of the actor-network approach by Yves Gingras: "Following scientists through society?—yes, but at arm's length!" *Cahiers d'Epistémologie* (Université de Québec à Montréal), no. 9203 (1992).

9. Michel Callon and Bruno Latour have warned against assuming in advance that we "know what society is made of." The comparative approach does not

begin from this assumption, but rather assumes that the nature of society depends upon time and place.

10. See, for example, Callon, "Society in the making," pp. 92–93; John Law and Michel Callon, "Engineering and sociology in a military aircraft project: A network analysis of technological change," *Social Problems* 35 (1988), no. 3, pp. 284–297; John Law, "Technology and heterogeneous engineering: The case of Portuguese expansion," in *The Social Construction of Technological Systems,* ed. Bijker et al.

11. The notion of problem transformations parallels the concept of "translation" used by Callon and Latour. See the references in note 8; see also the following: Callon, "Struggles and negotiations to define what is problematic and what is not: The socio-logic of translation," in *The Social Process of Scientific Investigation,* Sociology of the Sciences Yearbook 4 (1980), pp. 197–219; Callon, "Some elements of a sociology of translation: Domestication of the scallops and the fishermen of St. Brieuc Bay," in *Power, Action and Belief,* Sociological Review monograph 32 (Routledge, 1985). Some typical examples of problem transformations can be seen in the history of the electric light. At one point Thomas Edison transformed a technical desideratum to use a platinum filament into a socio-economic problem of organizing efforts to locate new platinum ore deposits. At another point, the economic problem of the high cost of copper was transformed into the technical problem of developing a high-resistance filament, and the problem of developing a high-resistance filament was in turn transformed into the problem of developing a spiral filament. See Robert Friedel and Paul Israel, *Edison's Electric Light: Biography of an Invention* (Rutgers University Press, 1986), pp. 55–56, 78–79, 80–81, 96, 106.

12. Studies concerned with technological knowledge represent a large and growing area of research within the history of technology with links to the debate on the relationship between science and technology. The following are some examples: Edwin T. Layton, Jr., "Technology as knowledge," *Technology and Culture* 15 (1974), no. 1, pp. 31–41; Walter Vincenti, *What Engineers Know and How They Know It: Analytical Studies from Aeronautical History* (Johns Hopkins University Press, 1990); Edward Constant, *The Origins of the Turbojet Revolution* (Johns Hopkins University Press, 1980); Ronald Kline, "Science and engineering theory in the invention and development of the induction motor, 1880–1900," *Technology and Culture* 28 (1987), no. 2, pp. 283–313; Eda Kranakis, "The French connection: Giffard's injector and the nature of heat," *Technology and Culture* 23 (1982), no. 1, pp. 3–38; David F. Channell, "The harmony of theory and practice: The engineering science of W. J. M. Rankine," *Technology and Culture* 23 (1982), no. 1, pp. 39–52.

13. In the literature on the history of technology there has generally been more emphasis on invention than on design. Edwin Layton was among the first to use design as an organizing concept; his approach has been taken up and extended by Walter Vincenti and Eugene Ferguson. See Layton, "American ideologies of science and engineering," *Technology and Culture* 17 (1976),

no. 4, pp. 688–701; Layton, "Science and engineering design," in *Bridge to the Future: A Centennial Celebration of the Brooklyn Bridge,* ed. M. Latimer et al. (New York Academy of Sciences, 1984); Vincenti, *What Engineers Know and How They Know It;* Eugene S. Ferguson, *Engineering and the Mind's Eye* (MIT Press, 1992). The concept of design is also central in the work of some historians of civil engineering, including David Billington—see his books *The Tower and the Bridge: The New Art of Structural Engineering* (Basic Books, 1983) and *Robert Maillart's Bridges: The Art of Engineering* (Princeton University Press, 1979).

14. For an introduction to this rapidly developing area of technology studies, see *The Social Shaping of Technology: How the Refrigerator Got Its Hum,* ed. D. MacKenzie and J. Wajcman (Open University Press, 1985); *The Social Construction of Technological Systems,* ed. W. Bijker et al. (MIT Press, 1987); *Shaping Technology/Building Society: Studies in Sociotechnical Change,* ed. W. Bijker and J. Law (MIT Press, 1992).

15. There are, however, important exceptions. For example, Louis Bucciarelli has explored the social character of the design process using ethnographic methods. See his book *Designing Engineers* (MIT Press, 1994). In "An ethnographic perspective on engineering design" Bucciarelli writes: "I see designing as all that goes on within the subculture of the firm. My frame is that of the ethnographer; my first premise, based upon what I have observed, is that designing is a social process. . . . A second premise is that all [daily activities within the firm] . . . are potentially important design acts; all may influence the way design proceeds and the ultimate form of the artifact. . . . A third premise is that all members of the firm and those they call upon outside the firm are potential contributors: individuals in marketing and production, purchasing and finance, as well as in engineering along with subcontractors, suppliers, and the customer." See Bucciarelli, "An ethnographic perspective on engineering design," *Design Studies* 5 (1984), no. 3, pp. 185–190. Also see Bucciarelli, "Engineering design process," in *Making Time: Ethnographies of High-Technology Organizations,* ed. F. Dubinskas (Temple University Press, 1988).

16. A body of literature focusing on technological and business practice is beginning to emerge. Three notable examples are Carolyn Cooper's *Shaping Invention: Thomas Blanchard's Machinery and Patent Management in Nineteenth-Century America* (Columbia University Press, 1991), which looks at the practice of managing technological innovation through the patent system; Friedel and Israel's *Invention of the Electric Light,* which looks in detail at the practice of designing and inventing in Edison's laboratory; and Judy McGaw's "Accounting for innovation: Technological change and business practice in the Berkshire County paper industry," *Technology and Culture* 26, no. 4 (1985), pp. 703–725. None of the studies I am aware of, however, have considered the broader historiographic implications of focusing on practice. In contrast, within the areas of history and sociology of science, the theoretical and historiographic implications of focusing on practice have been explicitly addressed.

See, for example, Frederic Lawrence Holmes, *Lavoisier and the Chemistry of Life: An Exploration of Scientific Creativity* (University of Wisconsin Press, 1985); Joan Hideko Fujimura, "Crafting science, transforming biology: The case of onco-gene research" (unpublished); Andy Pickering, "Knowledge, practice, and mere construction," *Social Studies of Science* 20 (1990), no. 4, pp. 682–728; Pickering, *Science as Practice and Culture* (University of Chicago Press, 1992); Timothy Lenoir, "Practice, reason, context: The dialogue between theory and experiment," *Science in Context* 2 (1988), no. 1, pp. 3–22; *The Uses of Experiment: Studies in the Natural Sciences*, ed. D. Gooding et al. (Cambridge University Press, 1989).

17. Particular ideologies may persist within a technical community for gen-erations. It is not my purpose here to enter the debate on the nature of ideology. It should be noted, however, that I use the term not merely in the sense of "false consciousness" but rather to refer more broadly to a set of biases, values, and perspectives that are characteristic of a particular social group and that shape their thoughts and actions. Useful introductions to the theories and debates concerning ideology include Clifford Geertz's "Ideology as a cultural system," in *Ideology and Discontent*, ed. D. Apter (Free Press of Glencoe, 1964), and Edward Shils's "Ideology: The concept and function of ideology," in *International Encyclopedia of the Social Sciences* (Macmillan, 1968), volume 7. I thank Steven Harris for bringing these references to my attention. Harris's Ph.D. dissertation, Jesuit Ideology and Jesuit Science: Religious Val-ues and Scientific Activity in the Society of Jesus, 1540–1773 (University of Wisconsin, 1988), also contains a valuable discussion of the concept of ideology.

18. John M. Staudenmaier, *Technology's Storytellers: Reweaving the Human Fabric* (MIT Press, 1985), p. 187.

Chapter 1

1. Source of birth date: Josiah V. Thompson of Uniontown, Pennsylvania, letters to Anna M. Wirtz, New Haven, Connecticut, December 15, 1931 and March 18, 1932, Llewellyn Edwards manuscripts, National Museum of Ameri-can History, Smithsonian Institution, Washington. Thompson, a wealthy in-dustrialist in Fayette County's iron industry with an interest in local history, visited and corresponded with descendants of Finley's three brothers, one of whom had also lived in Fayette County.

2. James G. Leyburn, *The Scotch-Irish: A Social History* (University of North Carolina Press, 1962), pp. 248–249; Charles W. Kepner, The Contributions Early Presbyterian Leaders Made in the Development of the Educational Institutions in Western Pennsylvania Prior to 1850, Ph.D. dissertation, Uni-versity of Pittsburgh, 1942, pp. 25–29; Henry Jones Ford, *The Scotch-Irish in America* (Princeton University Press, 1966), passim; Douglas Sloan, *The Scottish Enlightenment and the American College Ideal* (Columbia University Teachers' College Press, 1971), pp. 36–37.

3. According to Josiah Thompson, James Finley the bridge designer (henceforth "J.F.") was buried in the cemetery of the Laurel Hill Presbyterian parish. Independent local religious histories dating from the middle of the nineteenth century indicate that Samuel Finley (a brother of J.F.) was an elder of the Laurel Hill parish and that Samuel Finley came to the Laurel Hill parish from the East Nottingham parish in Maryland. See Joseph Smith, *Old Redstone; or, Historical Sketches of Western Presbyterianism* (Philadelphia: Lippincott, Grambo, 1854), pp. 285–287; David Elliott, *The Life of the Reverend Elisha Macurdy with an Appendix Containing Brief Notices of Various Deceased Ministers of the Presbyterian Church in Western Pennsylvania* (Allegheny: Kennedy & Brother, 1848), pp. 249–254; Franklin Ellis, *History of Fayette County, Pennsylvania* (Philadelphia: L. H. Everts, 1882). Early property maps of the region show that in North Union Township, where Laurel Hill parish was located, there was only one James Finley and only one Samuel Finley. See W. F. Horn, *The Horn Papers: Early Westward Movement on the Monongahela and Upper Ohio, 1765–1795* (Hagstrom, 1945), volume III, maps 5, 7, and 24. (These maps detail all the original properties of the Fayette County region surveyed and patented through the state land office.) See also Ellis, *History of Fayette County,* where J.F.'s property in North Union Township is mentioned specifically on p. 673. Although J.F. was not one of Rev. Finley's sons, he was very likely related to his namesake—Josiah Thompson does not suggest this, but the minister's genealogy leaves the possibility open. See Richard Webster, *A History of the Presbyterian Church in America* (Philadelphia: Joseph M. Wilson, 1857), pp. 610–611; *Encyclopaedia of the Presbyterian Church in the United States,* ed. A. Nevin (Philadelphia: Presbyterian Publishing Co., 1884), p. 234, s.v. Finley, Rev. James; *Pennsylvania: A History,* ed. G. Donehoo (New York: Lewis Historical Publishing Co., 1926), p. 96; Charles A. Hanna, *The Wilderness Trail; Or, the Ventures and Adventures of the Pennsylvania Traders on the Allegheny Path* (Putnam, 1911), volume II, p. 213.

4. James Veech, *The Monongahela of Old; or, Historical Sketches of South-Western Pennsylvania to the Year 1800* (Pittsburgh: n.p., 1858–1892), p. 160.

5. It has been suggested that Finley may have attended the College of Philadelphia—later the University of Pennsylvania (Emory L. Kemp, "Links in a chain: The development of suspension bridges, 1801–70," *The Structural Engineer* 57A (1979), August, p. 3)—but there is no evidence for this assertion. The school's enrollment records reveal that the only James Finley to attend that school in the eighteenth century was a James Edwards Burr Finley (1758–1819), but that Finley became a physician and lived in South Carolina (biographical file on James Edwards Burr Finley, University of Pennsylvania Archives). There is no evidence from enrollment records that J.F. attended Princeton, Harvard, or Yale. In the above-cited letter (Veech, *Monongahela of Old,* p. 160), Ephraim Douglass compares the levels of education of three men, including J.F., who had been elected to county offices, noting that one of them (Jonathan Rowland) has a "better English education" than either of the others. Since a person with a university or college education was considered at that time to be unusually well educated, the assertion that J.F. was not as well educated as Rowland seems to preclude a college education.

6. The "log colleges" were grammar-to-secondary schools organized by ministers at the parish or the county level. More than 65 had been established in the U.S. by 1800. According to Scotch-Irish and Presbyterian histories, there was one in virtually every parish. See J. G. Craighead, *Scotch and Irish Seeds in American Soil* (Philadelphia: Presbyterian Board of Publication, 1878), pp. 299–300; Barr Ferree, *Pennsylvania: A Primer* (New York: Leonard Scott, 1904), p. 23. See also Sloan, *Scottish Enlightenment and American College Ideal*, pp. 5–6, 36–88; Lawrence A. Cremin, *American Education: The Colonial Experience, 1607–1783* (New York, 1970), pp. 323–326, 378, 380, 402. The spread of Presbyterian education was associated with the migration of more than 225,000 Scots to the U.S. in the eighteenth century. A number of the Presbyterian academies eventually grew into colleges and universities, including Princeton and the University of Pennsylvania. If J.F. had attended a university, therefore, it almost certainly would have been one of these two Presbyterian universities. Rev. James Finley's brother Samuel was appointed president of the College of New Jersey (later Princeton University) in 1761.

7. The Nottingham Academy was initially directed by Rev. Samuel Finley. Since Rev. James Finley had served as pastor of East Nottingham parish since 1752, and since this parish joined with another nearby parish at about the time that Samuel Finley left, it may be that Rev. James Finley took over his brother's ministry and the directorship of the academy. Circumstantial evidence for this assertion can be seen by the fact that Rev. James Finley was said to have been the teacher of a Rev. James Dunlap who studied sometime between 1765 and 1771 (which was, moreover, just about the time that J.F. would have received his education). Further evidence can be seen in the fact that when Rev. James Finley moved to southwest Pennsylvania he became very active in promoting the development of education there. See "Finley, Samuel" in *Dictionary of American Biography* (hereafter *DAB*); Elliott, *Life of the Reverend Elisha Macurdy,*, pp. 249–254, 258–260; Ella Chalfant, *A Goodly Heritage* (University of Pittsburgh Press, 1955), pp. 114–115.

8. It was there that Rev. James Finley had received his education, under Rev. Samuel Blair. See Elliott, *Elisha Macurdy*, pp. 249–254.

9. Pecqua had been founded by Samuel Blair's cousin, Robert Smith. By 1769 (about the time J.F. would have begun his education), Smith had about 30 pupils. See Sloan, *Scottish Enlightenment and American College Ideal*, pp. 61, 168; Elliott, *Life of Elisha Macurdy*, pp. 249–254; Dwight Raymond Guthrie, *John McMillan, The Apostle of Presbyterianism in the West, 1752–1833* (University of Pittsburgh Press, 1952), pp. 80–81; Craighead, *Scotch and Irish Seeds in American Soil*, p. 306.

10. Sloan, *Scottish Enlightenment*, p. 61.

11. Quoted in Guthrie, *John McMillan*, p. 94.

12. J.F.'s property was first surveyed the day Fayette County's land office opened: April 3, 1769. J.F. took up ownership of the property on March 9 or

10, 1786, although he had been living in Fayette County before that date. *Horn Papers,* Map 24; Ellis, *History of Fayette County,* p. 673.

13. Among the others who moved were Rev. Finley and his eight sons. See Smith, *Old Redstone,* pp. 285–287; Elliott, *Life of the Rev. Elisha Macurdy,* pp. 249–254. Local presbyterian histories dating from the middle of the nineteenth century indicate that Rev. James Finley, minister of the East Nottingham parish, made trips to Western Pennsylvania in 1765, 1767, 1771, and 1772 for the Presbyterian synod. Extant letters written by Rev. James Finley tell of his desire to move to Western Pennsylvania (Rev. James Finley, Dunlap's Creek, Fayette County, March 13, 1783, letter to "a Gentleman in Westmoreland County," Society Collection, Pennsylvania Historical Society; Rev. James Finley, Cecil County, Maryland, April 28, 1783, letter to John Dickinson, President of the Supreme Executive Council of Pennsylvania, Society Collection, Pennsylvania Historical Society). Early property maps show that Rev. Finley bought land in Fayette County and in Westmoreland County in 1770 and 1771 (*Horn Papers,* volume III, maps 2 and 9). According to Elliott, Rev. James Finley moved permanently to Western Pennsylvania in 1783.

14. Lewis Clark Walkinshaw, *Annals of Southwestern Pennsylvania* (New York: Lewis Historical Publishing Co., 1939), volume II, pp. 188–189.

15. Solon J. Buck and Elizabeth Hawthorn Buck, *The Planting of Civilization in Western Pennsylvania* (University of Pittsburgh Press, 1939), pp. 149, 217–218.

16. The Laurel Hill Presbyterian church was built of logs, by the parishioners, in 1772. See Guthrie, *John McMillan,* p. 59.

17. U.S. Direct Tax of 1798: Tax Lists for the State of Pennsylvania (National Archives and Records Service, 1963), microfilm box 822, volume 22, Fayette County.

18. Lee Soltow, *Distribution of Wealth and Income in the United States in 1798* (University of Pittsburgh Press, 1989), p. 55.

19. Ibid., p. 245.

20. U.S. Direct Tax of 1798: Tax Lists for the State of Pennsylvania, Microfilm Box No. 822, volume 22, Fayette County.

21. Buck, *Civilization,* pp. 229–286, 318–348; Sidney Ratner, James H. Soltow, and Richard Sylla, *The Evolution of the American Economy: Growth, Welfare, and Decision Making* (Basic Books, 1979), pp. 105–108.

22. Buck, *Civilization,* p. 240; Alfred D. Chandler, Jr., *The Visible Hand: The Managerial Revolution in American Business* (Harvard University Press, 1977), p. 84.

23. Leland D. Baldwin, *Whiskey Rebels: The Story of a Frontier Uprising* (University of Pittsburgh Press, 1939), pp. 25–28; Edward Everett, "John Smilie,

forgotten champion of early Western Pennsylvania," *Western Pennsylvania Historical Magazine* (hereafter *WPHM*) 33 (1950), p. 87.

24. George W. Hughes, "The pioneer iron industry in western Pennsylvania," *WPHM* 14 (1931), pp. 207–224; Arthur Cecil Bining, "The rise of iron manufacturing in western Pennsylvania," *WPHM* 16 (1933), pp. 235–256; E. Earl Moore, "An introduction to the Holker Papers," *WPHM* 42 (1959), pp. 225–239.

25. The records for the U.S. Direct Tax of 1798 included inventories of slaves, but no slaves are listed for Fayette County.

26. Census of 1810.

27. Buck, *Civilization,* pp. 261–287 and 318–371.

28. Ibid.

29. The constitution was ultimately ratified by a vote of 46 to 23, a margin that contrasts sharply with the 7:2 ratio of the vote against ratification by the delegates from the western counties.

30. Buck, *Civilization,* pp. 454–487; Harry Marlin Tinkcom, *The Republicans and Federalists in Pennsylvania, 1790–1801* (Harrisburg: Pennsylvania Historical and Museum Commission, 1950), passim; Baldwin, *Whiskey Rebels,* passim; Everett, "John Smilie," pp. 77–89; "Gallatin, Abraham Alfonse Albert," *DAB.*

31. These were the major elected offices at the county level in the period when Finley served. The county commissioners were in charge of taxation, and the justices of the peace were "the cornerstone in the structure of administration of justice." See Buck, *Civilization,* pp. 433–441.

32. Ellis, *History of Fayette County,* pp. 150, 152–54, 162, 679; Veech, *Monongahela of Old,* pp. 142, 160, 170. The position of associate judge was equivalent to the position of justice of the peace. After 1790 in Fayette County the judges or justices were appointed for life. See Buck, *Civilization,* p. 435.

33. U.S. Direct Tax of 1798: Tax Lists for the State of Pennsylvania, microfilm box 822, volume 22, Fayette County, Union Township. This direct tax is often referred to as the "glass tax," because the tax was assessed on the number of glass window panes in the house. Only about 10 percent of the approximately 200 homes on the taxation list for Union Township had as many panes of glass as Finley's. I would like to thank Ronald A. Michael, Professor of Anthropology at California State College in Pennsylvania for the foregoing information concerning the glass tax. Since there is only one James Finley on the tax list for Union Township, there is no doubt that the house in question must have belonged to J.F. (Rev. James Finley did not settle in Union Township.)

34. For purposes of comparison, however, a two-story log home of exactly the same dimensions but with fewer and smaller windows and no outbuildings

was valued at $160. Thus, a substantial portion of the value of Finley's dwelling came from the extra buildings and the extra glass windows.

35. James Finley, Last Will and Testament, March 17, 1828, Register of Wills, Fayette County Courthouse, Uniontown.

36. Orphans Court Records, Fayette County Courthouse, Uniontown, entries for January Court, 1829, October Court 1838, June Court, 1839, December Court 1844, March Court 1848. After J.F.'s death, his three children, Mary, Elizabeth, and James, Jr., were put under the guardianship of Dr. Hugh Campbell. In 1838, in order to have additional funds for their upkeep and education, Dr. Campbell requested that the children's shares of the Finley property be sold, since the rental income they brought in was insufficient. The court and Mrs. Finley agreed, and the property was sold, including Mrs. Finley's share.

37. Soltow and Sylla, *Evolution of the American Economy,* p. 84.

38. Veech, *Monongahela of Old,* pp. 149–150.

39. Gaddis's log home (known as Fort Gaddis) still stands. See Ronald L. Michael and Ronald C. Carlisle, " A Log Settler's Fort/Home," *Pennsylvania Folklife* 25 (1976), no. 3, pp. 39–46; Ellis, *History of Fayette County,* p. 679.

40. Ellis, *History of Fayette County,* p. 153; Evelyn Abraham, "Isaac Meason, the first ironmaster west of the Alleghenies," *WPHM* 20 (1937), pp. 41–49; Carmel Caller, *Isaac Meason: The Man, Ironmaster and Businessman, His Mansion* (Connellsville Historical Society, 1975).

41. U.S. Direct Tax of 1798: Tax Lists for the State of Pennsylvania, microfilm box 822, roll 22, Fayette County. Meason resided in Tyrone Township. The tax lists show that Meason already in 1798 had large landholdings in several townships of Fayette County, several tracts of which included saw mills, grist mills, and furnaces.

42. "Gallatin," *DAB;* Buck, *Civilization,* pp. 307–308.

43. Everett, "John Smilie," *WPHM* 33 (1950), p. 84.

44. Ibid., pp. 77–89; Buck, *Civilization,* pp. 454–466.

45. Veech, *Monongahela of Old,* p. 142; Buck, *Civilization,* p. 465; Ellis, *History of Fayette County,* p. 673.

46. Russell J. Ferguson, *Early Western Pennsylvania Politics* (University of Pittsburgh Press, 1938), p. 117.

47. In 1796 a state senator from a neighboring county reported that there was only a very small Federalist party in Fayette County (Buck, *Civilization,* p. 476). An alternate explanation for Finley's election is that it represented a reaction against the Whiskey Rebellion.

48. Ephraim Douglass, letter to J.F., April 3, 1791, Gallatin manuscripts, 1791, no. 3, New York Historical Society.

49. "Gallatin," *DAB;* Albert Gallatin, *Report on the Subject of Roads and Canals* (1808; reprinted by Augustus M. Kelly in 1968), p. 62.

50. Abraham, "Isaac Meason," p. 43; Ellis, *History of Fayette County,* p. 250; Report on a petition for a bridge over Jacob's Creek at or near the residence of Isaac Meason, Clerk of Courts Office, Fayette County Courthouse, Uniontown. Meason consistently campaigned for transportation improvements that would aid his own business. In 1788 and 1794, for example, he petitioned for roads leading to his estate and to his iron works (Abraham, p. 43).

51. Lewis Clark Walkinshaw, *Annals of Southwestern Pennsylvania* (Lewis Historical Publishing Co., 1939), volume II, p. 189.

52. Buck, *Civilization,* p. 152.

53. Smith, *Old Redstone,* p. 383.

54. Elliott, *Elisha Macurdy,* pp. 249–254, 258–260; Guthrie, *John McMillan,* pp. 80–82.

55. Sloan, *Scottish Enlightenment,* pp. 103–145.

56. Guthrie, *John McMillan,,* p. 94; Elliott, *Elisha Macurdy,* pp. 258–260; Veech, *Monongahela of Old,* p. 105; Lois Mulkearn and Edwin V. Pugh, *A Traveler's Guide to Historic Western Pennsylvania* (University of Pittsburgh Press, 1954), p. 241. When Dunlap became principal of Jefferson College, Laurel Hill acquired a new minister: James Guthrie. Guthrie also became a trustee of the academy in Uniontown, along with J.F. See Smith, *Old Redstone,* p. 383; James Hadden, *A History of Uniontown, the County Seat of Fayette County, Pennsylvania* (Uniontown: James Hadden, 1913), p. 484.

57. William Frederick Worner, "The Strasburg Scientific Society," *Lancaster County Historical Society Papers* 25 (1921), no. 9, pp. 133–145.

58. Guthrie, *John McMillan,* pp. 48, 50, 88–96.

59. Quoted in ibid., p. 103.

60. Howard Miller, *The Revolutionary College: American Presbyterian Higher Education, 1707–1837* (New York University Press, 1976), p. 336, n. 51.

61. Buck, *Civilization,* pp. 376–378, 386–387.

62. H. M. Brackenridge, *Recollections of Persons and Places in the West* (Philadelphia: James Kay, Jr., 1834), pp. 53–54, 80.

63. Agnes Lynch Sarrett, *Through One Hundred and Fifty Years: The University of Pittsburgh* (University of Pittsburgh Press, 1937), pp. 7–9, 15–19.

64. Russell J. Ferguson, *Early Western Pennsylvania Politics* (University of Pittsburgh Press, 1938), p. 44.

65. Sarrett, *Through One Hundred and Fifty Years,* p. 35.

66. Ibid., pp. 35, 62; Edward Park Anderson, "The intellectual life of Pittsburgh, 1786–1836," *WPHM* 14 (1931), p. 22. When Nevill moved to Cincinnati, his library became the nucleus for the library of the Ohio Mechanics' Institute.

67. Sarrett, *Through One Hundred and Fifty Years,* p. 48.

68. Buck, *Civilization,* pp. 375–376; Miller, *Revolutionary College,* pp. 126, 133, 135, 148–149.

69. Buck, *Civilization,* p. 376. Aigster himself subsequently founded a company in Pittsburgh to manufacture sulphuric acid; see Edward Hahn, "Science in Pittsburgh, 1813–1848," *WPHM* 55 (1972), p. 69.

70. Hahn, "Science in Pittsburgh"; Sarrett, *Through One Hundred and Fifty Years,* p. 48; Buck, *Civilization,* p. 376.

71. Sarrett, *Through One Hundred and Fifty Years,* pp. 65–66, 72, 74–75, 105–106.

72. G. T. Fleming, *History of Pittsburgh and Environs* (American Historical Society, 1922), volume II, p. 621.

73. Ibid., p. 89; Hahn, "Science in Pittsburgh," pp. 70–72; Anderson, *Intellectual Life of Pittsburgh,* p. 62.

74. Isaac Meason and Albert Gallatin, who acquired considerable wealth from their industrial activities, are two examples. Meason had a more or less free reign in establishing an iron industry in Fayette County. There was a rapidly growing local market for iron, as well as a growing market further west, yet there was little need to worry about competition from the east, because the high cost of transporting iron over the Alleghenies had the same effect as a protective tariff. Moreover, since Meason was one of the first to set up a blast furnace in western Pennsylvania, he did not have much local competition either, at least in the beginning. Meason's enterprise was able to expand nearly into a vacuum, and similar conditions prevailed for other early enterprises as well, like Gallatin's glass works (Buck, *Civilization,* pp. 301–306; Bining, "Rise of iron manufacture," pp. 235–241).

75. Tinkcom, *Republicans and Federalists,* pp. 48–50; Whitfield J. Bell, "The scientific environment of Philadelphia, 1775–1790," *Proceedings of the American Philosophical Society* 92 (1948), March, pp. 6–14.

76. Reports on Petitions for Roads and Bridges, Clerk of Courts, Fayette County Courthouse, Uniontown; Ellis, *History of Fayette County,* pp. 250, 434–435, 620; Buck, *Civilization,* pp. 433–441.

77. Lance E. Davis et al., *American Economic Growth: An Economist's History of the United States* (Harper & Row, 1972), pp. 472–475; Arthur Cecil Bining, *The Rise of American Economic Life,* fourth edition (Scribner, 1964), pp. 179–182;

Curtis P. Nettels, *The Emergence of a National Economy, 1775–1815,* Economic History of the United States, volume 2 (Holt, Rinehart and Winston, 1962), pp. 251–255. On the Pennsylvania legislature's role with respect to internal improvements see Louis Hartz, *Economic Policy and Democratic Thought: Pennsylvania, 1776–1860* (Harvard University Press, 1948), pp. 37–69; George Rogers Taylor, *The Transportation Revolution, 1815–1860,* Economic History of the United States, volume 4 (Rinehart, 1951), pp. 15–28.

78. Brooke Hindle, *The Pursuit of Science in Revolutionary America, 1735–1789* (University of North Carolina Press, 1956), p. 373, n. 86.

79. Charles W. Peale, *An Essay on Building Wooden Bridges* (Philadelphia, 1797), p. iii. Peale also wrote about his plan to regional officials involved with bridge construction. See C. W. Peale, letter to Commissioners for building bridges over Jones Falls in Baltimore, November 4, 1795, Peale-Sellers Papers, American Philosophical Society Archives (hereinafter APS Archives). Peale's career and work are discussed in Sidney Hart's article "'To encrease the comforts of life': Charles Willson Peale and the mechanical arts," *Pennsylvania Magazine of History and Biography* 110 (1986), no. 3, pp. 323–357.

80. "Bridges," in *The Register of Arts,* ed. T. Fessenden (Philadelphia: C. & A. Conrad, 1808), pp. 311–312.

81. John Jones, letter, Indian River (Delaware), May 27, 1773, APS Archives. Jones, a judge in Delaware, was elected to the APS on January 21, 1774. In addition to his interest in bridges, Jones did agricultural experiments and invented several pieces of agricultural machinery, including a horse-drawn mowing machine. He also established a sawmill to make cedar shingles. Jones died around 1801. See the unpublished biographical sketch of Jones in the APS Archives; see also *A Catalogue of Instruments and Models in the Possession of the American Philosophical Society,* compiled by R. Multhauf (Philadelphia, 1961), pp. 40–43.

82. Thomas Gilpin, A Plan with an Estimation of the cost of a Chain Bridge over the River Schulkiln within the City of Philadelphia, Thomas Gilpin Letterbook, Maury Family Papers, University of Virginia Library. Gilpin had little formal education but acquired a knowledge of mathematics, practical navigation, surveying, natural history, and natural philosophy through self-study. Drawing inspiration (according to his son) from the career of Benjamin Franklin, Gilpin developed an interest in useful engineering projects; he frequently discussed plans for bridges, canals, and other structures at the American Philosophical Society, "which had become the great resort for designs of useful improvement." See "Memoir of Thomas Gilpin (Found Among the Papers of Thomas Gilpin, Jr.)," *Pennsylvania Magazine of History and Biography* 49 (1925), no. 4, pp. 289–328.

83. Joseph Needham et al., *Science and Civilization in China* (Cambridge University Press, 1971), volume IV, part 3, pp. 184–210.

84. Ibid.

85. Ibid.

86. Carl Christian Schramm, *Merkwürdigsten Brücken* (Leipzig: Ben Bernhard Christoph Breitkopf, 1735), pp. 59–61.

87. There is an engraving of the Mekong River chain bridge, dated 1782, in the Llewellyn Edwards manuscript collection at the National Museum of American History. The engraving was subsequently reproduced, slightly modified, on p. 458 of Robert Sears's book *Wonders of the World* (New York: E. Walker, 1847). Two other Asian chain bridges are described on pp. 54–59 of Samuel Turner's book *An Account of an Embassy to the Court of the Teshoo Lama in Tibet* (1800; reprinted in New Delhi by Manjusri in 1971).

88. For example, Thomas Pope, an American landscape gardener and amateur bridge architect, discussed Samuel Turner's descriptions on pp. 44–49 of *A Treatise on Bridge Architecture* (New York: A. Niven, 1811).

89. Jonathan Goldstein, *Philadelphia and the China Trade, 1682–1846* (Pennsylvania State University Press, 1978), pp. 2, 17, 73–77. William Chambers also mentioned Chinese chain bridges on p. 158 of his *Dissertation on Oriental Gardening* (London: W. Griffin, 1772).

90. Quoted in Goldstein, *Philadelphia and the China Trade*, p. 15.

91. John Jones, Indian River (Delaware), May 27, 1773, letter, APS Archives.

92. The APS does possess a bridge model constructed by John Jones; however, it does not coincide at all with the chain bridge described in Jones's proposal, and it was not received by the APS until 1786. See *A Catalogue of Instruments and Models in the Possession of the American Philosophical Society*, pp. 40–41.

93. Jones's proposal was discussed at a meeting of the APS on July 18, 1773, and the records of the proceedings reveal that Gilpin attended this session. (Gilpin was elected to the society in 1768.) See *Early Proceedings of the American Philosophical Society, 1744–1838* (Philadelphia: McCalla & Stavely, 1884), p. 80.

94. Thomas Gilpin, A Plan with an Estimation of the Cost of a Chain Bridge over the River Schulkiln within the City of Philadelphia, Thomas Gilpin Letterbook, Maury Family Papers, University of Virginia Library; "Memoir of Thomas Gilpin," *Pennsylvania Magazine of History and Biography* 49 (1925), p. 321.

95. "A statistical account of the Schuylkill permanent bridge," *Port Folio*, n.s. 5 (March 1808), p. 170.

96. Philosophical Hall, which housed the society, was just adjacent to Congressional Hall, where the Pennsylvania legislature met, and there was a degree of cooperation between the legislature and the society. Moreover, although Finley was not a member of the APS, Ephraim Douglass's letter asking him to make an inquiry at its headquarters suggests that Finley had contacts with it. See Ephraim Douglass, letter to James Finley, April 3, 1791, Gallatin manuscripts, 1791, no. 3, New York Historical Society.

97. Thomas Pope, *Treatise on Bridge Architecture*, pp. 171–173, 191. Pope also discussed South American suspension bridges and the Winch Bridge over the Tees in England. The latter was a small pedestrian chain bridge built (mainly for the use of miners) around 1741. It was analogous to those constructed in Asia and probably was inspired by them. Since Pope knew of these bridges, it is not unlikely that Finley learned of them. The sources most frequently cited on South American bridges and on the Winch Bridge were, respectively, the following: George Juan and Antonine de Ulloa, *Voyage Historique de l'Amérique Méridionale*, 2 vols. (Amsterdam and Leipzig: Arkste'e and Merkus, 1752); William Hutchinson, *The History and Antiquities of the County Palatine of Durham* (Carlisle, England: F. Jollie, 1794), volume III, p. 279. It is very unlikely that the Winch Bridge influenced Jones and Gilpin, because it appears to have been known to Americans, including Pope, solely through Hutchinson's work, which was not published until 1794.

98. Ellis, *History of Fayette County*, p. 250.

99. There are no known surviving copies of Finley's patent (June 1808). However, Finley published an article, "Description of the patent chain bridge," *Port Folio*, n.s. 3 (June 1810), pp. 441–453, which probably was an elaboration of his patent specification. All references in the following pages to Finley's patent are based on the *Port Folio* article. The date of Finley's patent, 1808, falls during a period of upsurge in patenting that began around 1806 and ended with the War of 1812. See Kenneth L. Sokoloff, "Inventive activity in early industrial America: Evidence from patent records, 1790–1846," *Journal of Economic History* 48 (1988), no. 3–4, pp. 813–850. See also Kenneth L. Sokoloff and B. Zorina Khan, "The democratization of invention during early industrialization: Evidence from the United States, 1790–1846," *Journal of Economic History*, 50 (1990), no. 1–2, pp. 363–378.

100. Finley, "Patent chain bridge," pp. 441–453.

101. Ibid., p. 451.

Chapter 2

1. James Finley, *A Description of the Chain Bridge Invented by Judge Finley* (Uniontown: W. Campbell, 1811), p. 12.

2. On Palmer see Bear Albright, "Palmer & Spofford: A heritage of building in the Merrimack River Valley," *Essex Institute Historical Collections* 122 (1986), no. 2, pp. 142–164; George B. Pease, "Timothy Palmer: Bridgebuilder of the eighteenth century," *Essex Institute Historical Collections* 83 (1947), pp. 97–111.

3. On Wernwag see Lee H. Nelson, *The Colossus of 1812: An American Engineering Superlative* (American Society of Civil Engineers, 1990).

4. On Peale see Sidney Hart, "'To encrease the comforts of life': Charles Willson Peale and the mechanical arts," *Pennsylvania Magazine of History and Biography* 110 (1986), no. 3, pp. 323–357.

5. For an excellent analysis of the American patent system as a context for the economic and social management of innovation see Carolyn Cooper, *Shaping Invention: Thomas Blanchard's Machinery and Patent Management in Nineteenth-Century America* (Columbia University Press, 1991). See also *Technology and Culture* 32 (1991), no. 4 (special issue on "Patents and Invention").

6. Ephraim Douglass, letter to James Finley, April 3, 1791, Gallatin manuscripts, New York Historical Society, 1791 no. 3.

7. Solon J. Buck and Elizabeth Hawthorn Buck, *The Planting of Civilization in Western Pennsylvania* (University of Pittsburgh Press, 1939), pp. 438–440.

8. Daniel Hovey Calhoun, *The American Civil Engineer: Origins and Conflict* (MIT Press, 1960), p. 22.

9. Peter Michael Molloy, Technical Education and the Young Republic: West Point as America's Ecole Polytechnique, 1802–1833 (Ph.D. dissertation, Brown University, 1975); Calhoun, *American Civil Engineer*, pp. 37–43.

10. See, e.g., Nelson, *Colossus of 1812.*

11. Calhoun, *American Civil Engineer*, pp. 8–53, 68–73, 92–93.

12. James Finley, "A description of the patent chain bridge," *Port Folio*, n.s. 3 (June 1810), p. 443.

13. For surveys of the Scottish Enlightenment and Scottish philosophy see Anand C. Chitnis, *The Scottish Enlightenment: A Social History* (Croom Helm, 1976); R. G. Cant, "The Scottish universities and Scottish Society in the eighteenth century," *Studies on Voltaire and the Eighteenth Century* 58 (1967), pp. 1953–1966; Douglas Sloan, *The Scottish Enlightenment and the American College Ideal* (Columbia University Teachers' College Press, 1971), pp. 1–35.

14. Quoted from Sloan, *Scottish Enlightenment*, p. 123.

15. Ibid., pp. 154–60.

16. Finley, "A description of the patent chain bridge," p. 448.

17. Finley, *A Description of the Chain Bridge*, p. 11.

18. Ibid., p. 12.

19. Finley, "A description of the patent chain bridge," p. 452.

20. Ibid.

21. Ibid.

22. Ibid.

23. Ibid., p. 447.

24. Finley's conclusion about the sag/span ratio at which the maximum cable tension equals the distributed load can be deduced mathematically from a

static analysis of a cable loaded uniformly along the horizontal. Such an analysis leads to 1/6.9 as the necessary ratio. Of course the validity of both this analysis and Finley's experiment rests on the following ideal assumptions: (1) the cables are perfectly flexible and negligible in weight compared to the load suspended from them; (2) all loads are perfectly static and uniformly distributed. See D. B. Steinman, *A Practical Treatise on Suspension Bridges, Their Design, Construction, and Erection,* second edition (Wiley, 1929), pp. 1–6, 12–13; Ferdinand P. Beer and E. Russell Johnston, Jr., *Vector Mechanics for Engineers: Statics and Dynamics,* third edition (McGraw-Hill, 1977), pp. 288–291. Real suspension bridges do not conform precisely to these criteria.

25. Finley, "A description of the patent chain bridge," p. 445. Designers of suspension bridges decades later utilized the sag/span ratio the same way as Finley. In 1872 the prominent American civil engineer John Trautwine published *The Civil Engineer's Pocket Book* (Philadelphia: Claxton, Remsen, and Haffelfinger), which included a "table of data required for calculating the main cables of suspension bridges." The first column of the table consisted of various sag/span ratios. Another column gave the ratio between the maximum overall tension of the cables and the "total suspended weight of the bridge and its load." For a sag/span ratio of 1/7, Trautwine found the latter ratio (tension/load) to be 1.01. For a sag/span ratio of 1/9, he found the tension/load ratio to be 1.23—a little less than that indicated by Finley's data.

26. Finley, "A description of the patent chain bridge," p. 443.

27. Ibid., pp. 444, 452–453. Finley used the first American edition: *Encyclopaedia: or, A Dictionary of Arts, Sciences, and Miscellaneous Literature* (Philadelphia: Thomas Dobson, 1798). The table to which Finley referred (volume XVIII, p. 10) summarized experiments by, among others, Petrus van Muschenbroek, one of the first researchers to carry out tensile tests to determine the breaking strength of wrought iron and other substances. See Stephen P. Timoshenko, *History of the Strength of Materials* (1953; Dover, 1983), p. 55. According to Harold Dorn, Robison's article, together with several others he wrote for the encyclopedia, constituted "the most systematic, learned, and potentially applicable and useful treatise on structural mechanics that had yet appeared in English." See pp. 183 and 187 of Dorn's Ph.D. dissertation, The Art of Building and the Science of Mechanics: A Study of the Union of Theory and Practice in the Early History of Structural Analysis (Princeton University, 1970).

28. John Jones, Indian River, Delaware, correspondence, May 27, 1773, APS Archives.

29. Thomas Gilpin, A Plan of a Chain Bridge, Thomas Gilpin Letterbook, Maury Family Papers, University of Virginia Library.

30. W. H. Boyer and Irving A. Jelly, "An early American suspension span," *Civil Engineering* 7 (1937), May, pp. 338–340.

31. Navier, *Rapport à Monsieur Becquey et mémoire sur les ponts suspendus,* second edition (Paris: Carilian-Goeury, 1830), p. 204.

32. Finley quoted a section from what he called "*Cyclopedia,*" part of which corresponds word for word to a passage from the article "Bridge" in *The Cyclopaedia; or, Universal Dictionary of the Arts, Sciences, and Literature,* ed. A. Rees (London: Longman, Hurst, 1802–1819). The other part of Finley's "quote" from Rees, however, was actually a condensation of data about iron bridges presented in the article. Yet there can be no doubt that this was the source that Finley drew from. Finley's article was published in 1810; the passage he quoted word for word mentioned the Teme bridge, constructed in 1795–1796. An examination of all the editions of encyclopedias published in English in this period and referred to either in Robert Collinson's *Encyclopaedias: Their History Throughout the Ages* (Hafner, 1964) or in the *National Union Catalogue of pre-1956 Imprints* reveals that only Rees's work contained the passage in question. This conclusion is, moreover, compatible with the fact that Rees's was among the most popular encyclopedias in America in the early nineteenth century. An American edition of it was published between 1805 and 1822. Finley could have used either edition.

33. Rees, "Bridge," n.p.; Dorn, Art of Building, pp. 116–117. There is evidence to indicate that Christopher Wren employed Hooke's solution to design the dome of St. Paul's Cathedral in London (Dorn, p. 116). However, the catenary principle could not be employed to build stone arch bridges, because it failed to allow for the fact that stonework had to be built up unevenly above the arch to make the roadway level.

34. Rees, "Bridge," n.p.

35. Dorn, Art of Building, pp. 129–133.

36. *The Papers of Thomas Jefferson,* ed. J. Boyd (Princeton University Press, 1956), volume 14, pp. 373–374, 562; Dorn, Art of Building, pp. 172–175. George Michael Danko suggests that Timothy Palmer employed the chain proportioning method as well; see pp. 138 and 141 of his Ph.D. dissertation, The Evolution of the Simple Truss Bridge, 1790–1850: From Empiricism to Scientific Construction (University of Pennsylvania, 1979).

37. Rees, "Bridge," n.p.

38. Finley, "A description of the patent chain bridge," p. 445.

39. Ibid., pp. 445–446.

40. Note that C-C′ is a single pulley. An extra load on side C will make the tensions at C and C′ momentarily different, until the pulley rotates enough to equalize them.

41. Finley, "A description of the patent chain bridge," pp. 446–447.

42. Ibid., p. 445.

43. See chapter 5 below.

44. Finley, *A Description of the Chain Bridge,* p. 12. Truss analysis emerged in the U.S. in the 1840s with the work of Squire Whipple and Hermann Haupt;

see Carl Condit, *American Building Art: The Nineteenth Century* (Oxford University Press, 1960), pp. 114–115, and Danko, Evolution of the Simple Truss Bridge, pp. 219–228. The method devised by Whipple provided a means of calculating the force on each member of a statically determinate truss and of determining whether the member was in tension or compression. Before this time, however, builders like Ithiel Town already thought of these structures as "frameworks with individual members in tension and compression," and they saw the need to position individual members so that these forces would be directed along their axes. See Gregory K. Dreicer, The Long Span: Intercultural Exchange in Building Technology: Development and Industrialization of the Framed Beam in Western Europe (Ph.D. dissertation, Cornell University, 1993), pp. 139–140.

45. Finley, *A Description of the Chain Bridge*, p. 11.

46. Ibid.

47. Finley, "A description of the patent chain bridge," p. 451.

48. Finley, *A Description of the Chain Bridge*, pp. 10–11.

49. Ibid., p. 10.

50. Ibid., p. 12.

51. Ibid., p. 11.

52. Ibid., pp. 11–12.

53. Ibid., pp. 12–13.

54. Finley, "A description of the patent chain bridge," pp. 448–449.

55. Ibid., pp. 442, 446, 448.

56. Ibid., pp. 448–450.

57. *Papers of Thomas Jefferson*, volume 13, p. 588; volume 14, pp. 561–564. For a discussion of Thomas Paine's work, see Emory Kemp, "Thomas Paine and his 'pontifical matters,'" *Newcomen Society Transactions* 49 (1977–78), pp. 21–40.

58. Finley, "A description of the patent chain bridge," p. 442.

59. The use of stirrups was not unique to Finley. Theodore Burr, a bridge builder contemporary to Finley, employed stirrups in his Delaware River bridge at Trenton (completed in 1806). Each span of the bridge was composed of three wooden arched ribs. Chains were suspended from the ribs; stirrups attached to these chains supported the transverse floor beams of the roadway. See Thomas Pope, *A Treatise on Bridge Architecture* (New York: Alexander Niven, 1811), p. 137. Lewis Wernwag, another bridge builder of the period, employed stirrups in his "Colossus" bridge over the Schuylkill in Philadelphia (completed in 1812). Wernwag's bridge was composed of three arched wooden ribs, each with a heavy truss railing. Wernwag used stirrups to help bind the arched ribs firmly to their respective railings. Each stirrup embraced a rib

(4 feet thick); the sides of each stirrup continued up to the top of the railings, where they were bolted to a kingpost (Danko, Evolution of the Simple Truss Bridge, pp. 53–57; Nelson, *Colossus,* pp. 28, 30, 33). It is not possible to say whether the stirrups of Finley, Burr, and Wernwag were independent innovations or whether there was borrowing.

60. Finley, "A description of the patent chain bridge," pp. 442, 449.

61. Ibid., p. 449.

62. Peale, *An Essay on Building Wooden Bridges,* pp. 7, 10–11.

63. Finley, "A description of the patent chain bridge," p. 446.

64. Ibid., pp. 442, 449.

65. Ibid., p. 446.

66. Ibid., pp. 442, 447–50.

67. Ibid., p. 450.

Chapter 3

1. Navier, *Rapport à Monsieur Becquey et mémoire sur les ponts suspendus,* second edition (Paris: Carilian-Goeury, 1830), p. 29.

2. A. P. Mills, "Tests of wrought iron links after 100 years in service," *Cornell Civil Engineer* 19 (1911), April, pp. 257–259; "Reinforcing the Newburyport suspension bridge," *Engineering Record* 42 (1900), October 6, p. 314; Malverd A. Howe, "An old chain suspension bridge," *Railroad Gazette* 32 (1900), January 5, p. 5.

3. "Reinforcing the Newburyport suspension bridge," p. 314; Howe, "Old chain suspension bridge," p. 5.

4. Mills, "Tests of wrought iron links," pp. 259–260.

5. Ibid., p. 257.

6. Howe, "Old chain suspension bridge," p. 5.

7. "Reinforcing the Newburyport suspension bridge," p. 315; Mills, "Tests of wrought iron links," pp. 261–262.

8. Jacob Blumer (1774–1830), son of a minister, was "well educated by his talented father and being of a mechanical turn of mind became a clockmaker" (Charles Rhoads Roberts, *History of Lehigh County, Pennsylvania* (Allentown: Lehigh Valley Publishing Co., 1914), volume 2, p. 118).

9. W. H. Boyer and Irving A. Jelly, "An early American suspension span," *Civil Engineering* 7 (1937), May, pp. 338–340.

10. Ibid.

11. Ibid., p. 339.

12. The four chain bridges known to have been built after 1812 included the three built by Blumer and the multiple-span Essex-Merrimack Bridge. It should be borne in mind, however, that information concerning the construction dates of Finley-type chain bridges is not always consistent. For example, Will H. Lowdermilk (*History of Cumberland (Maryland)* (Washington: James Anglin, 1875), p. 305) gives 1820 as the date for the construction of a Finley-type chain bridge at Cumberland, Maryland, yet Finley already mentions the construction of a chain bridge at Cumberland as of 1810.

13. The suspension bridge was reintroduced into America in the 1840s by Charles Ellet and John Roebling. Both Ellet and Roebling used wire cables rather than chains, following the practice of the French designer Marc Seguin, and both sought to apply the suspension principle to the construction of long-span bridges. See Carl Condit, *American Building Art: The Nineteenth Century* (Oxford University Press, 1960); Emory L. Kemp, "Ellet's contribution to the development of suspension bridges," *Engineering Issues—Journal of Professional Activities, Proceedings of the American Society of Civil Engineers* 99 (1973), no. PP3, pp. 331–351; Gene D. Lewis, *Charles Ellet, Jr.* (University of Illinois Press, 1968); David P. Billington, *The Tower and the Bridge: The New Art of Structural Engineering* (Basic Books, 1983), pp. 72–82; Arne Arthur Jakkula, "A history of suspension bridges in bibliographical form," *Texas Engineering Experiment Station Bulletin*, fourth series, 12 (1941), no. 7. Besides the work of Ellet and Roebling, smaller suspension bridges were built in the 1850s in northern California. These bear an outward resemblance to Finley's in their simplicity and lack of pretentiousness, but in fact the technological elements of which they were composed differed significantly. They used wire cables, cast-iron A-frame towers, and rolling saddles, and they were presumably not built as multiple spans. They accordingly did not represent a continuation of the Finley system, but rather a reworking of newer technological elements to create bridges that filled the same kind of market niche as that of the Finley bridges, but this time in the rugged, frontier areas of the far west. See Daniel L. Schodek, *Landmarks in American Civil Engineering* (MIT Press, 1987), pp. 111–112.

14. John Templeman, printed circular letter addressed to Nicholas Van Dyke, March 14, 1809, Longwood manuscripts, Eleutherian Mills Historical Library, group 4, box 10.

15. James Finley, *A Description of the Chain Bridge* (Uniontown: W. Campbell, 1811), p. 13.

16. *United States Gazette* (Philadelphia), January 1811.

17. Finley, "Patent chain bridge," *Port Folio*, p. 453.

18. *Standard History of Pittsburgh, Pennsylvania*, ed. E. Wilson (Chicago: H. R. Cornell, 1898), pp. 112–114; Pennsylvania General Assembly, Senate, Report

on Roads, Bridges, and Canals, March 23, 1822, table V; Richard T. Woley, *Monongahela, the River and Its Region* (Butler, Pennsylvania: Ziegler, 1937), pp. 54–55.

19. [John Robison], "Strength of materials," in *Encyclopaedia; or, A Dictionary of Arts, Sciences, and Miscellaneous Literature,* first American edition (Philadelphia: Thomas Dobson, 1798), volume 18.

20. W. Alexander and A. Street, *Metals in the Service of Man* (1944; revised editon: Penguin, 1954), pp. 92–93. See also Mario Salvadori, *Why Buildings Stand Up: The Strength of Architecture* (Norton, 1980), p. 63. Salvadori indicates that steel structures are not loaded beyond 60 percent of their elastic limit.

21. Navier (*Ponts suspendus,* second edition, pp. 133–135) found that the yield point of wrought iron was somewhere between one-third and one-half of its ultimate strength.

22. "Strength of materials," *Encyclopaedia* (Philadelphia, 1798), XVIII.

23. North American builders in the early nineteenth century combined the principles of the arch and the truss in various ways. In some cases, the truss itself was made to arch, so that it exerted thrust. (In a simple truss bridge, the truss is horizontal and produces no thrust.) In other cases, the arch and the truss were somehow interconnected so that the truss was either partly suspended from the arch or partly supported upon it. For the purpose of simplicity, I refer to all these combinations as "arch-truss" bridges. For a good survey of the most important arch-truss designs, see pp. 35–67 of Danko, Evolution of the Simple Truss Bridge.

24. In North America, Timothy Palmer is credited with introducing the practice of covering wooden bridges around 1806; however, covered wooden bridges were built in Switzerland in the eighteenth century. See Danko, Evolution of the Simple Truss Bridge, pp. 18–30, 35–67; Condit, *American Building Art,* pp. 78–89; Bear Albright, "Palmer and Spofford: A Heritage of Building in the Merrimack River Valley," *Essex Institute Historical Collections* 122 (1986), no. 2, pp. 142–164. Covered bridges also existed in Europe.

25. Danko, Evolution of the Simple Truss Bridge, pp. 20, 50.

26. Pennsylvania General Assembly, Senate, Report on Roads, Bridges, and Canals, March 23, 1822 (Philadelphia: C. Mowry, 1822), table V. The table listed the overall length between abutments of each bridge and the number of spans it comprised. A total of 25 bridges were listed; 21 were roofed and therefore must have been made of wood. The total length of each timber bridge was divided by the number of its spans in order to arrive at the average length of the spans. From these figures the average length of a wooden span was determined.

27. Lee H. Nelson, *The Colossus of 1812: An American Engineering Superlative* (American Society of Civil Engineers, 1990).

28. "A statistical account of the Schuylkill Permanent Bridge," *Port Folio*, n.s. 5 (1808), pp. 168–171, 182–187, 200–204.

29. Pennsylvania General Assembly, Report on Roads, Bridges, and Canals, table V; Robert Fletcher and J. P. Snow, "A history of the development of wooden bridges," *Transactions of the American Society of Civil Engineers* 99 (1934), pp. 325–332; Richard S. Allen, "Biographical sketch of Theodore Burr," unpublished manuscript, Library of Congress; Thomas Pope, *A Treatise on Bridge Architecture* (New York: Alexander Niven, 1811), p. 144.

30. John Templeman, printed circular letter addressed to Nicholas Van Dyke; Pennsylvania General Assembly, Report on Roads, Bridges, and Canals, table V and pp. 162–164; Mills, "Tests of wrought iron links," p. 257.

31. Lewis Wernwag, "Bridges," American Philosophical Society Library.

32. Ibid. See also Fletcher and Snow, "A history of the development of wooden bridges," pp. 328–329.

33. Condit, *American Building Art,* pp. 89–92, 114–15; Danko, Evolution of the Simple Truss Bridge, pp. 127–148, 219–228. Town's truss and its influence in Europe are analyzed in Gregory K. Dreicer, The Long Span. Intercultural Exchange in Building Technology: Development and Industrialization of the Framed Beam in Western Europe (Ph.D. dissertation, Cornell University, 1993), pp. 95–161.

34. Condit, *American Building Art,* pp. 89–92; Danko, Evolution of the Simple Truss Bridge, pp. 127–148.

35. Fletcher and Snow, "Development of wooden bridges," p. 333.

36. Danko, Evolution of the Simple Truss Bridge, pp. 148–158.

37. *United States Gazette* (Philadelphia), January 1811.

38. "Bridge," *Van Nostrand's Scientific Encyclopedia,* ed. D. Considine, fifth edition (Van Nostrand, 1976), pp. 362–363.

39. This bridge was built by Thomas Haven over the Merrimack River, near Newburyport. The total length of the bridge was about 1,000 feet (Condit, *American Building Art,* p. 167).

40. Finley's Statement of the Quantity of Iron for a Chain Bridge across the River Susquehanna, Stauffer Collection, Historical Society of Pennsylvania, volume 14, p. 989.

41. Finley, "Patent chain bridge," pp. 446–447. See also David B. Steinman, *A Practical Treatise on Suspension Bridges* (Wiley, 1922), p. 74.

42. Steinman, *Practical Treatise on Suspension Bridges,* pp. 12–14, 18–21, 72, 76, 82–83, 101.

43. Templeman, letter to Van Dyke, March 14, 1809, Eleutherian Mills Historical Library.

44. Besides the Palmerton Bridge, Blumer built (in 1813) a chain bridge with two full spans and two half-spans at Allentown, Pennsylvania. The total length of this bridge was 475 feet. See Asa Earl Martin and Hiram Herr Shenk, *Pennsylvania History Told by Contemporaries* (Macmillan, 1925), pp. 513–514. In 1824 Blumer built a chain bridge north of Allentown (Roberts, *History of Lehigh County*, I, pp. 380–381).

45. Lowdermilk, *History of Cumberland*, pp. 305–306.

46. Ibid., p. 306.

47. *United States Gazette*, January 1811. Clearly, as engineering works become more complex, it becomes increasingly important for the diverse activities of those involved in a project to be coordinated and directed toward a common end. Otherwise a workable technical system may never be achieved, whether it be a relatively simple suspension bridge or a very complex project such as a space mission. This was an important theme in Calhoun's book *The American Civil Engineer: Origins and Conflict* (MIT Press, 1960). He traced the emergence of the civil engineer as a scientifically educated professional working, not alone, but within an administrative hierarchy—a bureaucracy, if you will. And he cited several instances of complaints analogous to Finley's from other engineers in the early nineteenth century. For example, Charles B. Shaw, who held the post of State Engineer in Virginia, wrote in 1836: "An engineer who may have the misfortune to be scientific, is continually under necessity to vindicate his specifications from the self-claimed 'common sense' of superintendents anxious to execute some little change devised by themselves: much fruitless correspondence ensues, and in the mean time, and too late for remedy, the superintendent does what he thinks proper, to the injury of the commonwealth, in the substitution for the plans of their engineer, of some crudity of the 'practical man'; for common sense and practical views are by many deemed incompatible with study. Is not all learning but experience and collated facts? Why then are the extensive experiments of writers, many of them themselves practical, less valuable than the traditionary learning, or very limited personal experience of one man, be he ever so strong minded, yet unlearned in physical phenomena and their causes?" (Calhoun, *American Civil Engineer*, pp. 178–179) Louis Hartz's book *Economic Policy and Democratic Thought: Pennsylvania 1776–1860* (Harvard University Press, 1948) contains a valuable discussion of the early problems and limitations of administrative systems that evolved in Pennsylvania in conjunction with the growth of transportation (pp. 148–175, 289–297).

48. J. E. Gordon, *Structures: or, Why Things Don't Fall Down* (Da Capo, 1978), pp. 60–69; Mario Salvadori, *Why Buildings Stand Up: The Strength of Architecture* (McGraw-Hill, 1980), pp. 63–64.

49. Finley, "Patent chain bridge," pp. 442, 450.

50. "A note on early American suspension bridges," *Engineering News* 53 (1905), no. 11, pp. 269–270; Fred Perry Powers, "Historic bridges of Philadelphia," *Philadelphia Historical Society Papers* no. 11 (1917), pp. 300–305.

51. *History of Fayette County,* ed. F. Ellis (Philadelphia: L. H. Everts, 1882), p. 435.

52. Quoted from Mills, "Tests of wrought iron links," p. 260.

53. Boyer and Jelly, "Early American suspension span," pp. 338–340. Twice, however, a chain broke as a result of flooding: once in 1841 and once in 1857. The latter accident was due to the collapse of a dam—a canal boat swept down in the resulting flood struck the bridge and broke one of the chains.

54. Ellis, *History of Fayette County,* p. 435.

55. Finley, "Patent chain bridge," p. 449.

56. Boyer and Jelly, "Early American suspension span," p. 338.

57. Stephen P. Timoshenko, *History of Strength of Materials* (1953; Dover, 1983), pp. 62–66, 83–87.

58. See Nelson, *Colossus of 1812.* It appears that Wernwag and other builders of Finley's era relied on tests with scale models to help proportion their bridges.

59. Boyer and Jelly, "Early American suspension span," p. 339.

60. Personal communication, Hiroshi Tanaka, Professor of Civil Engineering, University of Ottawa.

61. Mills, "Tests of wrought iron links," passim.

62. Navier, *Ponts suspendus,* second edition, pp. 130–135.

63. Mills tested entire links and individual bars cut from links. I have used the data on the latter (Mills, "Tests of wrought iron links," p. 278).

64. Mills, "Tests of wrought iron links," p. 269.

65. Finley, "Patent chain bridge," p. 449.

66. Mills, "Tests of wrought iron links," pp. 263–264.

67. H. J. Hopkins, *A Span of Bridges* (Devon: David & Charles, 1970), p. 187; William Alexander Provis, *An Historical and Descriptive Account of the Suspension Bridge over the Menai Strait in North Wales* (London: Ibotson & Palmer, 1828).

68. Finley, in his *United States Gazette* article of January 1811, argued that wood varied at least as much in strength as iron.

69. Bruce Sinclair, *Philadelphia's Philosopher Mechanics: A History of the Franklin Institute, 1824–1865* (Johns Hopkins University Press, 1974), pp. 178, 183–184.

70. Condit, *American Building Art,* pp. 103–109, 115. In 1847, Squire Whipple still quoted 60,000 pounds per square inch as the tensile strength of iron;

however, unlike Finley, he realized that iron "can never be safely exposed in practice to more than a small proportion of these stresses, say from 1/6 to 1/4" (quoted from Condit, p. 115).

71. For example, Finley's Falls of Schuylkill bridge replaced a wooden span that collapsed after only 7 years (Templeman, printed circular), and Wernwag's "Colossus" was destroyed by fire in 1838 (Nelson, *Colossus of 1812,* p. 26). See table V of the Pennsylvania General Assembly's Report on Roads, Bridges, and Canals for evidence of many other disasters.

Chapter 4

1. *Un ingénieur des lumières: Emiland-Marie Gauthey,* ed. A. Coste et al. (Ecole des Ponts et Chaussées, 1993), p. 260. The engineer Gaspard Riche de Prony, in his biographical memoir on Navier ("Notice biographique de Navier," *Annales des Ponts et Chaussées: mémoires et documents* 7 (1837), no. 1, p. 5), states that Navier became an orphan at the age of 14, but this appears to be incorrect: Navier went to live with Gauthey in 1794, at the age of 9.

2. Navier, "Eloge historique de M. Gauthey," in Emiland Gauthey's *Traité de la construction des ponts,* ed. Navier (Paris: Firmin Didot, 1809), volume 1, pp. xi–xxxi.

3. Navier to le Baron Pasquier, Conseiller d'Etat, Paris, December 19, 1814, Archives Nationales, $F^{14}2289$.

4. Navier to le Baron Pasquier, Conseiller d'Etat, Paris, December 19, 1814, $F^{14}2289$, Archives Nationales; Prony, "Notice biographique sur Navier," pp. 6–7; Archives Nationales, Service de Navier, $F^{14}2289$.

5. There is no doubt that Gauthey gave Navier some training. In Dijon, he had trained another of his nephews to the extent that it was the nephew who took practical charge on a day-to-day basis of the construction of the Charolais Canal. The nephew died unexpectedly, however, and Gauthey took Navier under his wing. According to Navier, he assisted Gauthey in his work during his teenage years. See Navier to le Baron Pasquier, Conseiller d'Etat, Paris, December 19, 1814; Navier to Director General, Corps des Ponts et Chaussées, Chalon-sur-Saône, June 21, 1821; Archives Nationales, $F^{14}2289$.

6. Anne Blanchard, *Les ingénieurs du 'Roy' de Louis XIV à Louis XVI,* Collection du centre d'histoire militaire et d'études de défense nationale de Montpellier, no. 9 (Montpellier: Déhan, 1979), pp. 230–237; Anne Blanchard, *Dictionnaire des ingénieurs militaires, 1691–1791* (Montpellier: Université Paul Valéry, 1981).

7. Blanchard, *Les ingénieurs du 'Roy',* pp. 238–245.

8. Ibid., pp. 188–212; René Taton, "L'Ecole Royale du Génie de Mézières," in *Enseignement et diffusion des sciences en France au XVIIIᵉ siècle,* ed. R. Taton, second edition (Hermann, 1986), pp. 559–615.

9. Blanchard, *Les ingénieurs du 'Roy'*, p. 474. More information on the various state corps in the eighteenth century and on technical schools can be found in *Ecoles techniques et militaires au XVIIIᵉ siècle*, ed. R. Hahn and R. Taton (Hermann, 1986).

10. The Corps des Ponts et Chaussées is often said to have been founded in 1716, but in fact its roots go back further. See A. Brunot and R. Coquand, *Le Corps des Ponts et Chaussées* (Centre National de la Recherche Scientifique, 1982), pp. 3–15; Jean Petot, *Histoire de l'administration des Ponts et Chaussées, 1599–1815* (Marcel Rivière, 1958); E. J. M. Vignon, *Etudes historiques sur l'administration des voies publiques en France au XVIIᵉ et XVIIIᵉ siècle*, 3 volumes (Dunod, 1862).

11. In the historiography of American engineering, the process of professionalization has been conceptualized as a variant of the model offered by the liberal professions (medicine, law, etc.). See Edwin T. Layton, *The Revolt of the Engineers: Social Responsibility and the American Engineering Profession* (Case Western Reserve University Press, 1971); Daniel H. Calhoun, *The Civil Engineer, Origins and Conflict* (MIT Press, 1960). However, such a model is quite misleading for the case of French engineers of the state corps. Among other things, the ideology of the state engineering corps was significantly different from the kind of ideology associated with the liberal professions. In addition, in the French engineering corps most professional standards were imposed by the state; the ideal was loyal service to the state rather than the free and independent exercise of professional judgement; and the whole system was organized into an increasingly formal, military-style hierarchy, complete with military-like uniforms. Finally, the engineers of the state corps did not form independent professional societies, as American engineers did.

12. Perronet (1708–1794) worked from the age of 17 in the bureau of the chief architect of Paris. From 1737 to 1747 he served as an engineer of the Ponts et Chaussées in the region of Alençon, after which he became director of the Bureau des dessinateurs (soon renamed the Ecole des Ponts et Chaussées), where he remained until his death in 1794. Perronet also became the chief engineer of the corps, and he designed and built many bridges. A good biographical sketch of Perronet, with extensive references, can be found on pp. 346–349 and 356 of Antoine Picon's book *French Architects and Engineers in the Age of Enlightenment* (Cambridge University Press, 1992).

13. Brunot and Coquand, *Le Corps des Ponts et Chaussées*, pp. 3–37; Charles Coulston Gillispie, *Science and Polity in France at the End of the Old Regime* (Princeton University Press, 1980), pp. 479–498; Anne Querrien, "Ecoles et corps: Le cas des Ponts et Chaussées, 1747–1848," *Annales de la recherche urbaine*, no. 5 (1979), pp. 81–114; T.-A. Cotelle, *Esquisse historique sur l'institution des Ponts et Chaussées* (Paris: Paul Dupont, 1849); Alfred Picard, Note historique sur l'administration des travaux publics, manuscript collection, Ecole des Ponts et Chaussées.

14. Quoted from Gillispie, *Science and Polity*, p. 481.

15. Antoine Picon, *L'invention de l'ingénieur moderne: l'Ecole des Ponts et Chaussées, 1747–1851* (Ecole des Ponts et Chaussées, 1992), pp. 99–101.

16. The *corvée des routes* was a tax in the form of 6–30 days' mandatory labor per year. The tax, applied to peasants who lived within 5 or 10 miles of royal roads, was a major source of labor for road repair during the *ancien régime*.

17. On the role of corps engineers as an administrative elite in the eighteenth century and their involvement in directing *corvée* labor see Brunot and Coquand, *Le Corps des Ponts et Chaussées*, passim; H. Marcel, "Les débuts de Jean-Rodolphe Perronet ingénieur des Ponts et Chaussées de la généralité d'Alençon," *Annales des Ponts et Chaussées*, July-August 1948, pp. 419–440; *Un ingénieur des lumières*, ed. Coste et al., especially pp. 69–108. It is worth noting that a modified system of *corvée* labor was reintroduced and widely used in the nineteenth century, also under the direction of corps engineers. See A. Guillerme, *Corps à corps sur la route: les routes, les chemins et l'organisation des services au XIXe siècle* (Ecole des Ponts et Chaussées, 1984).

18. Quoted from Maurice Bonneviot, "Gauthey, le Chalonnais," in *Un ingénieur des lumières: Emiland-Marie Gauthey*, ed. Coste et al., p. 50.

19. Until the end of the eighteenth century there was no sharp dividing line between architects and engineers in France. They shared the same intellectual terrain, and their training and work practices often overlapped. On the evolution of the relationship between French architects and engineers during the eighteenth century, and on the rise of "a new division of labor" between them, particularly as manifested in their intellectual debates, see Picon's *French Architects and Engineers in the Age of Enlightenment*.

20. James Montgomery Oliver, The "Corps de Ponts et Chaussées," 1830–1848 (Ph.D. dissertation, University of Missouri, 1967), pp. 55–64.

21. Navier, "Eloge historique de M. Gauthey," pp. xi–xxxi; *Un ingénieur des lumières*, ed. Coste et al., passim.

22. Navier, "Eloge historique de Gauthey"; *Un ingénieur des lumières*, ed. Coste et al. p. 254; Turner W. Allen, The Highway and Canal System in Eighteenth-Century France (Ph.D. dissertation, University of Kentucky, 1953), volume 2, pp. 449–475, 521–531; Brunot and Coquand, *Le Corps des Ponts et Chaussées*, pp. 49–52; *Un ingénieur des lumières: Emiland-Marie Gauthey*, ed. Coste et al., p. 254. Navier incorrectly gives 1807 as the date of Gauthey's death.

23. Antoine Picon, "Entre ingénieur et architecte: L'itinéraire d'Emiland-Marie Gauthey," in *Un ingénieur des lumières*, ed. Coste et al., p. 207.

24. Ibid., pp. 207–236; see also Antoine Picon, "Les ingénieurs et l'idéal analytique à la fin du XVIIIe siècle," *Sciences et techniques en perspective* 13 (1987–1988), pp. 70–108; Antoine Picon, "Les ingénieurs et la mathématisation: L'exemple du génie civil et de la construction," *Revue d'histoire des sciences* 42 (1989), no. 1–2, pp. 155–172.

25. Jacques Desvigne, "La visite de l'église de Givry," and Anne Coste, "Eglise de Givry: les emprunts au modèle gothique," in *Un ingénieur des lumières,* ed. Coste et al. Gauthey's role as an engineer-architect emerges at numerous points in this volume. See especially pp. 207–236 of Picon's article "Entre ingénieur et architecte: L'itinéraire d'Emiland-Marie Gauthey."

26. Picon, "Entre ingénieur et architect," p. 223.

27. On the history of the Charolais Canal see Pierre Pinon, "Gauthey, les frères Brancion et le canal du Charolais: genèse d'un projet," in *Un ingénieur des lumières,* ed. Coste et al.; *Oeuvres de M. Gauthey,* ed. Navier (Paris: F. Didot, 1809–1816), volume 3, pp. 291–342; Navier, "Eloge de Gauthey"; Allen, Highway and Canal System in Eighteenth-Century France, volume 2, pp. 460–469.

28. Gauthey's estimates of the water supply needed and available for the Charolais Canal proved to be too low, with the result that the canal could not be used as many days of the year as predicted. See Patrice Notteghem, "Gauthey et la question de l'alimentation en eau du canal du Charolais," in *Un ingénieur des lumières,* ed. Coste et al.

29. Gauthey's statement, quoted from Pierre Pinon, "Gauthey, les frères Brancion et le canal du Charolais: Genèse d'un projet," in *Un ingénieur des lumières,* ed. Coste et al., p. 90.

30. Many studies within the history and sociology of science and technology deal, directly or indirectly, with the nature and importance of tacit knowledge. For example, Harry M. Collins ("The TEA set: Tacit knowledge and scientific networks," *Science Studies* 4 (1974), pp. 165–186) shows that written documents never provided all the knowledge and information needed to build a successful TEA laser; David P. Billington (*Robert Maillart's Bridges: The Art of Engineering* (Princeton University Press, 1979)) shows that some of Maillart's significant design innovations in reinforced-concrete bridges came from observing the patterns of cracks that emerged in previously built bridges; and Robert Friedel and Paul Israel (*Edison's Electric Light: Biography of an Invention* (Rutgers University Press, 1986)) show the many kinds of tacit knowledge that went into the invention of the light bulb.

31. Quoted from Edoardo Benvenuto and Massimo Corradi, "Gauthey et le *Traité de la construction des ponts* dans le cadre de la culture scientifique de son temps," in *Un ingénieur des lumières,* ed. Coste et al., pp. 153, 170.

32. On the role of observation in understanding arch behavior, see Billington, *Robert Maillart's Bridges.*

33. Benvenuto and Corradi, "Gauthey et le *Traité de la construction des ponts,*" pp. 165–168.

34. Ibid., p. 164.

35. Maurice Bonneviot, "Gauthey, le Chalonnais," in *Un ingénieur des lumières,* ed. Coste et al., pp. 47–48.

36. Benvenuto and Corradi, "Gauthey et le *Traité de la construction des ponts*," p. 171.

37. Ibid., pp. 153–154, 164–165, 171.

38. Antoine Picon, "Entre ingénieur et architecte: L'itinéraire d'Emiland-Marie Gauthey," in *Un ingénieur des lumières*, ed. Coste et al., p. 209.

39. The controversy over the Pantheon's dome has been analyzed extensively by historians. See the articles by Picon and by Benvenuto and Corradi cited above. See also Picon, *French Architects and Engineers in the Age of Enlightenment*, pp. 168–180. Navier's biographical memoir on Gauthey also discusses this debate.

40. On Patte see Picon, *Architects and Engineers in the Age of Enlightenment*, pp. 343–345.

41. Emiland-Marie Gauthey, *Mémoire sur l'application des principes de la mécanique à la construction des voûtes et des dômes* (Paris: C.-A. Jombert, 1771).

42. Emiland-Marie Gauthey, *Dissertation sur les dégradations survenues aux piliers du dôme du Panthéon français et sur les moyens d'y remédier* (Paris: H.-L. Perronneau, an VI).

43. See Picon, "Entre ingénieur et architecte."

44. Stephen P. Timoshenko, *History of Strength of Materials* (1953; Dover, 1983), p. 71.

45. There is a vast literature on the history and social significance of the Ecole Polytechnique. For a review essay and comprehensive bibliography of literature prior to 1987 see A. Fourcy, *Histoire de l'Ecole Polytechnique*, ed. J. Dhombres (Belin, 1987). On the very early years of the school during the revolutionary period see Janis Langins, *La république avait besoin de savants: Les débuts de l'Ecole Polytechnique: l'Ecole Centrale des Travaux Publics et les cours révolutionnaires de l'an III* (Belin, 1987); see also Terry Shinn, *Savoir scientifique et pouvoir social: L'Ecole Polytechnique, 1794–1914* (Presses de la Fondation Nationale des Sciences Politiques, 1980). Two important recent works edited by B. Belhoste et al. are *La France des X: deux siècles d'histoire*, ed. (Economica, 1995) and *La formation polytechnicienne, 1794–1994* (Dunod, 1994).

46. Maurice Crosland, *The Society of Arcueil* (Harvard University Press, 1967), pp. 192–196; Shinn, *Savoir scientifique et pouvoir social*, pp. 9–23; A Fourcy, *Histoire de l'Ecole Polytechnique*, ed. J. Dhombres (Belin, 1987), passim.

47. Y. Chicoteau, A. Picon, and C. Rochant, "Gaspard Riche de Prony ou le génie 'appliqué'," *Culture technique*, no. 12 (March 1984), pp. 176–177; Charles Coulston Gillispie, *Science and Polity*, pp. 479–498; Anne Querrien, "Ecoles et corps: le cas des Ponts et Chaussées, 1747–1848," manuscript, Bibliothèque de l'Ecole des Ponts et Chaussées.

48. Fourcy, *Histoire de l'Ecole Polytechnique*, p. 376.

49. Archives Nationales, Service de Navier, F¹⁴2289; Robert McKeon, "Navier, Claude-Louis-Marie-Henri," *Dictionary of Scientific Biography;* Fourcy, *Histoire de l'Ecole Polytechnique,* pp. 378–379.

50. René Taton, "Monge, Gaspard," in *Dictionary of Scientific Biography;* Fourcy, *Histoire de l'Ecole Polytechnique,* p. 256; Morris Kline, *Mathematical Thought from Ancient to Modern Times* (Oxford University Press, 1972), pp. 536–540, 565–569. On Monge see also René Taton, *L'oeuvre scientifique de Gaspard Monge* (Presses Universitaires de France, 1951).

51. René Taton, "Hachette, Jean Nicolas Pierre," in *Dictionary of Scientific Biography.*

52. Jean Itard, "Lacroix, Sylvestre François," in *Dictionary of Scientific Biography.*

53. Pierre Costabel, "Siméon-Denis Poisson: Aspect de l'homme et de son oeuvre," in *Siméon-Denis Poisson et la science de son temps,* ed. M. Métivier et al. (Palaiseau: Ecole Polytechnique, 1981), pp. 1–21.

54. Chicoteau et al., "Gaspard Riche de Prony," pp. 171–183; Robert M. McKeon, "Prony, Gaspard-François-Clair-Marie Riche de," in *Dictionary of Scientific Biography;* Note sur les services et les travaux de M. de Prony, Vendémiaire, An XIII (1805), manuscript collection, Ecole des Ponts et Chaussées, Paris. Navier's later contributions to hydrodynamic theory show that he was familiar with Prony's text.

55. Picon, *L'invention de l'ingénieur moderne,* p. 279.

56. According to Picon, during the first half of the nineteenth century lecture courses at the Ecole des Ponts et Chaussées were in session only four months of the year. In 1851 the duration of lecture courses was increased to six months. Picon argues that there was always a tension within the Ecole des Ponts et Chaussées resulting from the gap between the analytical ideal and the need to give students a more practical engineering training. See Picon, *L'invention de l'ingénieur moderne,* pp. 283–285, 389–400, 449–456.

57. Picon, *L'invention de l'ingénieur moderne,* p. 517.

58. See Picon, *L'invention de l'ingénieur moderne,* p. 447.

59. Eda Kranakis, "Hybrid careers and the interaction of science and technology," in *Technological Development and Science in the Industrial Age,* ed. P. Kroes and M. Bakker (Kluwer, 1992).

60. Archives Nationales, Service de Navier, F¹⁴2289.

61. Navier to Director General, Corps des Ponts et Chaussées, Chalon-sur-Saône, June 21, 1821, Archives Nationales, F¹⁴2289.

62. See "Les polytechniciens dans la vie intellectuelle parisienne au XIXᵉ siècle," in *Le Paris des polytechniciens: des ingénieurs dans le ville: 1794–1994,* ed. B. Belhoste et al. (Délégation à l'action artistique de la Ville de Paris, 1994),

pp. 53–57. The character and the systematic nature of the interactions among French engineers and mathematicians in Navier's era have been studied by I. Grattan Guinness in several works, including these: "The *ingénieur savant,* 1800–1830: A neglected figure in the history of French mathematics and science," *Science in Context* 6 (1993), no. 2, pp. 405–433; "Modes and manners of applied mathematics: The case of mechanics," in *The History of Modern Mathematics,* volume 2, ed. D. Rowe and J. McCleary (Academic Press, 1989); *Convolutions in French Mathematics, 1800–1840: From the Calculus and Mechanics to Mathematical Analysis and Mathematical Physics,* three volumes (Birkhäuser, 1990).

63. Prony, "Notice biographique de Navier," pp. 1–19; Yves Chicoteau, Antoine Picon, and Catherine Rochant, "Gaspard Riche-de-Prony ou le génie 'appliqué'," *Culture technique,* no. 12 (March 1984), pp. 171–183.

64. Navier to Sylvestre F. Lacroix, October 9, 1823, Lacroix manuscripts, Bibliothèque de l'Institut de France, 2396.

65. I. Grattan-Guinness, *Joseph Fourier, 1768–1830* (MIT Press, 1972), pp. ix–x; John Herivel, *Joseph Fourier: The Man and the Physicist* (Clarendon, 1975), pp. 128–129.

66. Charles Dupin, *Eloge de M. le Baron de Prony,* Chambre des Pairs, April 2, 1840, pp. 22–23, 32–33.

67. Herivel, *Fourier,* pp. 118, 129, 143.

68. Auguste A. Beugnot, *Vie de Becquey* (Paris: Firmin Didot, 1852), p. 265.

69. Although the French scientific community in Navier's era and later included a particularly high percentage of *polytechniciens,* it should not be thought that all graduates of the school had the same opportunity and interest to become part of the French scientific community. A somewhat different perspective on the school and its social role at the time Navier attended emerges when we look at what became of the students in Navier's class. Of the 96 students who graduated in his class, 61 (or 64 percent) went into the artillery corps, generally reaching the middle- to higher-level officer ranks (especially Chef de Bataillon, equivalent to the rank of Major). Of these 61, approximately one-third were killed in battle before 1815. What these statistics tell us (and the data for years immediately following 1802 follow roughly the same pattern) is that the *Ecole Polytechnique* in these years mainly produced not scientists but rather artillery officers for Napoleon's army. These statistics come from pp. 414–417 of Fourcy's *Histoire de l'Ecole Polytechnique.*

70. On the Société Philomathique see Maurice Crosland, *The Society of Arcueil: A View of French Science at the Time of Napoleon I* (Harvard University Press, 1967), pp. 169–179; Jonathan R. Mandelbaum, The "Société Philomathique de Paris" from 1788 to 1835: A Study in the Institutional History and Collective Biography of a Parisian Scientific Society (Ph. D. dissertation, Ecole des Hautes Etudes en Sciences Sociales, 1980).

71. On the history of the French Academy of Sciences see Roger Hahn, *The Anatomy of a Scientific Institution: The Paris Academy of Sciences, 1666–1803* (University of California Press, 1971); Maurice Crosland, *Science under Control: The French Academy of Sciences, 1795–1914* (Cambridge University Press, 1992). Crosland (pp. 171–172) extensively discusses the election system of the French Academy of Sciences and the elite social and intellectual status of academicians. He refers to them as forming part of a new "intellectual aristocracy" in France.

72. Navier to Sylvestre F. Lacroix, October 9, 1823, Lacroix manuscripts, Bibliothèque de l'Institut de France, 2396.

73. Navier to Director General, Corps des Ponts et Chaussées, Chalon-sur-Saône, June 21, 1821, Archives Nationales, $F^{14}2289$.

74. Louis L. Bucciarelli, "Poisson and the mechanics of elastic surfaces," in *Siméon-Denis Poisson*, ed. Métivier et al.; Timoshenko, *History of Strength of Materials*, pp. 119–122.

75. Cauchy began his professional career as a member of the Corps des Ponts et Chaussées, and did carry out engineering work; even more than Navier's, however, his research turned to mathematics. By 1828 he had reached the rank of *ingénieur-en-chef* in the corps, but he went into exile after the Revolution of 1830. Today Cauchy is known only as a mathematician. There is now a full-length biography of Cauchy: Bruno Belhoste's *Augustin-Louis Cauchy: A Biography* (Springer-Verlag, 1991). On Sophie Germain, see Louis Bucciarelli and N. Dworsky, *Sophie Germain: An Essay in the History of the Theory of Elasticity* (Reidel, 1980).

76. These norms and values emerge very clearly in Navier's vita and in letters he wrote to the Corps' directors in 1814, 1821, and 1830 requesting honors and promotions. See Archives Nationales, Service de Navier, $F^{14}2289$.

Chapter 5

1. Portions of chapters 5 and 6 have been published in the following articles: Eda Kranakis, "Navier's theory of suspension bridges," *Acta Historica Scientiarum Naturalium et Medicinalium* 39 (1987), pp. 247–258; Eda Kranakis, "The affair of the Invalides Bridge," *Jaarboek voor de Geschiedenis van Bedrijf en Techniek* 4 (1987), pp. 106–130. On Navier and suspension bridge technology see also Antoine Picon, "Navier and the introduction of suspension bridges in France," *Construction History* 4 (1988), pp. 21–34.

2. Thomas Telford, "Report, 1814, Respecting a Bridge over the River Mersey at Runcorn," manuscript collection, Ironbridge Gorge Museum, Ironbridge, Shropshire. For a detailed historical discussion of this proposal see Roland A. Paxton, "Menai Bridge (1818–1826) and its influence on suspension bridge development," *Transactions of the Newcomen Society* 49 (1977–78), pp. 89–95; Roland A. Paxton, "Menai Bridge 1818–1826: Evolution of design," in *Thomas Telford: Engineer*, ed. A. Penfold (Thomas Telford Ltd., 1980). Ideas

and plans for this suspension bridge continued to be modified and discussed until 1817.

3. Samuel Brown had earlier submitted a proposal for a suspension bridge at the Runcorn site.

4. Several wire suspension footbridges were also built in Scotland in 1816 and 1817, although these were actually cable-stayed bridges. See Robert Stevenson, "Description of bridges of suspension," *Edinburgh Philosophical Journal* 5 (1821), April–October, pp. 241–247.

5. The other main agent for the diffusion of suspension bridge technology from Britain to France was Marc Seguin, who visited France during the autumn of 1823. On Seguin see Michel Cotte, "Seguin et Cie (1806–1824): du négoce familial de drap à la construction du pont suspendu de Tournon-Tain," *History and Technology* 6 (1988), pp. 95–144; Michel Cotte, Le système technique des Seguin en 1824–35," *History and Technology* 7 (1990), no. 2, pp. 119–147. See also Marc Seguin (and brothers), "Notice sur les ponts en fil de fer," *Annales de l'industrie nationale et étrangère* 12 (1823), p. 291; Marc Seguin, *Des ponts en fil de fer*, second edition (Paris: Bachelier, 1826).

6. Navier, *Rapport à Monsieur Becquey et mémoire sur les ponts suspendus* (Paris: Imprimerie Royale, 1823). A second edition of this work (Paris: Carilian-Goeury, 1830), published after the removal of Navier's Invalides Bridge, has different pagination and includes a 50-page appendix concerning the *Pont des Invalides* but otherwise corresponds to the first edition.

7. Thomas Pope, *A Treatise on Bridge Architecture* (New York: Alexander Niven, 1811). R. A. Paxton states on p. 88 of his article "Menai Bridge (1818–1826) and its influence on suspension bridge development" that one of Finley's bridges had been described in 1808; however, the article in question (*Monthly Magazine*, November 1808, p. 344) was not cited by British engineers of the period, whereas Pope's work was cited extensively.

8. Pope's *Treatise* (p. 187) cited Finley's 1810 *Port Folio* article.

9. Charles Stewart Drewry, *A Memoir on Suspension Bridges* (London: Longman, Rees, 1832), pp. 17–18, 31–36; *Reports and Observations on the Patent Iron Cables Invented by Captain Samuel Brown*, ed. S. Brown (London: T. Sotheran, 1815); *Abridgements of Specifications Relating to Bridges, Viaducts, and Aqueducts, A.D. 1750–1866*, ed. B. Woodcroft (London: Office of the Commissioners of Patents for Inventions, 1868), pp. 19–21; Emory L. Kemp, "Samuel Brown: Britain's pioneer suspension bridge builder," *History of Technology* 2 (1977), pp. 8–12; Thomas Day, "Samuel Brown: His influence on the design of suspension bridges," *History of Technology* 8 (1983), pp. 61–90.

10. Since this model dates from around 1814, it is probable that it was made in conjunction with his proposal for a suspension bridge at Runcorn. Telford and Brown first met and had extensive technical discussions at the time of the Runcorn design competition. See Paxton, "Menai Bridge," p. 93.

11. Kemp, "Samuel Brown," pp. 8–16, 33–35; Drewry, *Memoir on Suspension Bridges,* pp. 39–41; Stevenson, "Description of bridges of suspension," pp. 247–252; Navier, *Ponts suspendus,* second edition (Paris: Carilian-Goeury, 1830), pp. 49–57 (subsequent references are to this edition); Pierre C. F. Dupin, *The Commercial Power of Great Britain* (London: Charles Knight, 1825), volume I, pp. 375–380.

12. Samuel Brown, "Description of the Trinity Pier of Suspension at New-haven, near Edinburgh," *Edinburgh Philosophical Journal* 6 (1822), October–April, pp. 22–28; Kemp, "Samuel Brown," pp. 16–33; Thomas Day, "Samuel Brown in North-East Scotland," *Industrial Archaeology Review* 7 (1985), no. 2, pp. 154–170.

13. The comment—made by Catherine Plymley, a sister of one of Telford's patrons—is quoted from J. B. Lawson's article "Thomas Telford in Shrews-bury: The metamorphosis of an architect into a civil engineer," in *Thomas Telford: Engineer,* ed. Penfold.

14. Ibid., pp. 1–22. On Telford's life see also Thomas Telford, *Life of Thomas Telford* (London: James and Luke G. Hansard and Sons, 1838); Alexander Gibb, *The Story of Telford* (London: Alexander Maclehose, 1935).

15. Thomas Telford, "Report, 1814."

16. Paxton, "Menai Bridge," p. 97. See also the account by Telford's chief assistant, William Alexander Provis: *An Historical and Descriptive Account of the Suspension Bridge Constructed over the Menai Strait* (London: Ibotson and Palmer, 1828).

17. These documents include a letter to Brown written by civil engineer George Buchanan on July 20, 1821, reproduced as an appendix in Day's "Samuel Brown"; Stevenson's "Description of bridges of suspension"; Peter Barlow's *Essay on the Strength and Stress of Timber* (London: J. Taylor, 1817); Davies Gilbert's article "On some properties of the catenarian curve with reference to bridges by suspension," *Quarterly Journal of Science, Literature, and the Arts* (1821), pp. 230–235; Telford's "Report, 1814"; and the published hearings for the Menai Bridge project (British Parliamentary Papers, House of Commons, Papers relating to building a Bridge over the Menai Strait near Bangor Ferry, Sess. 1819 (60), volume V, pp. 2–17; British Parliamentary Papers, House of Commons, Third Report from the Select Committee on the Road from London to Holyhead; Menai Bridge, Sess. 1819 (256), volume V, pp. 21–33.

18. The value of 370 tons refers to the total tension, which was distributed equally among the 12 cables supporting the bridge (Stevenson, "Description of bridges of suspension," pp. 254–255).

19. Day, "Samuel Brown," pp. 86–87. In 1824 Buchanan proposed a suspen-sion bridge to cross the River Esk at Montrose, but it was not built. See George Buchanan, "Report on the present state of the wooden bridge at Montrose,

and the practicability of erecting a suspended bridge of iron in its stead," *Edinburgh Philosophical Journal* 11 (1824), April–October, pp. 140–156.

20. Ibid., pp. 151–154.

21. There is some ambiguity here. When Buchanan's early formula is directly applied, using the data from the Union Bridge, it does produce a result of 900. His formula states that $T = X^2/8Y$, where T = tension, X = span, and Y = sag. For the Union Bridge, X is 432 feet and Y is 26 feet, which leads to a value for T of 900 tons when rounded off. However, further on in the letter Buchanan suggests that this tension of 900 tons would be produced by a load of 280 tons at the center of the span. If he meant by this a load of 280 tons in addition to the dead load of 100 tons, then Buchanan's results would still follow from his formula. The total load would then be 380 tons, and since the span of the bridge cables was around 360 feet, this would make a load per unit of span of a little over a ton. If Buchanan took the load parameter to be equal to one ton per foot, then he might see no reason to include it in the formula, since it would not affect the value of the final result. Nothing like this is explained in his letter to Brown, however.

22. In his 1824 paper Buchanan is also inconsistent in using a weight factor (i.e., a factor for the load per unit span borne by the cable). He introduces a factor S to denote the span of the cable, but later he suggests that this factor also denotes the weight of the bridge. Immediately thereafter, however, he explicitly introduces a weight factor, W, which he substitutes for S. He therefore rewrites his formula as $T_0 = nW/8$, where $n = 2x/y$ and $W = 2wx$. In these formulas T_0 is the tension at the cable's midpoint, w is the load per unit span borne by the cable, W is the total load borne by the cable, x is half the cable's span, and y is the cable's sag. Substituting for n and W, we find that his formula becomes $T_0 = wx^2/2y$—the standard formula for a cable loaded uniformly along the horizontal (parabola).

23. Stevenson, "Description of bridges of suspension," pp. 243–247.

24. In a cable-stayed bridge, the roadway is supported by a series of cables that radiate from the bridge towers to various points along the sides of the deck (i.e., the roadway).

25. Buchanan, "Report on the present state of the wooden bridge at Montrose," p. 153.

26. Stevenson, "Description of bridges of suspension," p. 247.

27. Navier, *Ponts suspendus*, p. 55.

28. Gilbert, "On some properties of the catenarian curve," pp. 230–235. Gilbert was also a member of parliament and president of the Royal Society.

29. This question had little relevance for engineering practice, since a cable with a very large sag—up to 190 feet for a 560-foot span—would require very long chains and very high towers.

30. Recall that the standard formula for a parabolic cable loaded uniformly along the horizontal gives $y = wx^2/2T_0$, where y is the sag, x is the span, w is the load per unit span, and T_0 is the tension at the cable's midpoint. Gilbert's formula gave $y = xz/2a$, where z is the length of the cable (which would not in practice be much longer than the cable's span) and where a represents a value that would give the tension at the cable's midpoint when multiplied by w. Gilbert did not actually include the factor w in the equation, however.

31. Gilbert, "Properties of the catenarian curve," p. 231.

32. Quoted from Day, "Samuel Brown," pp. 86–87.

33. The loads were not actually distributed uniformly over the cable's span; they "were disposed at the 1/4, 1/2, and 3/4 division of the distance over which [the wire] was stretched" (i.e., over the span). See Thomas Telford, "Report, 1814," p. 4.

34. Thomas Telford, "Report, plan, and estimate for building a bridge over the Menai Strait, near Bangor Ferry," in *Papers Relating to the Building of a Bridge over the Menai Strait, near Bangor Ferry*, British Parliamentary Papers, House of Commons, Sess. 1819 (60), volume V, p. 4.

35. Ibid., p. 30.

36. British Parliamentary Papers, House of Commons, Papers relating to building a Bridge over the Menai Strait near Bangor Ferry, Sess. 1819 (60), volume V; British Parliamentary Papers, House of Commons, Third Report from the Select Committee on the Road from London to Holyhead; Menai Bridge, Sess. 1819 (256), volume V. See also Navier, *Ponts suspendus*, pp. 36–49; Drewry, *Memoir on Suspension Bridges*, pp. 46–53.

37. In developing his ideas about suspension bridges prior to the Menai project, Telford had paid considerable attention to cable design. He had collected information on various chain designs for the Runcorn project and had recorded the details in a notebook. See Denis Smith, "The use of models in nineteenth century British suspension bridge design," in *History of Technology*, volume 2, ed. A. Hall and N. Smith (Mansell, 1977), p. 171.

38. Bryan Donkin, however, suggested that the permanent stretching of iron bars under loading, which gives them a higher tensile strength, could be used as a method of equalizing the tension in the kind of welded bar chain proposed by Telford. See British Parliamentary Papers, House of Commons, Papers relating to building a Bridge over the Menai Strait near Bangor Ferry, Sess. 1819 (60), volume V, p. 10.; Parliamentary Papers, House of Commons, Third Report from the Select Committee on the Road from London to Holyhead; Menai Bridge, Sess. 1819 (256), volume V.

39. British Parliamentary Papers, House of Commons, Third Report from the Select Committee on the Road from London to Holyhead; Menai Bridge, Sess. 1819 (256) volume V, pp. 23, 25, 28, 32. See also Buchanan, "Report on the present state of the wooden bridge at Montrose," p. 147.

40. Smith, "Use of models," pp. 169–214.

41. Ibid.; Chester H. Gibbons, "History of testing machines for materials," *Transactions of the Newcomen Society* 15 (1934–35), pp. 169–184; Stephen P. Timoshenko, *History of Strength of Materials* (1953; Dover, 1983), pp. 98–100; Navier, *Ponts suspendus,* pp. 235–248.

42. Drewry, *Memoir on Suspension Bridges,* pp. 154–157.

43. Smith explores the controversy on pp. 180–189 of "The use of models in nineteenth century British suspension bridge design."

44. Navier, *Ponts suspendus,* pp. 48–49; British Parliamentary Papers, House of Commons, Sess. 1823 (429), volume XIII.

45. Navier, *Ponts suspendus,* pp. 19–79.

46. Ibid., pp. 50, 54–57, 62–64, 67–77.

47. Marc Isambard Brunel (1769–1849) had an unusual career. Born in Normandy into a prosperous farm family, he strongly resisted his parents' efforts to prepare him for the priesthood. His parents eventually relented, and after studying drawing, perspective, and hydrography Brunel went to work on a French ship for 6 years. In 1792, during the French Revolution, he emigrated to the United States, where he became an architect and engineer, eventually winning the post of Chief Engineer of New York. In 1798 he emigrated to England—in part to marry an English woman he had met in France. He brought with him a plan to mechanize the production of pulley blocks. When he succeeded in this project, Brunel bought a sawmill and began a further project to mechanize boot production. He also invented dozens of other machines, and he designed bridges, tunnels, docks, and other engineering structures. In his work on suspension bridges, Brunel carried out experiments on the strength of iron. I know of no evidence to indicate whether he estimated cable tension empirically or by means of some mathematical technique. His suspension bridges, in any event, cleverly incorporated hangers whose length could be altered by means of adjustable nuts so as to make them correspond to the exact curve of the main cables. See, Richard Beamish, *Memoir of the Life of Sir Marc Isambard Brunel* (London: Longman, Green, 1862); L. T. C. Rolt, *Isambard Kingdom Brunel* (1957; Penguin, 1972), pp. 25–35. See also Navier, *Ponts suspendus,* pp. 67–77, 248.

48. Navier, *Ponts suspendus,* p. 49.

49. Ibid., pp. 39–40.

50. Ibid., p. 145.

51. Ibid., p. 126.

52. Ibid., p. 11.

53. Ibid.

54. Bernard Forest de Bélidor, *Architecture hydraulique,* ed. Navier (Paris: Firmin Didot, 1819), volume I, pp. 3–4.

55. Ibid., p. 278.

56. Ibid., p. 295.

57. Navier, *Ponts suspendus* (1823), pp. 81–92.

58. On the historical development of statics see Clifford Truesdell, *The Rational Mechanics of Flexible or Elastic Bodies, 1638–1788,* Leonhardi Euleri Opera Omnia, second series, volume XI, part 2 (Zurich: Orell Fussli, 1960), especially pp. 43–46, 50–53, and 64–88.

59. Navier, *Ponts suspendus,* pp. 187–192.

60. Ibid., pp. 54–55.

61. Ibid., pp. 93–101.

62. Ibid., p. 100.

63. Ibid., p. 52. Day indicates on p. 79 of "Samuel Brown" that Brown's patent called for fixed saddles, but contemporary descriptions of the Union Bridge, such as Navier's, describe rollers.

64. Provis, *Historical and Descriptive Account of the Menai Bridge,* pp. 38, 42.

65. Navier, *Ponts suspendus,* p. 72.

66. British Parliamentary Papers, House of Commons, Papers relating to building a Bridge over the Menai Strait near Bangor Ferry, Sess. 1819 (60), volume V, pp. 13–15.

67. According to Navier's classification, Brunel's tower-cable system could be treated as roughly equivalent to Telford's because they both allowed the cables to move back and forth with relatively little friction.

68. Navier, *Ponts suspendus,* pp. 101–112.

69. Truesdell, *Rational Mechanics,* pp. 377–380.

70. Navier, *Ponts suspendus,* p. 112.

71. Ibid., pp. 113–120.

72. Ibid., pp. 108–109, 113–120.

73. I thank Robert Gordon for helpful comments concerning the historical and metallurgical issues raised in this section.

74. For a popular exposition of the basic concepts and principles of the theory of elasticity see pp. 33–59 of J. E. Gordon's *Structures; or Why Things Don't Fall Down* (1978; Da Capo, 1981).

75. Timoshenko, *History of Strength of Materials,* pp. 18–21, 53, 74, 81–83, 92; Navier, *Résumé des leçons données à l'Ecole des Ponts et Chaussées sur l'application de la mécanique à l'établissement des constructions et des machines,* third edition, ed. A. Barré de Saint-Venant (Paris: Dunod, 1864), volume I, pp. xc–cvii; Thomas Young, *A Course of Lectures on Natural Philosophy and the Mechanical Arts* (London: Joseph Johnson, 1807), volume I, pp. 137, 151; volume II, p. 46; A. Duleau, *Essai théorique et expérimental sur la résistance du fer forgé* (Paris: Courçier, 1820), pp. v, 54. According to Truesdell (*Rational Mechanics,* p. 423), Euler formulated the concept of a modulus of elasticity equivalent to Young's as early as 1776. Nevertheless, it seems that Euler's result was overlooked by later researchers, because neither Duleau nor Navier nor Barré de Saint-Venant cited this contribution.

76. Emiland Gauthey, *Traité de la construction des ponts,* ed. Navier (Paris: Firmin Didot, 1809–1813), II, p. 145. Navier made this assertion in a lengthy note ("Sur l'évaluation de la force du fer"), one of many which he added to Gauthey's treatise.

77. The change in Navier's thinking, which he incorporated into his course on applied mechanics at the Ecole des Ponts et Chaussées in 1824, represented an important new direction in structural analysis. His student Barré de Saint-Venant recalled: " . . . in this . . . remarkable course of 1824, which [I] had attended . . . Navier brought together the theory of strength of materials and the theory of elasticity more plainly and completely than had his predecessors. . . . " (Navier, *Résumé des leçons données à l'Ecole des Ponts et Chaussées,* volume I, pp. lxxi, cv).

78. Duleau, *Résistance du fer forgé.*

79. Ibid., p. 54. The currently accepted value is 21,000 kg/mm^2. See Alan Cottrell, *An Introduction to Metallurgy,* second edition (Crane, Russak, 1975), p. 308.

80. Navier, *Ponts suspendus,* pp. 131–143.

81. Ibid., pp. 145–146.

82. British Parliamentary Papers, House of Commons, Third Report from the Select Committee on the Road from London to Holyhead; Menai Bridge, Sess. 1819 (256), volume V, p. 23.

83. Ibid., pp. 23, 26, 28.

84. Paxton, "Menai Bridge," p. 97.

85. Navier, *Ponts suspendus,* pp. 145–154.

86. Ibid., pp. 54–55.

87. Stevenson, "Description of bridges of suspension," p. 255.

88. Truesdell, *Rational Mechanics,* pp. 222–229, 234–315; Morris Kline, *Mathematical Thought from Ancient to Modern Times* (Oxford University Press, 1972), pp. 502–522.

89. One reason why eighteenth-century mathematicians failed to work out the method of constructing infinite trigonometric series to solve problems involving vibrating strings is that many doubted that the initial shape of a plucked string could be simulated in this way. The reasons for their doubt were linked to the ways in which mathematical functions were defined and understood at that time. See Truesdell, *Rational Mechanics,* pp. 244–300; Kline, *Mathematical Thought,* pp. 454–459, 502–514, 671–679.

90. Navier, *Ponts suspendus,* pp. 154–183.

91. Ibid., p. 68.

92. Marc Seguin, *Des ponts en fil de fer,* second edition (Paris: Bachelier, 1826), p. 40.

93. Ibid., p. 63.

94. Navier, *Ponts suspendus,* p. 34.

95. Ibid., p. 35.

96. Ibid., pp. 183–187.

97. Stevenson, "Description of bridges of suspension," p. 243.

98. Ibid., p. 245.

99. Navier, *Ponts suspendus,* p. 35.

100. Theodore von Kármán et al., *The Failure of the Tacoma Narrows Bridge* (Pasadena: [n.p.], 1941), p. i.

101. Friedrich Bleich et al., *The Mathematical Theory of Vibration in Suspension Bridges* (U.S. Government Printing Office, 1950), p. 11.

102. Ibid., p. 409.

103. Von Kármán et al., *Failure of the Tacoma Narrows Bridge,* pp. 20–34, V/1–V/15.

104. Von Kármán et al., *Failure of the Tacoma Narrows Bridge;* Bleich et al., *Mathematical Theory of Vibration in Suspension Bridges.*

105. See, e.g., Navier, "Détails historiques sur l'emploi du principe des forces vives dans la théorie des machines, et sur diverses roues hydrauliques," *Annales de Chimie et de Physique,* second series, 9 (1818), pp. 146–159.

106. A review of Navier's theory written in 1865 specifically criticized this lacuna. See E. Brissaud, Ponts suspendus: aperçu historique et comparatif, lithographed manuscript, Library of the Ecole des Ponts et Chaussées.

107. Stevenson, "Description of bridges of suspension," p. 255.

108. Navier, *Ponts suspendus*, p. 183.

109. Ibid., p. 183.

110. Navier's pathbreaking theory of elasticity was completed in 1821, and his memoir on fluid dynamics was completed at the end of 1822, the year following his first visit to Britain. See Navier, *Résumé des leçons données à l'Ecole des Ponts et Chaussées*, volume I, pp. lxii–lxv.

111. Navier, *Ponts suspendus*, p. 11.

112. Joan L. Richards, "Rigor and clarity: Foundations of mathematics in France and England, 1800–1840," *Science in Context* 4 (1991), no. 2, p. 298. Peter Barlow was an exception to this pattern; his 1817 *Essay on the Strength and Stress of Timber* does make use of Continental notation.

113. Navier, *Ponts suspendus*, p. 56.

114. Maurice Crosland, *Science under Control: The French Academy of Sciences, 1795–1914* (Cambridge University Press, 1992), pp. 119–123.

115. See, e.g., Navier, *Résumé des leçons*, p. lxix.

116. David B. Steinman, *A Practical Treatise on Suspension Bridges* (Wiley, 1922), pp. 1–18. Steinman did not cite Navier, but his analysis clearly did follow Navier's. Steinman even used the letter f to refer to the variable for cable sag. This letter was chosen by Navier for mnemonic reasons, because the French word for sag is *flèche*.

117. This was a conscious choice because Navier recognized that it could be made stiff and because he had the mathematical tools needed to consider the hypothesis of a stiffened deck.

118. Seguin, *Ponts en fil de fer*, p. 62.

119. Drewry, *Memoir on Suspension Bridges*, p. 209.

120. The first significant example was Roebling's Niagara suspension bridge, completed in 1855. On stiffness and the stiffening truss see James H. Cissel, "Stiffness as a factor in long span suspension bridge design," *Roads and Streets*, April 1941, pp. 64, 67–68, 70, 72; James Kip Finch, "Wind failures of suspension bridges; or evolution and decay of the stiffening truss," *Engineering News-Record* 126 (March 13, 1941), pp. 402–407. These works make clear, however, that not all builders showed equal concern for stiffness.

121. Sir Alfred Pugsley, *The Theory of Suspension Bridges*, second edition (Edward Arnold, 1968); Sven Olof Asplund, *On the Deflection Theory of Suspension Bridges* (Almqvist & Wiksell, 1943). For a recent historical account of the development of suspension bridge theory and its relationship to the evolution of stiffening trusses and problems of wind-induced oscillations see Stephen G. Buonopane and David P. Billington, "Theory and history of suspension bridge

design from 1823 to 1940," *Journal of Structural Engineering* 119 (1993), no. 3, pp. 954–977.

122. Pugsley, *Theory of Suspension Bridges,* p. 6; Steinman, *Practical Treatise on Suspension Bridges,* pp. 18–68.

123. Steinman, *Practical Treatise,* p. 12.

124. Ibid., p. 76.

125. Navier, *Ponts suspendus,* pp. 197–198.

126. Ibid., p. 64. Navier's assertion was based on an analysis he made of the equilibrium of cable-stayed bridges. For reasons of brevity, I did not include a summary of this part of Navier's theory (pp. 120–126 in the second edition).

127. Ibid., p. 7.

128. Seguin, *Ponts en fil de fer,* p. 61. See also John Augustus Roebling, "Memoir of the Niagara Falls and International Suspension Bridge," in *Papers and Practical Illustrations of Public Works of Recent Construction,* ed. J. Weale (London: John Weale, 1856), pp. 3–10.

129. Navier, *Ponts suspendus,* p. 11.

Chapter 6

1. The "technological look" of the design was due in part to an extensive use of iron, including iron framing for the towers.

2. Navier, *Rapport à M. Becquey et mémoire sur les ponts suspendus,* second edition (Paris: Bachelier, 1830), pp. 226–234.

3. Robert Stevenson, "Description of bridges of suspension," *Edinburgh Philosophical Journal* 5 (1821), April–October, p. 254.

4. British Parliamentary Papers, House of Commons, Third Report from the Select Committee on the Road from London to Holyhead; Menai Bridge, Sess. 1819 (256), volume V, p. 25.

5. This principle, whereby a change in pressure at one point in a closed body of water is transmitted throughout the entire body of water, is known as Pascal's Law.

6. Navier, *Ponts suspendus,* second edition, p. 228. (Subsequent references are to this edition.)

7. Ibid, pp. 226–234.

8. Ibid., p. 226.

9. Quoted from James M. Oliver, The "Corps des Ponts et Chaussées," 1830–1848 (Ph.D. dissertation, University of Missouri, 1967), pp. 175–176.

10. Navier, *De l'entreprise du pont des Invalides* (Paris: Firmin Didot, 1827), p. 8.

11. Navier, *Ponts suspendus*, p. 250. (These were the words of the executive council of the Corps des Ponts et Chaussées.)

12. Ibid., p. 255.

13. *Plan du pont extraordinaire dit des Invalides* (Paris, n.d.), pp. 1–2, 4.

14. Navier, *Invalides*, p. 8.

15. Ibid., pp. 7–8.

16. Ibid., p. 7.

17. Ibid., pp. 6, 22; Navier, *Ponts suspendus*, pp. 200–203.

18. Navier, *Invalides*, p. 4.

19. Ibid., pp. 3–4.

20. Navier, *Ponts suspendus*, p. 15.

21. Ibid., pp. 202–203, 208–215.

22. Ibid., p. 204.

23. British Parliamentary Papers, House of Commons, Third Report from the Select Committee on the Road from London to Holyhead; Menai Bridge, Sess. 1819 (256), volume V, pp. 26–27.

24. Ibid., p. 25.

25. James Finley, "A description of the patent chain bridge," *Port Folio*, n.s. 3 (1810), no. 6, p. 443.

26. Stevenson, "Bridges of suspension," p. 254.

27. Finley, "Patent chain bridge," pp. 442, 449.

28. W. H. Boyer and Irving A. Jelly, "An early American suspension span," *Civil Engineering* 7 (1937), no. 5, p. 339.

29. See James H. Cissel, "Stiffness as a factor in long-span suspension bridge design," *Roads and Streets* (April 1941), p. 70; David P. Billington, "History and esthetics in suspension bridges," *Journal of the Structural Division, Proceedings of the American Society of Civil Engineers* 103 (1977), no. ST8, p. 13. The pagination of the latter article refers to the reprint version of the paper.

30. Finley, "Patent chain bridge," pp. 440–453.

31. Navier, *Ponts suspendus*, pp. 50–53.

32. Ibid., pp. 200–202.

33. Ibid., p. 202.

34. Ibid., pp. 127–31, 202–203, 207–211, 298–300.

35. Navier, *Ponts suspendus*, pp. 211–215.

36. Ibid., pp. 215–217.

37. Ibid., pp. 217–220.

38. Ibid., pp. 220–226.

39. Ibid., pp. 67–71, 264–279, 281.

40. "Article publié par l'administration dans le Moniteur du 29 février 1828," *Journal du génie civil* 1 (1828), p. 446.

41. "Lettre au directeur du Journal du génie civil," *Journal du génie civil* 4 (1829), p. 678.

42. Navier, *Ponts suspendus*, pp. 249–253.

43. Ibid., p. 251.

44. Ibid., pp. 252–253.

45. Ibid., p. 252.

46. Ibid., pp. 249–254.

47. Ibid., p. 253.

48. "Article publié par l'administration," *Journal du génie civil* 1 (1828), p. 446; Auguste A. Beugnot, *Vie de Becquey* (Paris, 1852), pp. 184–190, 234–235.

49. Navier, *Ponts suspendus*, p. 253; *Pont des Invalides: mémoire à consulter* (n.p., 1826), p. 16.

50. Navier, *Invalides*, pp. 19–20, 25; Navier, "De l'exécution des travaux publics et des concessions," *Annales des Ponts et Chaussées: mémoires* 1 (1832), pp. 1–31.

51. Navier, *Ponts suspendus*, p. 251.

52. The Menai Bridge was not an entrepreneurial undertaking; as noted earlier, it was financed by the state and administered by a committee of the House of Commons. See Navier, *Invalides*, p. 17.

53. Navier, *Ponts suspendus*, pp. 253–255.

54. *Pont des Invalides: mémoire à consulter* (n.p., 1826), passim.

55. Navier, *Ponts suspendus*, pp. 255, 258; "Le pont de fer," *La pandore*, November 25, 1826, pp. 2–3.

56. J. Hopkins, *A Span of Bridges* (Devon: David and Charles, 1970), pp. 202–203.

57. Navier, *Ponts suspendus*, pp. 45, 66–68, 282, 284–294.

58. Ibid., pp. 263–278.

59. Ibid., pp. 279–280.

60. Ibid., p. 280. See also Archives Nationales F^{14}2289 and "Note relative à la disposition des chaînes du pont des Invalides," *Journal du génie civil* 2 (1829), pp. 181–185.

61. Navier, *Ponts suspendus*, pp. 258–263.

62. Navier, letter to Becquey, July 1, 1826, Archives Nationales F^{14}2289.

63. Navier, letter to Becquey, July 24, 1826, Archives Nationales F^{14}2289.

64. Navier, *Ponts suspendus*, pp. 262–263; Prony, "Notice biographique de Navier," *Annales des Ponts et Chaussées: mémoires et documents* 7 (1837), no. 1, p. 12.

65. "The New York anchorage of the East River bridge," *Scientific American*, 34 (1876), January 8, pp. 15–16. See also David B. Steinman, *A Practical Treatise on Suspension Bridges* (Wiley, 1922), pp. 77, 92, 106–107, 119, 161.

66. Charles Stewart Drewry, a British contemporary of Navier who discussed this anchorage design in 1832, came to a similar conclusion. See pp. 100–101 of his *Memoir on Suspension Bridges* (London: Longman, Rees, 1832).

67. Navier, *Invalides*, p. 14; Navier, *Ponts suspendus*, pp. 127–131, 263, 299–300.

68. Navier, *Ponts suspendus*, p. 129.

69. Ibid., pp. 300–301; Navier, *Invalides*, pp. 10–11, 18; Prony, "Notice biographique de Navier," pp. 12–14; Barré de Saint-Venant, "Notice sur les ouvrages de Navier," in Navier, *Résumé des leçons . . . de la résistance des corps solides*, ed. Barré de Saint-Venant (Paris, 1864), p. lxvii.

70. James Kip Finch, "Wind failures of suspension bridges," *Engineering News-Record* 126 (March 13, 1941), pp. 402–407.

71. The hinge was included to make the structure statically determinate (so that forces in the structure could be calculated theoretically), but it had practical disadvantages. David Steinman explained: "Intermediate hinges are troublesome and expensive details, particularly in long spans; besides augmenting the deflections, they cause sudden reversals of shear under moving load, and constitute a point of weakness and wear in the structure. . . . Furthermore, a center hinge conduces to a serious distortion of the cable from the ideal parabolic form, with a resulting overloading of some of the hangers. In the case of the Brooklyn Bridge, the center hinge or slip joint has caused excessive bending stresses in the cable at that point, and the breaking of the adjacent suspenders; 120 suspenders near the hinge had to be replaced by larger ropes." (*Practical Treatise on Suspension Bridges*, p. 81)

72. "Article publié par l'administration," *Journal du génie civil* 1 (1828), p. 452.

73. *Pont des Invalides: mémoire à consulter,* passim; Navier, *Invalides,* pp. 1, 9, 11, 16–17; Navier, *Ponts suspendus,* p. 301; *Résistance des corps solides,* pp. lxvii-lxviii; Prony, p. 14.

74. *Pont des Invalides: mémoire à consulter,* passim.

75. Ibid.; *Plan du pont extraordinaire dit des Invalides,* pp. 1–4.

76. Navier, *Invalides,* pp. 19–24; Prony, "Notice biographique de Navier," p. 14; Navier, *Résistance des corps solides,* pp. lxvi-lxviii; *Plan du pont extraordinaire dit des Invalides,* pp. 1–2, 4.

77. William Alexander Provis, *An Historical and Descriptive Account of the Suspension Bridge Constructed Over the Menai Strait* (London: Ibotson and Palmer, 1828), pp. 3–5; Gene D. Lewis, *Charles Ellet, Jr.* (University of Illinois Press, 1968), pp. 121–126; David Jacobs and Anthony E. Neville, *Bridges, Canals, and Tunnels* (Van Nostrand, 1968), p. 79.

78. Alan Trachtenberg, *Brooklyn Bridge: Fact and Symbol,* second edition (University of Chicago Press, 1979), p. 35; "New York Anchorage of East River Bridge," p. 16.

79. "Note relative à la disposition des chaînes du pont des Invalides," *Journal du génie civil* 2 (1829), p. 182.

80. This theme will be explored further in Part II.

81. "Ponts en chaînes et en fils de fer," *Moniteur de l'industrie,* 4 (1829), p. 100.

82. Honoré de Balzac, *The Village Rector* (Boston, 1900), p. 242.

83. *An Account of Various Suspension Bridges, Particularly of Captain S. Brown's, Near Berwick-upon-Tweed* (n.p., [ca. 1826]), p. 51, Edwards manuscripts, National Museum of American History. Parts of this article repeat, verbatim, sections from Robert Stevenson, "Description of Bridges of Suspension," so Stevenson may have been its author.

84. "Pont des Invalides," *Le pandore* (9 September 1826).

85. Navier, "Le pont de fer," *Le pandore,* No. 1286 (25 November 1826), p. 3.

86. "Article publié par l'administration," *Journal du génie civil* 1 (1828), pp. 451–452; "Lettre au directeur du Journal du génie civil," *Journal du génie civil* 4 (1829), pp. 677–680; Archives Nationales F^{14}10205; Navier, *Résistance des corps solides,* ed. Saint-Venant, pp. lxvii-lxviii; "Pont suspendu des Champs-Elysées," *Journal du génie civil* 8 (1830), pp. 144–149; Alexandre Corréard, "Observations sur le pont suspendu des Champs-Elysées," *Journal du génie civil* 8 (1830), pp. 149–154.

87. Corréard, "Observations sur le pont suspendu des Champs-Elysées," p. 152.

88. Ibid., pp. 149–154. For more descriptions of the other bridges see pp. 102–107 of Charles Stewart Drewry's *Memoir on Suspension Bridges* (London: Longman, Rees, et al, 1832) and Archives Nationales F^{14}10205.

89. Archives Nationales F^{14}2289.

90. Navier, letter to Bérard, October 25, 1830, Archives Nationales F^{14}2289.

91. Prony, "Notice biographique de Navier," p. 13.

92. See, for example, Mellet and Henry, *L'arbitraire administratif des Ponts et Chaussées* (Paris, 1835); P.D. Martin, *Description du pont suspendu, construit sur la Garonne, à Langon* (Paris, 1832), passim.

93. See, for example, Cecil O. Smith, Jr., "The longest run: Public engineers and planning in France," *American Historical Review* 95 (1990), no. 3, pp. 657–692; Arthur Louis Dunham, *The Industrial Revolution in France, 1815–1848* (Exposition Press, 1955), pp. 49–84, 399–407.

94. "Les travaux publics aux Etats-Unis d'Amérique: Analyse d'un rapport de M. Malézieux," *Annales du génie civil*, second series, 4 (1875), p. 460.

95. For more on Becquey's ideas and policies see pp. 661–665 of Smith, "Longest run."

96. See, e.g., Smith, "Longest run."

97. See p. 13 of "Des chemins de fer aux États-Unis d'Amérique," *Journal de l'industriel et du capitaliste* 2 (1836). This article used a comparison of American and French railroads as a basis for an attack on the Corps des Ponts et Chaussées. When the author referred to the "sterile monuments" created by the Corps, he did not specifically mean the Pont des Invalides; however, the latter was representative of the type of undertaking he was criticizing.

Introduction to Part II

1. Dennis Hart Mahan, *An Elementary Course of Civil Engineering, for the Use of the Cadets of the United States Military Academy* (New York: Wiley and Putnam, 1837); Charles Storrow, *A Treatise on Water-Works for Conveying and Distributing Supplies of Water* (Boston: Hilliard & Gray, 1835); Peter A. Ford, "Charles S. Storrow, civil engineer: A case study of European training and technological transfer in the antebellum period," *Technology and Culture* 34 (1993), no. 2, pp. 271–299; Todd Shallot, "Building waterways, 1802–1861: Science and the United States Army in early public works," *Technology and Culture* 31 (1990), no. 1, pp. 18–50.

2. Stephen P. Timoshenko, *History of Strength of Materials* (1953; Dover, 1983).

3. T. M. Charlton, *A History of Theory of Structures in the Nineteenth Century* (Cambridge University Press, 1982), pp. 175–186. Other histories that could also have been cited include the following: Hunter Rouse and Simon Ince,

History of Hydraulics (1957; Dover, 1963); D. S. L. Cardwell, *From Watt to Clausius: The Rise of Thermodynamics in the Early Industrial Age* (Cornell University Press, 1971).

4. In 1800 France had a population about 4.5 times that of the United States; by 1820 the ratio was only a little over 3:1; between 1860 and 1870 the U.S. overtook France in population.

5. Thomas Parke Hughes, *Elmer Sperry: Inventor and Engineer* (Johns Hopkins University Press, 1971); Thomas Parke Hughes, *American Genesis: A Century of Invention and Technological Enthusiasm* (Viking, 1989); Reese V. Jenkins, *Images and Enterprise: Technology and the American Photographic Industry, 1839–1925* (Johns Hopkins University Press, 1975); Edwin T. Layton, "Scientific technology, 1845–1900: The hydraulic turbine and the origins of American industrial research," *Technology and Culture* 20 (1979), January, pp. 64–89; Kendall Birr, *Pioneering in Industrial Research* (Washington: Public Affairs Press, 1957); Kendall Birr, "Science in American industry," in *Science and Society in the U.S.*, ed. D. Van Tassel and M. Hall (Dorsey, 1966); Howard R. Bartlett, "The development of industrial research in the United States," in *Industrial Research: A National Resource* (Government Printing Office, 1941); Leonard S. Reich, *The Making of American Industrial Research: Science and Business at GE and Bell, 1876–1926* (Cambridge University Press, 1985); David Noble, *America by Design: Science, Technology, and the Rise of Corporate Capitalism* (Oxford University Press, 1977); Matthew Josephson, *Edison* (McGraw-Hill, 1959); Paul Israel, *From Machine Shop to Industrial Laboratory: Telegraphy and the Changing Context of American Invention, 1830–1920* (Johns Hopkins University Press, 1992); Robert Friedel and Paul Israel, *Edison's Electric Light: Biography of an Invention* (Rutgers University Press, 1988); W. Bernard Carlson, *Innovation as a Social Process: Elihu Thomson and the Rise of General Electric, 1870–1900* (Cambridge University Press, 1991); Bruce Sinclair, *Philadelphia's Philosopher Mechanics: A History of the Franklin Institute, 1824–1865* (Johns Hopkins University Press, 1974); Alfred D. Chandler, *The Visible Hand: The Managerial Revolution in American Business* (Harvard University Press 1977), p. 375.

6. Terry Shinn, "The genesis of French industrial research, 1880–1940," *Social Science Information* 19 (1980), no. 3, pp. 607–640. With respect to the use of American patents, Shinn found, for example, that St.-Gobain used American patents for manufacturing window glass and plate glass, while the electrical companies CEM and CGE relied on patents from the Edison company and later from General Electric.

7. Robert H. Thurston, "The modern mechanical laboratory, especially as in process of evolution in America," *Society for the Promotion of Engineering Education Proceedings* 8 (1900), p. 334. On the importance of the Conservatoire des Arts et Métiers in the first half of the nineteenth century see Alexandre Herlea, "Préliminaires à la naissance des laboratoires publics de recherche industrielle en France," *Culture technique* no. 18 (March 1988), pp. 220–231; Alexandre Herlea, "Advanced technology, education, and industrial research laboratories in 19th century France," in *Technological Education—Technological Style* (San Francisco Press, 1986).

8. There are admittedly many problems in using patent statistics in the present context. Perhaps most important, the validity of this measure rests on the assumption that patents are based on empirico-inductive research oriented toward industrial and design problems. Although a patent by definition embodies some design-oriented research, we cannot necessarily assume that this research was empirico-inductive in nature. In principle a patent could be the result of mathematical research, though in practice most were not. In addition, many were based simply on "cut and try." This measure is also problematic in the sense that it does not take into account the differences between the two patent systems that affected the rates of patenting. For example, a larger percentage of French patents were awarded to foreigners. Throughout most of the nineteenth century, between 5 percent and 11 percent of French patents were awarded to foreigners, as opposed to only about 3–4 percent of American patents. This situation changed, however, after 1870.

9. Yves Plasseraud and François Savignon, *Paris 1883: Genèse du droit unioniste des brevets* (Paris: Litec, 1983), p. 27.

10. Brooke Hindle and Steven Lubar, *Engines of Change: The American Industrial Revolution, 1790–1860* (Smithsonian Institution Press, 1986), pp. 78–79.

11. "Statistique de l'industrie," *Le génie industriel* 18 (1859), p. 92.

12. Plasseraud and Savignon, *Paris 1883: Genèse du droit unioniste des brevets*, pp. 27, 32–44, 70, 76–82, 91–93, 113.

13. Unfortunately, some authors seem unable to recognize that it is quite valid to study trends at the national level, and that doing so is not necessarily the same as either having a nationalist bias or holding a racialist theory of "national character." See, notably, Gregory K. Dreicer, The Long Span. Intercultural Exchange in Building Technology: Development and Industrialization of the Framed Beam in Western Europe (Ph.D. dissertation, Cornell University, 1993).

Chapter 7

1. André Thépot, "Les ingénieurs du Corps des Mines," *Culture Technique,* no. 12 (March 1984), pp. 55–61.

2. A. Brunot and R. Coquand, *Le Corps des Ponts et Chaussées* (Paris: Editions du Centre National de la Recherche Scientifique, 1982); André Guillerme, *Corps à corps sur la route: Les routes, les chemins et l'organisation des services au XIXième siècle* (Paris: Ecole des Ponts et Chaussées, 1984).

3. John Hubbel Weiss, *The Making of Technological Man: The Social Origins of French Engineering Education* (MIT Press, 1982), p. 61.

4. Guillerme, *Corps à corps sur la route*, p. 101.

5. Antoine Picon, *L'invention de l'ingénieur moderne: L'Ecole des Ponts et Chaussées, 1747–1851* (Ecole des Ponts et Chaussées, 1992), p. 418.

6. Terry Shinn, "Des corps de l'Etat au secteur industriel: Genèse de la profession d'ingénieur, 1750–1920," *Revue française de sociologie* 19 (1978), p. 42. For a somewhat different version of this, in English, see "From 'corps' to 'profession': The emergence and definition of industrial engineering in modern France," in *The Organization of Science and Technology in France, 1808–1914,* ed. R. Fox and G. Weisz (Cambridge University Press, 1980).

7. Picon, *L'invention de l'ingénieur moderne,* pp. 591–592.

8. Shinn, "Des corps de l'état au secteur industriel," p. 42; Thépot, "Les ingénieurs du Corps des Mines," p. 55.

9. Aspects of France's social structure in the nineteenth century are examined in many works. Several examples available in English follow: Roger Magraw, *France 1815–1914: The Bourgeois Century* (Fontana, 1987); Theodore Zeldin, *France 1848–1945: Ambition and Love* (Oxford University Press, 1979); Louis Chevalier, *Laboring Classes and Dangerous Classes in Paris During the First Half of the Nineteenth Century* (Princeton University Press, 1973).

10. Alfred Beillard, *L'enseignement technique et les travailleurs* (Besançon: Imprimerie Franc-Comtoise, 1888), p. 9.

11. The overview of French technical education presented here leaves out regional schools, in part because of a lack of sufficient information about them. Charles R. Day examines some of these schools in "The development of higher primary and intermediate technical education in France, 1800 to 1870," *Historical Reflections/Réflexions historiques* 3 (1976), no. 2, pp. 49–67. See also "Ecole central lyonnaise pour l'industrie et le commerce," *Annales du génie civil* 6 (1867), pp. 592–596 and "Ecole industrielle et commerciale de Rouen," *Journal du génie civil* 14 (1847), pp. 509–516. See also Thérèse Charmasson, *L'enseignement technique de la Révolution à nos jours,* volume 1 (Economica, 1987). I have not considered the role of the Conservatoire des Arts et Métiers, because it did not have a program of technical training leading to a formal degree; on this institution see Jacques Payen, "The role of the Conservatoire Nationale des Arts et Métiers in the development of technical education up to the middle of the 19th century," *History and Technology* 5 (1988), pp. 95–138.

12. I use the name that corresponds to the period I am discussing, in part for the sake of accuracy and in part because each name of the school was associated with a different curriculum.

13. For a description of this school see Frederick B. Artz, *The Development of Technical Education in France, 1500–1850* (MIT Press, 1966), pp. 228–229.

14. Adeline Daumard, "Les élèves de l'Ecole Polytechnique de 1815 à 1848," *Revue d'histoire moderne et contemporaine* 5 (1958), July–September, pp. 226–234.

15. Georges Ribeill, *La révolution ferroviaire: la formation des compagnies des chemins de fer en France (1823–1870)* (Belin, 1993), p. 304.

16. Charles R. Day, *Education for the Industrial World: The Ecoles d'Arts et Métiers and the Rise of French Industrial Engineering* (MIT Press, 1987), pp. 78, 81–82, 84.

17. Shinn, "From 'corps' to 'profession'," p. 198; Andrew J. Butricia, "The Ecole Supérieure de Télégraphie and the beginnings of French electrical engineering education," *IEEE Transactions on Education* E30 (1987), no. 3, pp. 121–129; Terry Shinn, "Des sciences industrielles aux sciences fondamentales: La mutation de l'Ecole Supérieure de Physique et de Chimie," *Revue française de sociologie* 22 (1981), pp. 167–182; *Soixantenaire de l'Ecole Supérieure d'Electricité, 1894–1954* (Paris: Imprimerie Chaix, 1955).

18. For a perspective on the *grandes écoles* and the French engineering corps that is similar in several respects to that presented here see Peter Lundgreen's article "Engineering education in Europe and the U.S.A., 1750–1930: The rise to dominance of school culture and the engineering professions," *Annals of Science* 47 (1990), pp. 33–75.

19. Henry Barnard, *Systems, Institutions, and Statistics of Scientific Instruction, Applied to National Industries in Different Countries* (New York: E. Steiger, 1872), p. 414.

20. Terry Shinn, *Savoir scientifique et pouvoir social: L'Ecole Polytechnique, 1794–1914* (Paris: Presses de la Fondation Nationale des Sciences Politiques, 1980), pp. 185.

21. Picon, *L'invention de l'ingénieur moderne*, p. 409.

22. The French science faculties also began to take up this role during the last quarter of the nineteenth century. See Terry Shinn, "The French science faculty system, 1808–1914: Institutional change and research potential in mathematics and the physical sciences," *Historical Studies in the Physical Sciences* 10 (1979), pp. 271–332; George Weisz, *The Emergence of Modern Universities in France, 1863–1914* (Princeton University Press, 1983), pp. 162–195; Mary Jo Nye, *Science in the Provinces: Scientific Communities and Provincial Leadership in France, 1860–1930* (University of California Press, 1986).

23. Weiss, *The Making of Technological Man*, pp. 57–87, 245–256.

24. Day, *Education for the Industrial World*, p. 190.

25. Ibid., pp. 138–142.

26. Ibid., p. 141.

27. Alexander Dallas Bache, *Report on Education in Europe to the Trustees of the Girard College for Orphans* (Philadelphia: Lydia R. Bailey, 1839), p. 556.

28. Ibid., p. 559.

29. Ibid., pp. 542–561.

30. Christian Licoppe, "Physique et chimie à l'Ecole Polytechnique (1795–1850)," in *La France des X: deux siècles d'histoire*, ed. B. Belhoste et al. (Economica, 1995), pp. 273–281.

31. Barnard, *Systems, Institutions, and Statistics of Scientific Instruction*, pp. 410–411.

32. Michel Atten, "La physique en souffrance, 1850–1914," in *La formation polytechnicienne, 1794–1994*, ed. B. Belhoste et al. (Dunod, 1994), p. 241. Atten indicates that no laboratory work in physics was done between 1850 and 1880. However, Christian Licoppe ("Physique et chimie à l'Ecole Polytechnique," p. 278) says that in 1851 students did four experiments (*manipulations*) per semester.

33. Société des Amis de l'Ecole Polytechnique, *L'Ecole Polytechnique* (Gauthier-Villars, 1932), p. 90.

34. Picon, *L'invention de l'ingénieur moderne*, pp. 284–285.

35. Bache, *Report on Education in Europe*, pp. 561–562.

36. Ministère des Travaux Publics, *Programmes de l'enseignement intérieur de l'Ecole des Ponts et Chaussées* (Paris: Imprimerie Nationale, 1875), manuscript 19952, Ecole des Ponts et Chaussées, Paris.

37. Ministère des Travaux Publics, *Programmes de l'enseignement intérieur de l'Ecole des Ponts et Chaussées* (Corbeil: J. Crété, 1889), manuscript 24941, Ecole des Ponts et Chaussées, Paris.

38. Picon, *L'invention de l'ingénieur moderne*, pp. 469–505. Navier published his lectures as a textbook: *Résumé des leçons données à l'Ecole des Ponts et Chaussées sur l'application de la mécanique à l'établissement des constructions et des machines* (Paris: Firmin Didot, 1826). This work went through many editions. A particularly noted edition was edited by Navier's student Adhemar Barré de Saint-Venant: Navier, *Résumé des leçons données à l'École des Ponts et Chaussées sur l'application de la mécanique à l'établissement des constructions et des machines*, third edition (Paris: Dunod, 1864).

39. Picon, *L'invention de l'ingénieur moderne*, p. 512.

40. A. J. C. Defontaine, Nouvelle organisation de l'Ecole Royale des Ponts et Chaussées, 1839, Ecole des Ponts et Chaussées (henceforth EPC) manuscript 2208; Programmes de l'enseignement, 1867 and 1875, EPC manuscript 19952; Programmes de l'enseignement, 1888, EPC manuscript 24941; Admission des élèves externes aux cours de l'école, 1864, EPC manuscript 2209/c.98.

41. Ministère de l'Agriculture, du Commerce, et des Travaux Publics, *Programmes de l'enseignement intérieur de l'Ecole des Ponts et Chaussées* (Paris: Thunot, 1867), pp. 77–91, box 19952, manuscript collection, EPC.

42. Barnard, *Systems, Institutions, and Statistics of Scientific Instruction*, pp. 422–423.

43. Picon, *L'invention de l'ingénieur moderne*, pp. 396–397, 416–418.

44. Ministère des Travaux Publics, *Laboratoires de l'Ecole Nationale des Ponts et Chaussées* (Paris: Imprimerie Nationale, 1891).

45. Ministère des Travaux Publics, *Laboratoire et atelier expérimental du nouveau dépôt de l'Ecole des Ponts et Chaussées*, manuscript 137356, Bibliothèque Historique de la Ville de Paris. This document reproduces an article published in *Annales des Ponts et Chaussées* 1 (1871).

46. Ministère des Travaux Publics, *Laboratoires de l'Ecole Nationale des Ponts et Chaussées*.

47. The title of this course was "ateliers de constructions mécaniques."

48. Louis Aguillon, *L'Ecole des Mines de Paris: notice historique* (Paris: Dunod, 1889), pp. 228–237.

49. Bache, *Report on Education in Europe*, pp. 562–563.

50. "Laboratoire des essais à l'Ecole des Mines," *Revue industrielle* 3 (1873), p. 379. I do not know to what extent this laboratory was used by students, but in the years 1868–1872 nearly 600 analyses were carried out per year on average.

51. Barnard, *Systems, Institutions, and Statistics of Scientific Instruction*, pp. 463–474; Weiss, *Making of Technological Man*, p. 170. Weiss discusses the curriculum of the *Ecole Centrale* as it developed up to about 1840. Barnard reviews the school's curriculum for the years 1837 and 1868.

52. Weiss, *Making of Technological Man*, pp. 123–145; Barnard, *Systems, Institutions, and Statistics*, pp. 464–474; "Rapport fait au conseil de perfectionnement de l'Ecole Centrale des Arts et Manufactures, le 12 Juillet 1830," *Annales de l'industrie française et étrangère* 6 (1830), pp. 190–198. The latter two sources contain relevant information about student projects and laboratory work that is not discussed by Weiss.

53. Barnard, *Systems, Institutions and Statistics*, pp. 463–474.

54. Day, *Education for the Industrial World*, pp. 71–74, 78–79, 96–99 ; Barnard, *Systems, Institutions, and Statistics*, pp. 451–458.

55. Barnard, *Systems, Institutions, and Statistics*, pp. 458–460.

56. Day, *Education for the Industrial World*, pp. 69–107; "Renseignements sur les écoles professionnelles en France: Ecoles Impériales d'Arts et Métiers," *Annales du génie civil* 3 (1864), pp. 582–586.

57. For an overview of changes in the French engineering profession in the first half of the nineteenth century, see John Weiss, "Bridges and barriers: Narrowing access and changing structure in the French engineering

profession, 1800–1850," in *Professions and the French State, 1700–1900*, ed. G. Geison (University of Pennsylvania Press, 1984), pp. 15–65.

58. Guillerme, *Corps à corps sur la route*, pp. 83–109, 125–160.

59. Ibid., pp. 83–86.

60. Day, *Education for the Industrial World*, p. 192

61. Eugène Flachat, "Discours—Société des Ingénieurs Civils, Séance du 10 janvier 1862," *Annales du génie civil* 1, part 2 (1862), p. 7.

62. Proposition de loi portant réorganisation du Corps des Ponts et Chaussées, 1881, EPC manuscript 15842/c.836, pp. 4, 8. On the functions of the *conducteurs* see also A. Corréard, "De la nécessité de réorganiser le corps des Ingénieurs des Ponts et Chaussées," *Journal du génie civil* 15 (1847), p. 94; Assemblée Nationale, Rapport fait par le Citoyen Stourm, 17 Novembre 1848, EPC manuscript 2156/c.96, pp. 15–16; Guillerme, *Corps à corps sur la route*, pp. 105–106.

63. An *arrondisement* is a subdivision of a municipality.

64. Observations . . . par les Inspecteurs Généraux et Divisionnaires des Ponts et Chaussées . . . relatif à un nouveau mode de recrutement des ingénieurs, 1848, EPC manuscript 2153/c.97.

65. Corréard, "De la nécessité de réorganiser le corps des Ingénieurs des Ponts et Chaussées," p. 111. See also Guillerme, *Corps à corps sur la route*, pp. 139–140.

66. "Considerations sur les conducteurs," *Journal du génie civil* 10 (1831), p. 26.

67. Ibid., pp. 25–30; Corréard, "De la nécessité de réorganiser le corps des Ingénieurs des Ponts et Chaussées," pp. 92–111; James M. Oliver, The Corps des Ponts et Chaussées, 1830–1848 (Ph.D. dissertation, University of Missouri, 1967), pp. 70–77; Citoyen Stourm; "Administration des Ponts et Chaussées," *Journal de l'industriel et du capitaliste* 3 (1837), January, p. 58; Porée, Notes sur l'organisation du Corps . . . des Ponts et Chaussées, 1948, EPC manuscript 2154/c. 96; "Nouveau mode d'admission pour le grade de conducteur des Ponts-et-Chaussées," *Journal des chemins de fer* 6 (1847), pp. 621–623. See also a further group of documents concerning proposed laws to allow conducteurs to be admitted into the corps hierarchy: EPC manuscripts. 2151/c.94, 2153/c.97, 2155/c.97, 2158/c.97 2159/c.97, 2161/c.97.

68. EPC manuscripts. 2167/c.97, 15842/c.836, 10184/c.550; "Décret réglant les conditions d'admission des conducteurs dans le corps des Ingénieurs des Ponts et Chaussées," *La propagation industrielle* 3 (1868), pp. 80–81; A. Brunot and R. Coquand, *Le Corps des Ponts et Chaussées* (Paris: Editions du Centre National de la Recherche Scientifique, 1982), pp. 241–245, 296–302, 343–344, 385–394, 433–435, 695–698, 715–719, 725–732. On the struggle between

conducteurs and corps engineers see also Guillerme, *Corps à corps sur la route;* John H. Weiss, "The lost baton: The politics of intraprofessional conflict in nineteenth-century French engineering," *Journal of Social History* 16 (1982), no. 1, pp. 1–19. Weiss's data show that the *conducteurs* tended to be recruited from lower social strata than the corps engineers.

69. Day, *Education for the Industrial World*, p. 194.

70. Georges Ribeill, *La révolution ferroviaire: la formation des compagnies des chemins de fer en France (1823–1870)* (Belin, 1993), p. 310.

71. Quoted from James M. Edmonson, From Mécanicien to Ingénieur: Technical Education and the Machine-Building Industry in Nineteenth-Century France (Ph.D. dissertation, University of Delaware, 1981), pp. 400–401. A brief discussion of Poulot's career is presented in Day, *Education for the Industrial World*, pp. 94–96.

72. Shinn, *Savior scientifique et pouvoir social*, pp. 36–37, 94–97, 185.

73. Dominique Barjot, "Les entrepreneurs polytechniciens: l'exemple des travaux publics (1883–1974)," in *La France des X: deux siècles d'histoire*, ed. B. Belhoste et al. (Economica, 1995).

74. Edmonson's data must be used with care, because they were compiled mainly from the files of those who were awarded the French Legion of Honor. This meant that large and well-known machine shops were probably over-represented in his listing. It was mainly the largest shops that drew *centraliens*. Thus, Edmonson's data may overstate the role of *centraliens* in the entire machine-building sector; nevertheless, they show clearly the small role played by graduates of the Ecole Polytechnique and the *écoles d'application*. See Edmonson, From Mécanicien to Ingénieur, pp. 402–404, 585.

75. Day, *Education for the Industrial World*, pp. 106–107; "Des ingénieurs en France," *Journal des chemins de fer* 4 (1845), pp. 316–317; "Note sur la nécessité de former en France une société portant la dénomination: Institut des Ingénieurs Civils de France," *Journal du génie civil* 13 (1846), pp. 479–487; *Observations . . . par la Société des Ingénieurs Civils sur le projet de décret relatif au mode de recrutement du Corps des Ponts et Chaussées* (Paris: Imprimerie centrale de Napoléon Chaix, 1848); "Les ingénieurs de l'état et les ingénieurs civils," *Chronique industrielle* 2 (1879), February 16, pp. 41–43; Ribeill, *La révolution ferroviaire*, pp. 323–334.

76. Emile Thomas, *Histoire des ateliers nationaux* (Paris: Michel Lévy, 1848), pp. 42–43.

77. "Société des Ingénieurs Civils," *La propagation industrielle* 3 (1868), p. 28.

78. On the history of this society see Bruno Jacomy, "A la recherche de sa mission: la Société des Ingénieurs Civils," *Culture technique*, no. 12 (March 1984), pp. 209–219.

79. Weiss, *The Making of Technological Man*, pp. 85–86.

80. Charles Rodney Day, "Des ouvriers aux ingénieurs?" *Culture technique,* no. 12 (March 1984), pp. 281–291.

81. Georges Ribeill, "Les associations d'anciens élèves d'écoles d'ingénieurs des origines à 1914," *Revue française de sociologie* 27 (1986), pp. 317–338.

82. Terry Shinn, "From 'corps' to 'profession': The emergence and definition of industrial engineering in modern France," in *The Organization of Science and Technology in France, 1808–1914,* ed. R. Fox and G. Weisz (Cambridge University Press, 1980), pp. 203–204.

83. Day, *Education for the Industrial World,* pp. 106–107.

84. Shinn, "From 'corps' to 'profession'," p. 204. Unfortunately, little historical research has been done on another French institution that perhaps also served as something of a professional society in the nineteenth century: the Société d'Encouragement pour l'Industrie Nationale. On the origins and early history of this organization see Pietro Redondi, "Nation et entreprise: La Société d'Encouragement pour l'Industrie Nationale, 1801–1815," *History and Technology* 5 (1988), pp. 193–222.

Chapter 8

1. Alan Brinkley et al., *American History: A Survey,* eighth edition (McGraw-Hill, 1991), p. xxxvii; Dominique and Michèle Frémy, *Quid 1986* (Paris: Robert Laffont, 1985), pp. 504, 514; B. R. Mitchell, *European Historical Statistics, 1750– 1975,* second revised edition (Facts on File, 1981), p. 30.

2. Frederick Jackson Turner, *The Frontier in American History* (1920; Holt, Rinehart and Winston, 1962), pp. 2–3.

3. These are only two dimensions of the many critiques of the Turner thesis. For an overview of the debates see William Cronon, "Revisiting the vanishing frontier: The legacy of Frederick Jackson Turner," *Western Historical Quarterly* 18 (1987), pp. 157–176. See also Richard Slotkin, *The Fatal Environment: The Myth of the Frontier in the Age of Industrialization, 1800–1890* (Atheneum, 1985); *American Frontier and Western Issues: A Historiographical Review,* ed. R. Nichols (Greenwood, 1986); Walter Nugent, *Structures of American Social History* (Indiana University Press, 1981). For an important recent analysis of the interactions between urban and rural areas in the expanding American west, see William Cronon, *Nature's Metropolis: Chicago and the Great West* (Norton, 1991).

4. Another major limitation of Turner's thesis is that it does not adequately recognize the extent to which frontier societies are shaped by the economic and political systems of the states in which they are located or to which they are most closely linked. It is not merely expansion into a wilderness that explains the character of a frontier society, but rather expansion within a particular social, economic and political context. In the slave states, for example, westward expansion occurred within an economic and legal context that transplanted the slave system and the values that went with it into new areas.

American slavery did entail a system of rigid social hierarchy, and this was reproduced in the process of westward expansion. See Eugene D. Genovese, *The Political Economy of Slavery: Studies in the Economy and Society of the Slave South* (Vintage Books, 1965); Nugent, Structures of American Social History, pp. 94–103.

5. James Veech, *The Monongahela of Old* (Pittsburgh: n.p. 1858–1892), pp. 149–150.

6. Omer Buyse, *Méthodes américaines d'éducation générale et technique* (Paris: Charleroi, 1908), pp. 12–13. See also Alexis de Tocqueville, *Democracy in America* (Knopf, 1945), volume II, pp. 161–162. Carl Bridenbaugh discusses the comparatively high social standing of artisans in colonial America on pp. 155–181 of *The Colonial Craftsman* (University of Chicago Press, 1950).

7. Margo Stein, Working to Rule: The Social Origins of French Engine Drivers (Ph.D. dissertation, Harvard University, 1977), pp. 119–120, 382.

8. Theodore D. Weld, *First Annual Report of the Society for Promoting Manual Labor in Literary Institutions* (New York: S. W. Benedict, 1833), quoted in Charles Alpheus Bennett, *History of Manual and Industrial Education up to 1870* (Peoria: Manual Arts Press, 1926), p. 191.

9. James Earle Ross, *Democracy's College: The Land-Grant Movement in the Formative Stage* (Arno, 1969), pp. 21–26.

10. Buyse, *Méthodes américaines*, p. 658.

11. A sense of the character and activities of the Army Corps of Engineers can be gained from Martin Reuss's article "Andrew A. Humphreys and the development of hydraulic engineering: Politics and technology in the Army Corps of Engineers, 1850–1950," *Technology and Culture* 26 (1985), pp. 1–33. See also "Rodman, Thomas J.," *Dictionary of American Biography*. On the Engineer Corps of the U.S. Navy see Monte Calvert, *The Mechanical Engineer in America, 1830–1910* (Johns Hopkins University Press, 1967), passim. See also "Isherwood, Benjamin Franklin" in *Dictionary of American Military Biography* and the chapter on Isherwood in *Captains of the Old Steam Navy: Makers of the American Naval Tradition, 1840–1880*, ed. J. Bradford (Naval Institute Press, 1986).

12. During the second half of the nineteenth century, graduates of the Ecole Centrale came to play a larger role in the mechanical and metallurgical industries. This was linked with a rise in the status of this industrial sector. See James Edmonson, From Mécanicien to Ingénieur: Technical Education and the Machine Building Industry in Nineteenth-Century France (Ph.D. dissertation, University of Delaware, 1981), pp. 373–438.

13. Calvert, *Mechanical Engineer in America*, passim.

14. "William Sellers," *Journal of the Franklin Institute* 159 (1905), May, pp. 365–381; "Sellers, William," *Dictionary of American Biography;* Calvert, *Mechanical Engineer in America*, pp. 8–10.

15. "Sellers, Coleman," *Dictionary of American Biography*. See also "Sellers, Coleman," *National Cyclopaedia of American Biography*.

16. Coleman Sellers, "The Worcester County Free Institute of Industrial Science," *Journal of the Franklin Institute* 65 (1873), no. 4. p. 221.

17. Jeanne McHugh, *Alexander Holley and the Makers of Steel* (Johns Hopkins University Press, 1980), pp. 1–42; Calvert, *Mechanical Engineer in America*, passim.

18. Edwin T. Layton, Jr., *The Revolt of the Engineers: Social Responsibility and the American Engineering Profession* (Case Western Reserve University, 1971), pp. 25–46.

19. Layton, *Revolt of the Engineers*, pp. 25–52; Bruce Sinclair, *A Centennial History of the American Society of Mechanical Engineers, 1880–1980* (University of Toronto Press and ASME, 1980); A. Michael McMahon, *The Making of a Profession: A Century of Electrical Engineering in America* (IEEE Press, 1984), pp. 31–43.

20. Layton, *Revolt of the Engineers*, pp. 29–30.

21. "Engineering schools of the United States XXVII," *Engineering News and American Railway Journal* 28 (October 20, 1892), p. 375.

22. "Engineering schools of the United States XXV," *Engineering News and American Railway Journal* 28 (October 13, 1892), p. 354.

23. Henry F. May, *The Enlightenment in America* (Oxford University Press, 1976), pp. 344, 348.

24. Paul Kurtz, "American philosophy," in *Encyclopedia of Philosophy*.

25. "Engineering Schools of the United States III," *Engineering News and American Railway Journal* 27 (1892), April 2, p. 318.

26. Daniel H. Calhoun, *The American Civil Engineer, Origins and Conflict* (Harvard University Press, 1960), pp. 24–53.

27. Calvert, *Mechanical Engineer in America*, passim; Layton, *Revolt of the Engineers*, p. 4; Leonard S. Reich, *The Making of American Industrial Research: Science and Business at GE and Bell, 1876–1926* (Cambridge University Press, 1985), p. 18. According to Reich college-trained engineers were in the majority in the U.S. by the mid 1880s; according to Layton they remained in the minority until the twentieth century.

28. These numbers do not include military engineers; in the case of France they include only those who entered the Corps des Ponts et Chaussées (about 20%) or the Corps des Mines (about 5%); they do not include the approximately 5% of *polytechniciens* who became civil engineers in the private sector.

29. "The Engineering Schools of the United States III," *Engineering News and American Railway Journal* 27 (1892), April 2, pp. 318–319; Antoine Picon, *L'Invention de l'ingénieur moderne: L'Ecole des Ponts et Chaussées, 1747–1851* (Paris:

Ecole des Ponts et Chaussées, 1992), pp. 733–738; Terry Shinn, *Savoir scientifique et pouvoir social: L'Ecole Polytechnique, 1794–1914* (Paris: Presses de la Fondation Nationale des Sciences Politiques, 1980), p. 185. Calhoun's *American Civil Engineer* gives statistics for the supply of civil engineers produced by West Point that are much lower than those given in the first article noted above. Calhoun claims that the number of practicing civil engineers trained at West Point was only 15 in 1830.

30. Edward S. Farrow, *West Point and the Military Academy*, second edition (New York: Orange Judd, 1879), pp. 17–18; Edward S. Farrow, *West Point and the Military Academy*, third edition (New York: Military-Naval Publishing Co., 1895), pp. 23–25.

31. Roswell Park, *A Sketch of the History and Topography of West Point and the U.S. Military Academy* (Philadelphia: Henry Perkins, 1840), pp. 101–102.

32. West Point's curriculum cannot be compared with the Ecole Polytechnique's alone, because the former covered four years and constituted a complete engineering education whereas the latter covered two years and represented only a partial education. Therefore, I have compared it with the Ecole Polytechnique's curriculum plus the curriculum of one of the *écoles d'application*—specifically, the Ecole des Ponts et Chaussées, which was a three-year school.

33. "Engineering Schools of the United States IV," *Engineering News and American Railway Journal* 27 (1892), April 9, pp. 342–343.

34. Park, *History and Topography of West Point*, pp. 104–106. See also James L. Morrison Jr., *"The Best School in the World": West Point, the Pre-Civil War Years, 1833–1866* (Kent State University Press, 1986), pp. 87–101, 160–163.

35. Morrison, *"Best School in the World,"* pp. 93–98. In the period 1854–1861 West Point was reorganized to have a five-year curriculum in which greater emphasis was placed on military engineering and applied military training, including artillery, gunnery, ordnance, equitation, and marksmanship. The military engineering course included still more extensive fieldwork, and the gunnery course included a period of summer fieldwork in which students "manufactured gunpowder, rockets, and flares in the ordnance laboratory." (ibid., pp. 114–125; quote from p. 119)

36. Alexander Dallas Bache, who disapproved of the fact that *polytechniciens* did not have to do such work (see chapter 7 above), was a West Point graduate.

37. Thomas E. Griess, Dennis Hart Mahan: West Point Professor and Advocate of Military Professionalism, 1830–1871 (Ph.D. dissertation, Duke University, 1968), pp. 354, 356.

38. Hugh T. Reed, *Cadet Life at West Point* (Chicago: the author, 1896), pp. 109, 111.

39. Park, *History and Topography of West Point*, p. 119.

40. Ibid., p. 120.

41. Morrison, *"Best School in the World,"* pp. 61–62, 155–159.

42. Terry S. Reynolds, "The education of engineers in America before the Morrill Act of 1862," *History of Education Quarterly* 32 (1992), no. 4, pp. 465–465; James Gregory McGivern, *First Hundred Years of Engineering Education in the United States (1807–1907)* (Gonzaga University Press, 1960), pp. 42–45.

43. Brooke Hindle, *The Pursuit of Science in Revolutionary America, 1735–1789* (University of North Carolina Press, 1956).

44. The Franklin Institute, founded in 1824 in Philadelphia and devoted to "the mechanic arts," organized public lectures, systematic technological research, and a drawing school, and became a leading forum for American technologists. Its *Journal of the Franklin Institute*, established in 1826, became one of the most prestigious American engineering journals in the nineteenth century. See Bruce Sinclair, *Philadelphia's Philosopher Mechanics: A History of the Franklin Institute, 1824–1865* (Johns Hopkins University Press, 1974).

45. Carl Bode, *The American Lyceum: Town Meeting of the Mind* (Oxford University Press, 1974).

46. Bennett, *History of Manual and Industrial Education,* pp. 326–327.

47. Quoted from Carl Bode, *The American Lyceum: Town Meeting of the Mind* (Oxford University Press, 1956), p. 11.

48. Ibid., p. 12.

49. Ibid., pp. 121, 185–187; Dirk J. Struik, *Yankee Science in the Making* (Collier, 1962), pp. 268–269.

50. Quoted from Bennett, *History of Manual and Industrial Education,* p. 349.

51. Quoted from ibid., p. 348.

52. Asa Fitch diary, volume E (November 1, 1826–July 23, 1827), p. 26, Sterling Memorial Library, Yale University. I thank my former student Ann M. Plane for calling my attention to this source. The following account of Eaton and the Rensselaer Institute draws on her senior essay, Popular Science, Industry, and Social Class: Amos Eaton and the Rise of Technical Education, 1820–1840 (Yale University, 1985).

53. Quoted from Plane, Popular Science, Industry, and Social Class, p. 29.

54. On Common Sense philosophy in American science see George H. Daniels, *American Science in the Age of Jackson* (Columbia University Press, 1968).

55. Quoted from John C. Greene, *American Science in the Age of Jefferson* (Iowa State University Press, 1984), pp. 179, 181.

56. Ibid., pp. 181–182.

57. Prospectus of the "Preparation Branch" of the Rensselaer School, 1826, reproduced in "Engineering Schools of the United States VI," *Engineering News and American Railway Journal* 27 (1892), April 21, p. 412.

58. Ibid.

59. Fitch diary, volume C (April 25–July 17, 1826), p. 11; volume D (July 18–October 31, 1826), pp. 21–58, 143.

60. Ray Palmer Baker, *A Chapter in American Education* (Scribner, 1924), pp. 24–25.

61. Plane, Amos Eaton and the Rise of Technical Education, pp. 10, 15.

62. Ibid., pp. 31–32.

63. "Engineering Schools of the United States XVI," *Engineering News and American Railway Journal* 28 (1892), August 4, p. 112.

64. Notice of Rensselaer Institute, 1835, reproduced in "Engineering Schools of the United States VI," *Engineering News and American Railway Journal* 27 (1892), April 21, p. 412.

65. Terry S. Reynolds, "The education of engineers in America before the Morrill Act of 1862," *History of Education Quarterly* 32 (1992), no. 4, pp. 459–482.

66. "Entrance requirements for engineering colleges," *Society for the Promotion of Engineering Education Proceedings* 4 (1896), pp. 103–104. (This journal will hereafter be cited as *SPEE Proceedings*.) The regionalism of American technical education can be seen in data compiled between 1923 and 1929 by the Society for the Promotion of Engineering Education. In a survey of more than 4,000 students from 32 institutions, they found that 83.7% went to an engineering school in their own state. See "A study of engineering students at the time of entrance to college," in *Report of the Investigation of engineering Education, 1923–1929*, volume I (Pittsburgh: Society for the Promotion of Engineering Education, 1930), pp. 188–189.

67. McGivern, *First Hundred Years of Engineering Education*, pp. 58–61.

68. "Engineering Schools of the United States XVI," *Engineering News and American Railway Journal* 28 (1892), August 4, p. 114; Lawrence P. Grayson, "A brief history of engineering education in the United States," *Engineering Education* 68 (1977), no. 3, p. 249. It is admittedly an oversimplification to lump all the polytechnics together, because several had peculiarities that made them different from Rensselaer. Stevens trained only mechanical engineers until the end of the nineteenth century; it even organized electrical engineering as a branch of mechanical engineering. WPI was unique in its emphasis on shop training, which effectively put students through a controlled apprenticeship, and this model was taken up by Rose Polytechnic. Moreover, some of the polytechnics, such as MIT, were top-flight schools that recruited widely,

whereas others, including Brooklyn Polytechnic, were second-rate schools that recruited only locally. Yet they all had formal degree programs and gave what they held to be a comprehensive engineering training. Before the end of the nineteenth century, for example, WPI had become a broad-based engineering school, with programs in civil and electrical engineering as well as mechanical engineering. (McGivern, *First Hundred Years,* pp. 98–102.)

69. Quoted from David F. Noble, *America by Design: Science, Technology, and the Rise of Corporate Capitalism* (Oxford University Press, 1977), p. 30.

70. Several links can be traced from advocates of the manual labor and lyceum movements to advocates of the land-grant college movement. One link already mentioned was the People's College movement. Another link was a plan for an "industrial university" formulated by Jonathan Baldwin Turner in 1850. In this university, Turner counseled, "let . . . experiments be made in all . . . interests of agriculture and mechanic or chemical art, mining, merchandise, and transportation by water and by land, and daily practical and experimental instruction given to each student in attendance in his own chosen sphere of research or labor in life. Especially let the comparative merits of all labor-saving tools, instruments, machines, engines, and processes be thoroughly and practically tested and explained. . . . " Turner was subsequently consulted in the formulation of the Morrill Act. (Ross, *Democracy's College,* pp. 37–38, 51–55; Bennett, *History of Manual and Industrial Education,* pp. 355–359, 366–373.)

71. "Engineering Schools of the United States XVI," *Engineering News and American Railway Journal* 28 (1892), August 4, p. 112

72. McGivern, *First Hundred Years,* pp. 132–133.

73. The U.S. Naval Academy, however, maintained its importance as a school for engineers.

74. Turner, *Frontier in American History,* p. 283.

75. Existing histories of American technical education have not drawn attention to this fact. For example, McGivern's *First Hundred Years of Engineering Education in America* largely ignores the significance of workshop training in American technical schools in the last half of the nineteenth century and does not adequately address the issue of the scale and role of laboratory training. The way McGivern lists school curricula, moreover, makes it difficult to determine which courses even had laboratory or workshop content. Other historical overviews also suggest that laboratory training of American engineering students began to be implemented only in the second half of the nineteenth century, thus overlooking the early curricula of schools such as Rensselaer. (See, e.g., Grayson, "Brief history," p. 252.)

76. Bennett, *Manual and Industrial Education,* pp. 360–362.

77. Samuel C. Prescott, *When MIT Was "Boston Tech," 1861–1916* (Technology Press, 1954), pp. 54–60, 98.

78. These shops were initially run by master mechanic John Sweet, who taught practical shop courses to the engineering students. See Calvert, *Mechanical Engineer in America*, p. 87. Workshop training began to be specifically listed in Cornell's curriculum in 1874. See William Wickenden, "A comparative study of engineering education in the United States and in Europe," in *Report on the Investigation of Engineering Education, 1923–1929*, p. 820.

79. Calvert, *Mechanical Engineer in America*, passim; "The technical education of the mechanical engineer," *Journal of the Franklin Institute* 102 (1876), September, pp. 145–150. In the twentieth century the amount of shop work began to decline, particularly with the growth of cooperative programs in which students spent time working in industry. See SPEE, *Investigation of Engineering Education, 1923–1929*, volume I, bulletins 11 and 12. For the history of one of those cooperative programs see W. Bernard Carlson, "Academic entrepreneurship and engineering education: Dugald C. Jackson and the MIT-GE Cooperative Engineering Course, 1907–1932," *Technology and Culture* 29 (1988), no. 3, pp. 536–567.

80. Buyse, *Méthodes américaines*, p. 9. Buyse felt that shop work exemplified the spirit of American education, in which the training of the mind and the training of the hand were rarely separated. In American schools, he explained, "physical action precedes or accompanies the act of thinking; the most abstract branches of learning, for us, are presented in a material and concrete form, and require, in order to be assimilated, as much manual dexterity as intellectual vigor. . . . The supreme form of action, manual training, which is universally practiced in [American] schools, is an exercise in moral strength. All the education allies physical, muscular effort with the assimilation of ideas."

81. Harry E. Smith, "Drawing, shop work and engineering laboratory practice in engineering colleges," *Society for the Promotion of Engineering Education Proceedings* 7 (1899), pp. 176–180. With WPI excluded, the average is 678 hours.

82. C. L. Crandall, "A general course in civil engineering," *Society for the Promotion of Engineering Education Proceedings* 3 (1895), pp. 268–274.

83. W. H. Schuerman, "Is not too much time given to merely manual work in the shops?" *Society for the Promotion of Engineering Education Proceedings* 4 (1896), p. 341.

84. "Engineering Schools of the United States XXXIV," *Engineering News and American Railway Journal* 28 (1892), December 1, p. 519.

85. Frances C. Caldwell, "A comparative study of the electrical engineering courses given at different institutions," *SPEE Proceedings* 7 (1899), pp. 127–129.

86. Harry E. Smith, "Drawing, shop work and engineering laboratory practice in engineering colleges," *SPEE Proceedings* 7 (1899), pp. 176–180.

87. Charles H. Benjamin, "The true place of drawing and shopwork in engineering colleges," *SPEE Proceedings* 3 (1895), p. 131.

88. C. W. Marx, "Amount and kind of shop work required in a mechanical engineering course," *SPEE Proceedings* 2 (1894), pp. 207–208.

89. Ibid., p. 132.

90. John H. Barr, "Teaching machine design," *SPEE Proceedings* 2 (1894), pp. 243–234.

91. L. S. Randolph, "The economic element in technical education," *SPEE Proceedings* 3 (1895), pp. 183–184.

92. Henry S. Munroe, "Courses in mining engineering," *Engineering News and American Railway Journal* 28 (1892), November 3, p. 416. On the development of various sorts of fieldwork in American technical education see also McGivern, *First Hundred Years,* pp. 127–128.

93. Buyse, *Méthodes américaines,* pp. 668–669.

94. Robert H. Thurston, "The modern mechanical laboratory especially as in process of evolution in America," *SPEE Proceedings* 8 (1900), pp. 337–338.

95. "The Purdue University locomotive on grades," *Engineering News and American Railway Journal* 28 (1892), September 22, p. 278. For some other examples of American technical-school laboratories see the illustrated article by Robert Rosenberg, "The origins of EE education: A matter of degree," *IEEE Spectrum* 21 (1984), no. 7, pp. 60–68.

96. Prescott, *When M.I.T. Was "Boston Tech,"* pp. 132–139.

97. Buyse, *Méthodes américaines,* pp. 668–681.

98. The laboratory facilities of the French Science Faculties are discussed in Mary Jo Nye's *Science in the Provinces: Scientific Communities and Provincial Leadership in France, 1860–1930* (University of California Press, 1986). See also Ministère des Travaux Publics, *Laboratoires de l'Ecole Nationale des Ponts et Chaussées: note sur leurs origines, leurs installations, les appareils et méthodes d'essai employés et leurs travaux* (Paris: Imprimerie Nationale, 1891).

99. American engineering educators in the Society for the Promotion of Engineering Education extensively discussed the role and importance of laboratories. See, for example, Gaetano Lanzo, "The organization and conduct of engineering laboratories," *SPEE Proceedings* 2 (1894), pp. 149–161; Lanzo, "The class room and the laboratory in their mutual adjustment to the end of the most efficient instruction," *SPEE Proceedings* 6 (1898), pp. 37–48; R. C. Carpenter, "Engineering laboratory courses," *SPEE Proceedings* 5 (1897), pp. 26–39.

100. Buyse, *Méthodes américaines,* p. 10.

101. J. B. Johnson, "Discussion," *SPEE Proceedings* 6 (1898), p. 43; George F. Swain, "Comparison between American and European methods in engineering education," *SPEE Proceedings* 1 (1894), pp. 85, 96–97, 103; Robert H.

Thurston, "The modern mechanical laboratory especially as in process of evolution in America," *SPEE Proceedings* 8 (1900), pp. 331–358.

102. De Volson Wood, presidential address, *SPEE Proceedings* 2 (1894), p. 29.

103. Harry E. Smith, "Drawing, shop work and engineering laboratory practice in engineering colleges," *SPEE Proceedings* 7 (1899), p. 179.

104. Buyse, *Méthodes américaines*, pp. 661–664.

105. John H. Barr, "Teaching machine design," *SPEE Proceedings* 2 (1894), p. 239. Experimental work in American technical schools in the twentieth century is examined in Bruce Seely's "Research, engineering, and science in American engineering colleges: 1900–1960," *Technology and Culture* 34 (1993), no. 2, pp. 344–386.

Chapter 9

1. Terry Shinn ("From 'corps' to 'profession': The emergence and definition of industrial engineering in modern France," in *The Organization of Science and Technology in France, 1808–1914*, ed. R. Fox and G. Weisz (Cambridge University Press, 1980), pp. 187, 207) states that "the functions exercised by engineers in the state corps were almost totally lacking in scientific and engineering content," and that "their work did not entail research or the development of new techniques." Shinn argues that the state engineers were little more than elite administrators. In other cases they have been portrayed more as scientists; some entries in the *Dictionary of Scientific Biography* fall into this category. Neither of these characterizations captures their role adequately, however.

2. For an overview of the activities of Ponts et Chaussées engineers see A. Brunot and R. Coquand, *Le Corps des Ponts et Chaussées* (Paris: Editions du Centre National de la Recherche Scientifique, 1982). My analysis is also based on research on the corps engineers' role in the development of suspension bridge technology in France. In this context I have often found that their work—even their administrative work—demanded up-to-date technical knowledge. For example, when a suspension bridge collapsed at Angers in 1850, members of the corps (who were part of an investigative committee) helped to prepare a report evaluating the probable causes of the collapse—yet this was clearly a task that could not have been accomplished by individuals with no engineering experience. See Rapport de la commission d'enquête . . . pour rechercher les causes et les circonstances qui ont amené la chute du pont suspendu de la Baisse-Chaine, 1850, manuscript 3130, Ecole des Ponts et Chaussées).

3. For details of the administrative tasks of engineers in the Corps des Mines see André Thépot, "Les ingénieurs du Corps des Mines," *Culture Technique*, no. 12 (March 1984), pp. 55–61.

4. See, e.g., Edwin T. Layton, Jr., *The Revolt of the Engineers: Social Responsibility and the American Engineering Profession* (Case Western Reserve University Press, 1971); Alfred D. Chandler, *The Visible Hand: The Managerial Revolution in*

American Business (Harvard University Press 1977); Thomas P. Hughes, *Networks of Power* (Johns Hopkins University Press, 1983).

5. Brunot and Coquand, *Le Corps des Ponts et Chaussées;* Société des Amis de l'Ecole Polytechnique, *L'Ecole Polytechnique* (Gauthier-Villars, 1932).

6. Brunot and Coquand, *Le Corps des Ponts et Chaussées,* p. 545.

7. Institut de France, *Index Biographique des Membres et Correspondants de l'Académie des Sciences* (Paris, 1954). The careers of members in the mechanics section of the Académie des Sciences were checked through a number of sources, including *Dictionary of Scientific Biography;* Brunot and Coquand, *Le Corps des Ponts et Chaussées;* and Société des Amis de l'Ecole Polytechnique, *L'Ecole Polytechnique.*

8. Brunot and Coquand, pp. 471–473; T. M. Charlton, *A History of Theory of Structures in the Nineteenth Century* (Cambridge University Press, 1982), pp. 51–57, 94–105.

9. *L'Ecole Polytechnique,* pp. 442–444; Brunot and Coquand, pp. 459–461, 481–483.

10. *L'Ecole Polytechnique,* pp. 404–406.

11. Ibid., pp. 408–409, 410–411.

12. Ibid., pp. 377–379.

13. Ibid., pp. 331–333, 144–145, 392–397. On the work of Piobert see also Alexandre Herlea, "Préliminaires à la naissance des laboratoires publics de recherche industrielle en France," *Culture Technique,* no. 18 (March 1988), p. 230.

14. *L'Ecole Polytechnique,* pp. 159–161; Stephen P. Timoshenko, *History of Strength of Materials, with a Brief Account of the History of Theory of Elasticity and Theory of Structures* (1953; Dover, 1983), pp. 244–245; Isaac Todhunter and Karl Pearson, *A History of the Theory of Elasticity and of the Strength of Materials from Galilei to Lord Kelvin* (1893; Dover, 1960), volume 2, part 1, pp. 330–350. The latter contains a detailed discussion of Phillips's theoretical analysis of springs.

15. Alphonse Léon, *Des changements projetés dans le mode de recrutement du Corps des Ponts et Chaussées* (Paris: Carilian-Goeury, 1849), pp. 9–10.

16. See Observations . . . par les Inspecteurs Généraux et Divisionnaires des Ponts et Chaussées . . . relatif à un nouveau mode de recrutement des ingénieurs, 1848, manuscript 2153/c.97, Ecole des Ponts et Chaussées, pp. 4–6.

17. Assemblée Nationale, Rapport Fait par le Citoyen Stourm, 17 Novembre 1848, manuscript 2156/c.96, Ecole des Ponts et Chaussées, p. 9. For other examples of how the *ingénieurs civils* regarded theory see "De l'organisation des conducteurs des Ponts et Chaussées," *Journal du génie civil* 15 (1847),

pp. 85–91; Eugène Karr, "Sur les privilèges," *Annuaire de la Société des Anciens Elèves des Ecoles d'Arts et Métiers* 2 (1849), pp. 49–54.

18. Paul de Rousiers, *La vie américaine: ranches, fermes, et usines* (Paris: Firmin-Didot, [n.d., ca. 1896]), p. 275.

19. Charles Coulston Gillispie, *The Montgolfier Brothers and the Invention of Aviation* (Princeton University Press, 1983), pp. 159–177. Seguin's work is also examined in two recent publications by Michel Cotte: "Seguin et Cie (1806–24): du négoce familial de drap à la construction du pont suspendu de Tournon-Tain," *History and Technology* 6 (1988), pp. 95–144; "Le système technique des Seguin en 1824–25," *History and Technology* 7, no. 2 (1990), pp. 119–147. See also Dominique Barjot, "From Tournon to Tancarville: The contribution of French civil engineering to suspension bridge construction, 1824–1959," *History and Technology* 6 (1988), pp. 177–201.

20. Marc Seguin, *De l'influence des chemins de fer et de l'art de les tracer et de les construire* (Paris: Carilian-Goeury, 1839), pp. ix–x.

21. Ibid., pp. 28–29. Seguin made similar criticisms in an even more hostile tone in his *Mémoire sur les causes et sur les effets de la chaleur, de la lumière, et de l'électricité* (Paris: A. Tramblay, 1865), pp. 15–18. The latter was specifically an attack against abstruse mathematics in physics, but it clearly reflects his sentiments about mathematics in engineering.

22. Marc Seguin, *Des ponts en fil de fer*, second edition (Paris: Bachelier, 1826), pp. 4–7, 18–19, 33–34, 75–105.

23. Ibid., p. 114.

24. Ibid., pp. 72–73; Dominique Amouroux and Bertrand Lemoine, "L'Age d'or des ponts suspendus en France," *Annales des Ponts et Chaussées*, n.s., no. 19 (1981), pp. 53–63. The kind of approach taken by Seguin was mirrored later in the nineteenth century by yet another bridge designer, François Hennebique (1843–1921), who began his career as an apprentice to a stonemason. In 1892 Hennebique patented a system for building with reinforced concrete, and the company he established built thousands of reinforced concrete bridges, factories, water towers, and other structures. David Billington writes in his book *The Tower and the Bridge: The New Art of Structural Engineering* (Basic Books, 1983) (pp. 148–151) that "Hennebique and his colleagues . . . did not feel bound by formulas; as forms proved successful they experimented further, and faced with economic competition, tried lighter and lighter forms."

25. Robert Fox, "Contingency or mentality? Technical innovation in France in the age of science-based industry," in *Technological Education—Technological Style*, ed. M. Kranzberg (San Francisco Press, 1986), p. 62; John H. Weiss, *The Making of Technological Man: The Social Origins of French Engineering Education* (MIT Press, 1982), p. 76; Charles R. Day, "The making of mechanical engineers in France: The Ecoles d'Arts et Métiers, 1803–1914," *French Historical Studies* 10 (1978), no. 3, p. 443.

26. Gillispie, *Montgolfier Brothers,* p. 164.

27. *Annales de l'industrie française et étrangère* 3 (1829), pp. 64–68.

28. Brunot and Coquand, *Le Corps des Ponts et Chaussées,* p. 407.

29. Ibid., pp. 405 and 407.

30. Quoted from Gillispie, *Montgolfier Brothers,* pp. 176–177.

31. Marc Seguin, letter to Henry Desgrand, September 9, 1845, Marc Seguin manuscripts, Archives of the Academy of Sciences, Paris. The Academy of Sciences possesses photocopies of the letters. The originals are in the possession of a descendant of Henry Desgrand. Portions of the letters have been published in Claude Bechetoille's book *Marc Seguin: grand savant méconnu* (Largentière: Humbert & Fils, 1975).

32. Marc Seguin to Henry Desgrand, April 16, 1838, February 2, 1839, and September 9, 1845, Archives of Academy of Sciences, Paris.

33. Marc Seguin to Henry Desgrand, December 7, 1860, Archives of Academy of Sciences, Paris.

34. Marc Seguin to Henry Desgrand, September 30, 1850, Archives of Academy of Sciences, Paris.

35. Jules Guillemin, "Benôit Fourneyron—notice biographique," *Bulletin de la Société de l'Industrie Minérale* 12 (1867), pp. 551–552.

36. Eugène Flachat, "Discours—Société des Ingénieurs Civils, séance du 10 janvier 1862," *Annales du génie civil* 1, part 2 (1862), p. 7.

37. James M. Edmonson, From Mécanicien to Ingénieur: Technical Education and the Machine-Building Industry in Nineteenth-Century France (Ph.D. dissertation, University of Delaware, 1981), pp. 178–181, 200–205; Charles R. Day, *Education for the Industrial World: The Ecoles d'Arts et Métiers and the Rise of French Industrial Engineering* (MIT Press, 1987), pp. 214–215.

38. Jacques Eugène Armengaud, "Préface du tome premier de la deuxième édition de la *Publication industrielle,*" in *Traité pratique des moteurs hydrauliques et à vapeur* (Paris: [by author], 1844), p. vii.

39. Jacques Eugène Armengaud, "Notice biographique sur Eugène Pihet," *Le génie industriel* 32 (1866), pp. 265–266. If Armengaud unnecessarily slighted Pihet for not being a Lagrange, he received similar treatment in turn. When Armengaud was awarded a medal by the Société d'Encouragement pour l'Industrie Nationale for his *Publication industrielle,* the citation stated that "although . . . the attention of the author might tend more toward application than toward theoretical discussions, . . . one can cite justly, with great praise, excellent articles that are particularly useful because they give more easily applicable results" (quoted from Edmonson, From "Mécanicien" to "Ingénieur," p. 180). Whether intended or not, the citation implied that Armengaud's journal would have been even more praiseworthy if it had

focused more on theoretical discussions than on applications—a curious point of view for an institution whose ostensible mission was to encourage practical, industrial development.

40. Armengaud, "Notice sur Eugène Pihet," pp. 265–275.

41. Ordonnance du roi portant création d'une Ecole de Mineurs à Saint-Etienne, 1816, manuscript 2198/c.96, Ecole des Ponts et Chaussées; Frederick B. Artz, *The Development of Technical Education in France, 1500–1850* (MIT Press, 1966), pp. 228–229.

42. Guillemin, "Benôit Fourneyron," pp. 533–562; B. Fourneyron, "Mémoire sur l'application en grand dans les mines et manufactures, des turbines hydrauliques ou roues à palettes courbes de Bélidor," *Bulletin de la Société d'Encouragement pour l'Industrie Nationale* 33 (1834), pp. 3–17, 49–61, 85–96.

43. Billington, *The Tower and the Bridge*, p. 64.

44. Bertrand Lemoine, *La tour de Monsieur Eiffel* (Gallimard, 1989), p. 24.

45. Quoted from Billington, *The Tower and the Bridge*, p. 65.

46. Navier, *De l'entreprise du pont des Invalides* (Paris: Firmin Didot, 1827), p. 15.

47. For a fuller analysis of Giffard's work see Eda Kranakis, "The French connection: Giffard's injector and the nature of heat," *Technology and Culture* 23 (1982), no. 1, pp. 3–38.

48. Even *conducteurs* did theoretical research. One by the name of Godillot published a work on the theory of bending of beams: *Etude sur le calcul de la résistance des poutres*. Another, F. Birot, wrote a treatise on the "theory and practice" of engineering: *Exposé théorique et pratique de l'art de l'ingénieur*. Both works are cited in "Noms des ingénieurs, des professeurs, et des industriels qui ont voulu promettre leur collaboration pour la rédaction des Annales du génie civil," *Annales du génie civil* 1, part 1 (1862), pp. vi–vii.

49. Henri Giffard to Dumas, July 21, 1878, Archives of Academy of Sciences, Paris.

50. James M. Oliver, The Corps des Ponts et Chaussées, 1830–1848 (Ph.D. dissertation, University of Missouri, 1967), p. 79.

51. C. Thirion, "Du role des ingénieurs civils dans l'industrie privée," *Le génie industriel* 27 (1864), p. 69; also published in *La propagation industrielle* 1 (1865–66), p. 209.

52. P. Janet, *Leçons d'électrotechnique générale*, fourth edition (Paris: Gauthier-Villars, 1920), volume 3, pp. 1–83, 126–186; François Cahen, "L'évolution en France de l'enseignement de l'électrotechnique des courants forts," in *Soixantenaire de l'Ecole Supérieure d'Electricité* (Paris: Imprimerie Chaix, 1955), pp. 10–16; Terry Shinn,"Raisonnement déductif: formation et déformation," *Bulletin de la Société Française de Physique*, n.s. 42 (1981), p. 34; Terry Shinn, "Des corps

de l'état au secteur industriel: genèse de la profession d'ingénieur, 1750–1920," *Revue française de sociologie* 19 (1978), pp. 3–56. One other, indirect, way in which engineers in the private sector emulated the values and traditions of the state engineers was by sending their sons to the *grandes écoles*. The recent work of Charles Day on the history of the Ecoles d'Arts et Métiers includes data from a three-generational study comprising *gadzarts* who graduated between 1830 and 1870, their fathers, and their sons and sons-in-law. Information on the educational attainments of 100 out of 201 sons and sons-in-law revealed that 41% also attended one of the Ecoles d'Arts et Métiers, but that 17.4% attended either the Ecole Polytechnique or the Ecole Centrale. See Day, *Education for the Industrial World,* pp. 183–134.

53. Robert Fox, "Science, industry, and the social order in Mulhouse, 1798–1871," *British Journal for the History of Science* 17 (1984), pp. 127–169.

54. In fact, nearly every one of the provincial corps engineers who published work about either suspension bridges or hydraulics (my areas of research) was a member of one or more regional scientific societies. For example, N.-R.-D. Le Moyne, P. D. Martin, and Endrès, all of whom wrote on the theory of suspension bridges, were members, respectively, of the Société des Lettres, Sciences et Arts de Metz; the Société Royale Littéraire d'Arras and the Société Royale Litteraire de Boulogne sur Mer; and the Académie des Sciences, Inscriptions & Belles-Lettres de Toulouse. J. F. D'Aubuisson de Voisins, who did work in hydraulics, was Secrétaire Perpétuel de l'Académie des Sciences, Inscriptions et Belles-Lettres de Toulouse; Lermier, a hydraulic theorist, was also a member of this academy and of the Académie Royale des Sciences, Belles-Lettres et Arts de Bordeaux. Arthur Morin, who also did research in hydraulics, was a member of the Académie Royale de Metz; Ordinaire de Lacolonge, who did research on the theory of turbines, was a member of four provincial societies: the Acadámy des Sciences de Bordeaux, the Société Industrielle de Mulhouse, the Société d'Emulation de Doubs, and the Académie de Metz. My research suggests, in short, that there was some tendency to recreate the scientific-technical culture of the state engineering corps at the regional level.

55. Edwin T. Layton, Jr., "European origins of the American engineering style of the nineteenth century," in *Scientific Colonialism,* ed. N. Reingold and M. Rothenberg (Smithsonian Institution, 1987), p. 156. See also Layton, "American ideologies of science and engineering," *Technology and Culture* 17 (1976), October, pp. 691–693.

56. Edwin T. Layton, Jr., "Mirror image twins: The communities of science and technology in nineteenth-century America," *Technology and Culture* 12 (1971), October, pp. 562–580.

57. Edmonson, From "Mécanicien" to "Ingénieur," pp. 350–351. Le Chatelier was a *polytechnicien* who worked alternately in the state engineering corps and in the private sector.

58. Alfred R. Wolff, "The value of the study of the mechanical theory of heat," *Journal of the Franklin Institute,* third series, 8 (1881), January, p. 34. Wolff's career is examined in chapter 1 of Gail Cooper's *Man-Made Weather: Air-Conditioning Engineers and Their Vision of the New Technology, 1902–1960* (Johns Hopkins University Press, in press).

59. Coleman Sellers, "Scientific method in mechanical engineering," *Journal of the Franklin Institute,* third series, 90 (1885), December, p. 424.

60. Emile Malézieux, *Travaux publics des Etats-Unis d'Amérique en 1870* (Paris: Dunod, 1873), p. 546.

61. Quoted from Layton, "European origins of the American engineering style," pp. 117, 159.

62. Quoted from Layton, "American ideologies of science and technology," p. 694.

63. Layton, "European origins of the American engineering style," pp. 163–164.

64. Coleman Sellers, *American Supremacy in Applied Mechanics* (New York: Engineering Magazine Co., 1892), pp. 14–19.

65. Wolff, "Value of the study of the mechanical theory of heat," p. 35.

66. This was basically the philosophy of Benjamin Isherwood and of Coleman Sellers. Sellers asserted, e.g., that "the wonderful progress of modern times is due wholly to the method that has been pursued of grouping facts in proper order." See Sellers, "Scientific method in mechanical engineering," p. 425; Layton, "European origins of the American engineering style," p. 157. This philosophy also prevailed among American scientists during the first half of the nineteenth century. See George Daniels, *American Science in the Age of Jackson* (Columbia University Press, 1968).

67. Thomas Edison and Elmer Sperry are two others.

68. Henry Ford, *My Life and Work,* ed. S. Crowther (London: William Heinemann, 1922), pp. 23–24.

69. Ibid., 22–32.

70. Bruce Sinclair, *Philadelphia's Philosopher Mechanics: A History of the Franklin Institute, 1824–1865* (Johns Hopkins University Press, 1974), pp. 134–194.

71. Edwin T. Layton, "Scientific technology, 1845–1900: The hydraulic turbine and the origins of American industrial research," *Technology and Culture* 20 (1979), January, pp. 70–87. See, however, Arthur Morin's *Expériences sur . . . les turbines* (Metz: Thiel, 1838) and his *Hydraulique,* second edition (Paris: Hachette, 1858). Morin, a corps engineer, conducted an important series of tests on the Fourneyron turbine in 1837, and in *Hydraulique* he discusses further tests which he and others did on a variety of turbines.

72. Arthur T. Safford and Edward P. Hamilton, "The American mixed-flow turbine and its setting," *Transactions of the American Society of Civil Engineers* 85 (1922), p. 1269.

73. Frederick Winslow Taylor, *The Principles of Scientific Management* (1911; Norton, 1967), pp. 104–114.

74. Dugald C. Jackson, "The equipment for electrical engineering laboratories," *Society for the Promotion of Engineering Education Proceedings* 2 (1894), p. 224.

75. Edwin T. Layton, Jr., "Millwrights and engineers, science, social roles, and the evolution of the turbine in America," in *The Dynamics of Science and Technology,* ed. W. Krohn et al. (Reidel, 1978).

76. Safford, "American mixed-flow turbine," p. 1269; Layton, "Hydraulic turbine," p. 76.

77. Its invention made Henri Giffard a millionaire.

78. Whereas American libraries generally had dozens of injector trade catalogs, I was unable to find even one such catalog in any French library. I did, however, find one early trade brochure of the Flaud company for sale in an antiquarian bookstore specializing in science and technology in Paris. It is, of course, difficult to draw conclusions from "non-evidence," but it does seem clear that something was happening in injector design and manufacturing in the U.S. that was not happening in France.

79. It would be too lengthy to list all these sources here, but they are listed in the bibliography. For an initial guide to this literature and to the history of the injector scc pp. 3–38 of Kranakis, "The French connection."

80. Dudebout and Croneau, *Appareils accessoires des chaudières à vapeur* (Paris: Gauthier-Villars, 1895), p. 87.

81. Articles and discussions on engineering laboratories published in the *Society for the Promotion of Engineering Education Proceedings* make clear that injectors were standard apparatus in the laboratories of engineering school.

82. See, e.g., Omer Buyse, *Méthodes américaines d'éducation générale et technique* (Paris: Charleroi, 1908), pp. 672–673; George N. Nissenson, *Practical Treatise on Injectors* (New York: [author], 1890), pp. 20–21; "Is it economical to use injectors upon locomotives, and if so, to what extent?" *American Railway Master Mechanics Association Proceedings* 9 (1876), pp. 31–52; "Report of the Committee on Science and the Arts on the steam injector and ejector of J. H. Irwin," *Journal of the Franklin Institute* 109 (1880), February, pp. 106–113, 386–391; D. S. Jacobus, "Comparative efficiency of injectors and steam pumps," *Stevens Indicator* 4 (1888), April, pp. 112–120; Committee on Science and the Arts, Tests of Injectors, 1895, Franklin Institute Archives, Philadelphia; Karl Andreu, "Tests of the efficiency of an injector," *Engineering News* 36, no. 3 (July 16, 1896), p. 39; George Henry Trautmann, "A comparative test of steam

injectors," *Bulletin of the University of Wisconsin,* Engineering Series, 2, no. 2 (1897), pp. 89–122; Francis Tucker and Frank Wilmer Lawrence, "A comparison of the efficiencies of pumps and injectors: report of a series of tests made between January and May 1897 at the Department of Dynamic Engineering, University of Pennsylvania," manuscript, University of Pennsylvania Library. This conclusion is also based on examination of dozens of injector trade catalogs, many of which contain illustrations and descriptions of company research facilities.

83. "William Sellers," *Journal of the Franklin Institute* 159 (1905), May, pp. 372–373; Strickland L. Kneass, *Practice and Theory of the Injector* (Wiley, 1899); Strickland L. Kneass, "Discharge of steam through orifices," *Proceedings of the Engineers' Club of Philadelphia* 8 (1891), July, pp. 176–187.

84. Villiers, "Résultats des expériences faites sur l'injecteur Giffard," *Annales des mines,* fifth series, 17 (1860), pp. 357–366; Deloy, "Expériences sur l'injecteur automoteur de M. Giffard," *Annales des mines,* fifth series, 17 (1860), pp. 301–317; Résal, "Recherches théoriques sur les effets mécaniques de l'Injecteur automoteur de M. Giffard," *Annales des mines,* sixth series, 1 (1862), pp. 575–605; "Injecteur automoteur des chaudières à vapeur: système du chemin de fer du Nord," manuscript, Conservatoire des Arts et Métiers, Paris. The fact that French libraries and archives contain no injector trade catalogs also suggests a rather weak tradition of injector design and testing in France, although it may also reflect a sense that trade catalogs were not worth saving. However, American libraries have preserved thousands of trade catalogs.

85. "Injecteur Giffard—type de la marine construit par Flaud," *Revue industrielle* 5 (July 7, 1875), pp. 246–247.

86. Henri Giffard, *Notice théorique et pratique sur l'injecteur automoteur* (Paris: H. Flaud, 1860), pp. 20, 22.

87. Kneass, *Practice and Theory of the Injector;* Kneass, "Discharge of steam through orifices."

88. Sellers trade brochure, 1861, National Museum of American History.

89. Flaud trade brochure, ca. 1860, in the author's possession. (This kind of reference would have appeared out of place in an American trade catalog.)

90. Combes, "Rapport . . . sur les perfectionnements apportés à l'injecteur Giffard, par M. Turck," *Bulletin de la Société d'Encouragement pour l'Industrie Nationale,* second series, 11 (1864), May, p. 265; Giffard, *Notice théorique et pratique sur l'injecteur automoteur,* p. 20.

91. Deloy, "Expériences sur l'injecteur automoteur de M. Giffard," *Annales des mines,* fifth series, 17 (1860), pp. 301–317.

92. Maurice Demoulin, *Traité pratique de la locomotive* (Paris: Librairie Polytechnique, 1898), p. 113.

93. Kneass, *Practice and Theory of the Injector,* pp. 20–21.

94. This conclusion is based on analysis of the descriptive literature on injectors published in the U.S., France, and Britain during the nineteenth century. It is very easy to trace the internationally important injector designs through this literature.

95. Demoulin, *Traité pratique de la locomotive*, p. 113, Kneass, *Practice and Theory of the Injector*, p. 93.

96. Yearly Indexes of Patents Issued, Institut national de la propriété industrielle, Paris; Alexandre Gouilly, "Analyse de l'oeuvre de Henri Giffard," *Mémoires et comptes rendus de la Société des Ingénieurs Civils* [n.v.] (September 1888), pp. 486–487.

97. U.S. Patent Office, *Index of Patents Issued, 1790–1880;* Yearly Indexes of Patents Issued, 1880–1925; "Kneass, Strickland Landis," *National Cyclopedia of Biography.* Kneass, incidentally, always kept at his desk a complete set of all injector patents issued in the U.S.; this suggests the extent to which his research was oriented toward patenting.

98. A variety of economic and legal factors may affect patenting activity, of course. Yet, in my view, on the basis of a comparative analysis of the French and American patent systems, these are not sufficient to explain the wide discrepancy between the number of patents taken out by Giffard and Flaud and the number taken out by Sellers and Kneass (or, more generally, by French vs. American injector manufacturers). Giffard and Flaud could also have hired someone to do injector research (as Sellers hired Kneass), which no doubt would have led to further patents. The American evidence suggests that important improvements in injector design—and hence important patents— came from extensive and sustained testing and research. Thus, I would argue that Flaud and Giffard did not get many patents because they did not do a lot of empirical research.

99. One American engineering educator felt that trade catalogs were great learning aids for students: "One most valuable part of the equipment [for instruction in design] is a collection of trade catalogs, and any school can have these. Among such publications of the present day are the best treatises on machine design." (John H. Barr, "Teaching machine design," *Society for the Promotion of Engineering Education Proceedings* 2 (1894), p. 250.

100. John H. Kouwenhoven, "The designing of the Eads Bridge," *Technology and Culture* 23 (1982), pp. 535–568; Howard S. Miller, *The Eads Bridge* (University of Missouri Press, 1979); Billington, *The Tower and the Bridge*, pp. 112–117.

101. See, e.g., Terry Shinn, "Reactionary technologists: The struggle over the Ecole Polytechnique, 1880–1914," *Minerva* 22 (1984), no. 3–4, pp. 329–345; Coleman Sellers, Course of Lectures in the Department of Engineering Practice, Lecture Notes—Observations Made in Europe Bearing on Questions of Water-Wheels ([n.p.]; reprint: *Stevens Indicator*, April 1892), p. 4.

Conclusion to Part II

1. Coleman Sellers, "Technical education: The mechanic arts abroad and at home," *Journal of the Franklin Institute*, third series, 89 (1885), January, p. 23.

2. There is an enormous body of literature addressing the question of the reality and the causes of the technological and economic "retardation" of France relative to other industrializing countries during the nineteenth century. Among the classic and more recent works in this domain which I have found useful are the following: R. Aldrich, "Late-comer or early starter? New views on French economic history," *Journal of European Economic History* 16 (1987), pp. 89–100; Rondo E. Cameron, *France and the Economic Development of Europe, 1800–1914* (Princeton University Press, 1961); Rondo Cameron and Charles E. Freedeman, "French economic growth: A radical revision," *Social Science History* 7 (1983), pp. 3–30; J. H. Clapham, *The Economic Development of France and Germany, 1815–1914* (Cambridge University Press, 1936); S. B. Clough, "Retardative factors in French economic development in the 19th and 20th centuries," *Journal of Economic History* 6, supplement (1946), pp. 91–210; F. Crouzet, "England and France in the eighteenth century: A comparative analysis of two economic growths," in *The Causes of the Industrial Revolution*, ed. R. M. Hartwell (Methuen, 1967); N. F. R. Crafts, "Economic growth in France and Britain, 1830–1910: A review of the evidence," *Journal of Economic History* 44 (1984), pp. 49–67; Tom Kemp, *Economic Forces in French History* (Longman, 1972); Caglar Keyder and P. K. O'Brien, *Economic Growth in Britain and France, 1780–1914: Two Paths to the 20th Century* (Allen and Unwin, 1978); Charles P. Kindleberger, *The Economic Growth of France and Britain, 1851–1950* (Harvard University Press, 1964); David Landes, *The Unbound Prometheus: Technological Change and Industrial Development in Western Europe from 1750 to the Present* (Cambridge University Press, 1969); J. Marczewski, "The take-off hypothesis and French experience," in *Economics of Take-Off into Sustained Growth*, ed. W. Rostow (Macmillan, 1963); Maurice Lévy-Leboyer, "La croissance économique en France au xixe siècle," *Annales: économie, société, culture* 3 (1968), no. 4, pp. 788–807. For an entry point to this literature from the perspective of history of science and technology, see Robert Fox, "Contingency or mentality? Technical innovation in France in the age of science-based industry," in *Technological Education — Technological Style*, ed. M. Kranzberg (San Francisco Press, 1986).

General Conclusion

1. An important exception is John Weiss's book *The Making of Technological Man: The Social Origins of French Engineering Education* (MIT Press, 1982). Some standard histories of American engineering education, notably James Gregory McGivern's *First Hundred Years of Engineering Education in the United States, 1807–1907* (Gonzaga University Press, 1960), say very little about the content of curricula and the apportionment of time. Yet recent research has begun to analyze curricula more critically. See, for example, Bruce Seely, "Research,

engineering, and science in American engineering colleges," *Technology and Culture* 34 (1993), no. 2, pp. 344–386.

2. New technical schools that were established in the last two decades of the nineteenth century in France do appear, however, to have concentrated significantly on laboratory training, and some (such as the Ecole Spéciale des Travaux Publics, established in 1892 to train *conducteurs*) also introduced workshop training. On the Ecole Spéciale des Travaux Publics see William E. Wickenden, "A comparative study of engineering education in the United States and in Europe," in *Report of the Investigation of Engineering Education, 1923–1929,* volume I (University of Pittsburgh, 1930), p. 769.

3. Bridge engineers now pay great attention to torsional vibrations too. In the context of Navier's formulation of the wind problem, however, the point is that he ignored evidence of the possibility of significant vertical oscillations. The data to which he had access did not clearly address the possibility of torsional oscillations.

Bibliography

Manuscripts and Special Collections

Allen, Richard S. "Biographical sketch of Theodore Burr." Unpublished manuscript, Library of Congress.

Archives Nationales, Paris. F^{12}1299A, 1299B, 2197, 2323–5, 2328–9, 6747–9; F^{13}713, 751, 1011–12; F^{14}771, 835, 1138, 1177, 2289, 3185, 3186, 3190, 10205.

Assemblée Nationale, Session 1850. Projet de loi tendant à apporter des modifications au mode de rectrutement des ingénieurs des Ponts et Chaussées. Manuscript 2158/c.97, Ecole des Ponts et Chaussées, Paris.

Assemblée Nationale. Rapport fait par le Citoyen Stourm . . . relatif à des changements dans l'organisation du Corps des Conducteurs des Ponts et Chaussées . . . , 17 Novembre 1848. Manuscript 2156/c.96, Ecole des Ponts et Chaussées, Paris.

Bresse, J. A. C. Cours de mécanique apliquée. 2ième partie: hydraulique. 1858. Manuscript 23039, Ecole des Ponts et Chaussées, Paris.

Brissaud, E. Ponts suspendus; aperçu historique et comparatif, 1865. Manuscript 10249/c.541, Ecole des Ponts et Chaussées, Paris.

Carvallo. Essai sur la théorie de l'injecteur Giffard, 1859. Manuscript 2373, Ecole des Ponts et Chaussées, Paris.

Charras et Latrade. Assemblée nationale, Session 1850. Proposition tendant à apporter des modifications au mode de recrutement des ingénieurs des Ponts et Chaussées, 1850. Manuscript 2159/c.97, Ecole des Ponts et Chaussées, Paris.

Crozet, M. Description du pont suspendu construit sur le Drac, près de Grenoble. 1829. Manuscript 4696, Ecole des Ponts et Chaussées, Paris.

Defontaine, C. Projet d'organisation de l'Ecole des Ponts et Chaussées, 1839. Manuscript 2208, Ecole des Ponts et Chaussées, Paris.

Detours, Auguste Mie, and Delbetz. Proposition relative aux emplois d'ingénieur des Ponts et Chaussées, 1849. Manuscript 2155/c.97, Ecole des Ponts et Chaussées, Paris.

Douglass, Ephraim. Uniontown, Pennsylvania, April 3, 1791, letter to James Finley, Philadelphia. Albert Gallatin manuscripts, New York Historical Society.

Duleau, A. J. C. Leçons sur les ponts suspendus faites à l'Ecole des Ponts et Chaussées en 1827–8. Manuscript 4560, Ecole des Ponts et Chaussées, Paris.

Ecole Centrale des Arts et Manufactures. Notes sur le cours d'hydraulique, 1850–51. Ecole Centrale des Arts et Manufactures.

Edwards, Llewellyn. Papers. National Museum of American History, Smithsonian Institution.

Etats de Correspondence. Archives de la Chambre de Commerce, Paris.

Finley, James. Last Will and Testament, probated March 17, 1828. Register of Wills, Fayette County Courthouse, Uniontown, Pennsylvania.

Finley, James. Statement of the Quantity of Iron for a Chain Bridge Across the River Susquehanna. Stauffer Collection, Historical Society of Pennsylvania.

Finley, Rev. James. Correspondance. Society Collection, Historical Society of Pennsylvania.

Fitch, Asa. Manuscript diary. Sterling Memorial Library, Yale University.

Flaud Trade Brochure, ca. 1860. In the author's possession.

Giffard, Henri. Correspondance. Archives de l'Académie des Sciences, Paris.

Giffard, Henri. Papers. Musée de l'Air, Paris.

Gilpin, Thomas. Letterbook. Maury Family manuscripts, University of Virginia Library.

Grévy, Jules and Raynal. Chambre de Députés, Session de 1884. Projet de loi portant réorganisation du Corps des Ponts et Chaussées. Manuscript 15842/c.836, Ecole des Ponts et Chaussées, Paris.

Jones, John. Indian River Delaware. Correspondence, May 27, 1773. American Philosophical Society Archives.

Kermaingaut. Mémoires diverses sur les ponts suspendus. Manuscript 268, Ecole des Ponts et Chaussées, Paris.

Lagrené et Prudhomme. Assemblée nationale, Session 1850. Amendements à la proposition de MM. Charras et Latrade . . . , 1850. Manuscript 2160/c.97, Ecole des Ponts et Chaussées, Paris.

Latrade et al. Chambre des députés, Session 1882. Proposition de loi relative au Corps des Ponts et Chaussées. Manuscript 15842/c.836, Ecole des Ponts et Chaussées, Paris.

Ménard-Dorian. Chabre de Députés, Session de 1884. Rapport fait au nom de la commission chargée d'examiner les propositions de lois de M. Cantagrel . . . , de M. Latrade . . . , et de M. Jean David. . . . Manuscript 15842/c.836, Ecole des Ponts et Chaussées, Paris.

Ministère des Travaux Publics. Programmes de l'enseignement intérieur de l'Ecole des Ponts et Chaussées, 1875. Manuscript 19952, Ecole des Ponts et Chaussées, Paris.

Navier, C. L. H. M. Correspondance. Lacroix manuscripts, Bibliothèque de l'Institut de France.

Navier, C. L. H. M. Rapport sur le cours de mécanique appliquée . . . 1830. Ecole des Ponts et Chaussées, Paris.

Observations . . . par les Inspecteurs Généraux et Divisionnaires des Ponts et Chaussées . . . relatif à un nouveau mode de recrutement des ingénieurs, 1848. Manuscript 2153/c.97, Ecole des Ponts et Chaussées, Paris.

Orphan's Court Records, 1828–1848. Fayette County Courthouse, Uniontown, Pennsylvania.

Patent Records, France, Service Nationale de la Propriété Industrielle. (Injectors, suspension bridges, turbines.)

Patent Tubular Iron Bridge. Zaccheus Collins manuscripts, 1810–11. Historical Society of Pennsylvania.

Peale-Sellers papers. American Philosophical Society.

Pennsylvania Railroad, Test Department. Sellers Exhaust Steam Injector, Altoona, Pennsylvania, May 16, 1938, Typescript, Hagley Museum and Library, Wilmington, Delaware.

Pièces relatives aux ponts suspendus. Manuscript 2438, Ecole des Ponts et Chaussées, Paris.

Programme des cours professés à l'Ecole des Ponts et Chaussées en 1848 et 1849. Manuscript 19952, Ecole des Ponts et Chaussées, Paris.

Proposition de loi relative au Corps des Ponts et Chaussées. Paris: Imprimerie du Journal Officiel, 1881. Manuscript 15842/c.836, Ecole des Ponts et Chaussées, Paris.

Reports on Petititions for Roads and Bridges. Clerk of Courts. Fayette County Courthouse, Uniontown, Pennsylvania.

Schuylkill-Lancaster Bridge Co. Correspondence. Historical Society of Pennsylvania.

Seguin, Marc. Correspondence. Manuscript 24707, Bibliothèque Nationale, Paris.

Seguin, Marc. Correspondence and Miscellaneous Papers, Archives de l'Académie des Sciences, Paris.

Soult de Dalmatie. Assemblée nationale, Session 1850. Rapport . . . sur la proposition de MM. Charras et Latrade . . . , 1850. Manuscript 2161/c.97, Ecole des Ponts et Chaussées, Paris.

Telford, Thomas. Report, 1814, Respecting a Bridge over the River Mersey at Runcorn. Ironbridge Gorge Museum, England.

Templeman, John. Georgetown, D.C., March 14, 1809, printed circular letter addressed to Nicholas Van Dyke. Group 4, Box 10, Longwood manuscripts, Hagley Museum and Library, Wilmington, Delaware.

Tests of Injectors, 1895, Committee on Science and the Arts, Franklin Institute Archives, Philadelphia.

Trade Catalogue Collection. Goss Library, Purdue University.

Trade Catalogue Collection. Franklin Institute, Philadelphia.

Trade Catalogue Collection. National Museum of American History, Smithsonian Institution.

Tucker, Francis and Frank Wilmer Lawrence. "A Comparison of the Efficiencies of Pumps and Injectors: Report of a Series of Tests Made Between January and May 1897 at the Department of Dynamic Engineering, University of Pennsylvania." Unpublished manuscript, University of Pennsylvania Library.

Wernwag, Lewis. Printed sheet on bridges. American Philosophical Society.

William Sellers & Co. Visitors Register, 1861–1947. Acc. 1466, v. 2, Hagley Museum and Library, Wilmington, Delaware.

Periodicals

Annales de la Société d'Economie Politique, 1846–1887.

Annales de l'industrie française et étrangère, 1828–1830.

Annales des mines, 1858–1900.

Annales des Ponts et Chaussées, 1831–1900.

Annales du génie civil, 1862–1880.

Annales industrielles, 1869–1890.

Année industrielle, 1889–1892.

Annuaire de la Société des Anciens Elèves des Ecoles d'Arts et Métiers, 1848–1870.

Bulletin de la Société des Inventeurs et de Protecteurs de l'Industrie, 1846–47.

Bulletin de la Société d'Encouragement pour l'Industrie Nationale, 1858–1900.

Bulletin de la Société Industrielle de Mulhouse, 1858–1900.

Bulletin des sciences technologiques, 1824–1831.

Chronique industrielle, 1878–1880.

Comptes rendus de l'Académie des Sciences, 1858–1900.

Le constitutionnel, 1826.

Cosmos, 1852–1860.

Engineer, 1858–1885.

Engineering News, 1875–1882.

Engineering News and American Railway Journal, 1888–1900.

Le génie industriel, 1851–1871.

Le globe, 1826.

L'industriel, 1826–1830.

Journal de l'industriel et du capitaliste, 1836–37.

Journal des chemins de fer, des mines, et des travaux publics, 1842–1847.

Journal des usines, 1841–1847.

Journal du génie civil, 1828–1847.

Journal of Railway Appliances, 1883–1895.

Journal of the Franklin Institute, 1826–1900.

Mechanic's Magazine, 1858–1890.

Mémoires et comptes rendus de la Société des Ingénieurs Civils, 1858–1890.

La pandore, 1826.

Pennyslvania Correspondent and Farmers' Advertiser (Doylestown, Pennsylvania), 1810.

La propagation industrielle, 1865–1870.

La quotidienne, 1826.

Locomotive Enginering, 1888–1900.

Revue industrielle, 1870–1880.

Scientific American, 1858–1900.

Society for the Promotion of Engineering Education Proceedings, 1893–1900.

Transactions of the American Society of Civil Engineers, 1867–1900

Transactions of the American Society of Mechanical Engineers, 1880–1900.

United States Gazette (Philadelphia), 1811.

Van Nostrand's Eclectic Engineering Magazine, 1869–1878.

Primary Sources: United States

"Advantages and disadvantages of lifting and non-lifting injectors." *Proceedings of the the American Railway Association,* division 5 (mechanical) (1921–22): 352–357.

Andreu, Karl. "Tests of the efficiency of an injector." *Engineering News* 36 (July 16, 1896): 39.

Baker, George. "Locomotive injectors." *Scientific American Supplement* 38, no. 987 (December 1, 1894): 15770–15771.

Baker, Ira O. "Engineering education in the United States at the end of the century." *Society for the Promotion of Engineering Education (SPEE) Proceedings* 8 (1900): 11–27.

Barr, John H. "Teaching machine design." *SPEE Proceedings* 2 (1894): 236–252.

Benjamin, Charles H. "The true place of drawing and shopwork in engineering colleges." *SPEE Proceedings* 3 (1895): 126–136.

Bissell, G. W. "Mechanical laboratory work at Ames, Iowa." *SPEE Proceedings* 2 (1894): 217–220.

Bleich, Friedrich, et al. *The Mathematical Theory of Vibration in Suspension Bridges.* Washington: U.S. Government Printing Office, 1950.

Brackenridge, Henri Marie. *Recollections of Persons and Places in the West.* Philadelphia: James Kay, Jr., 1834.

"Bridges." In *Register of Arts,* ed. Thomas G. Fessenden. Philadelphia: C. & A. Conrad, 1808.

Buyse, Omer. *Méthodes américaines d'éducation générale et technique.* Paris: Charleroi, 1908.

Caldwell, Frances C. "A comparative study of the electrical engineering courses given at different institutions." *SPEE Proceedings* 7 (1899): 127–129.

Capraro, Paul. "The injector and some requirements for its successful operation." *The Engineer* 44, no. 19 (October 1, 1907): 905–907.

Carpenter, R. C. "Engineering laboratory courses." *SPEE Proceedings* 5 (1897): 26–39.

Carpenter, R. C. "The injector." *Cassier's Magazine* 1 (1892): 211–216, 277–283, 278–383.

Chevalier, Michel. *Histoire et description des voies de communication aux Etats-Unis.* Paris: C. Gosselin, 1840.

Colvin, Fred H. "Plain talks on the injector." *Locomotive Engineering* 11 (1898): 64–65, 77–78, 143–144, 252–253.

Crandall, C. L. "A General Course in Civil Engineering." *SPEE Proceedings* 3 (1895): 268–274.

Creighton, W. H. P. "Methods of instruction." *SPEE Proceedings* 5 (1897): 104–113.

Early Proceedings of the American Philosophical Society, 1744–1838. Philadelphia: McCalla & Stavely, 1884.

Edwards, Emory. *Modern American Locomotive Engines; Their Design, Construction, and Management.* Philadelphia: Henry Carey Baird, 1888.

Elbridge, Gerry, Jr. *Diary.* New York: Brentano's, 1926.

Elliott, David. *The Life of the Reverend Elisha Macurdy with an Appendix Containing Brief Notices of Various Deceased Ministers of the Presbyterian Church in Western Pennsylvania.* Allegheny, Pennsylvania: Kennedy & Brother, 1848.

Emery, Charles. "The Giffard injector." *Journal of the Franklin Institute* 81 (January 1866): 52–55.

Encyclopaedia: or A Dictionary of Arts, Sciences, and Miscellaneous Literature, first American edition. Philadelphia: Thomas Dobson, 1798.

"Engineering schools of the United States, I–XII." *Engineering News and American Railway Journal* 27 (January–June 1892): 277–278, 294–296, 318–319, 342–5, 371–373, 412–414, 433, 459–461, 514–516, 541–543, 589–590, 660–661.

"Engineering schools of the United States, XIII–XXXV." *Engineering News and American Railway Journal* 28 (July–December 1892): 6, 28–29, 65–66, 87–89, 111–114, 139–140, 161–162, 186–187, 207–208, 210–211, 231–233, 256, 268–269, 277–278, 302, 327–328, 354–355, 375–376, 401–402, 414–417, 437–438, 471–472, 488–489, 518–520, 546–547.

"Entrance requirements for engineering colleges." *SPEE Proceedings* 4 (1896): 103–104.

Farrow, Edward S. *West Point and the Military Academy,* second edition. New York: Orange Judd, 1879.

Farrow, Edward S. *West Point and the Military Academy,* third edition. New York: Military-Naval Publishing Co., 1895.

Finley, James. "Description of the patent chain bridge." *Port Folio,* n.s. 3 (June 1810): 441–453.

Finley, James. *A Description of the Chain Bridge Invented by Judge Finley.* Uniontown, Pennsylvania: W. Campbell, 1811.

Ford, Henry. *My Life and Work.* London: William Heinemann, 1922.

Gallatin, Albert. *Report on the Subject of Roads and Canals.* 1808; New York: Augustus M. Kelley, 1968.

Hood, Onzi. "An apprenticeship system in college shops." *SPEE Proceedings* 7 (1899): 62–70.

Howe, Malverd A. "An old chain suspension bridge." *Railroad Gazette,* 32 (January 5, 1900): 5.

"Is it economical to use injectors upon locomotives, and if so, to what extent?" *American Railway Master Mechanics Association, Proceedings* 9 (1876): 31–52.

Jackson, Dugald C. "The equipment for electrical engineering laboratories." *SPEE Proceedings* 2 (1894): 221–235.

Jacobus, D.S. "Comparative efficiency of injectors and steam pumps." *Stevens Indicator* 4 (April 1888): 112–120.

King, R. S. "Selecting an injector." *The Engineer* 44, no. 19 (October 1, 1907): 904–905.

Kirkman, Marshall M. *Locomotive Appliances.* New York: World Railway Publishing Co., 1906.

Kneass, Strickland L. "The development of the injector." *Proceedings of the Engineers' Club of Philadelphia* 10, no. 1 (1893): 91–99.

Kneass, Strickland L. "Discharge of steam through orifices." *Proceedings of the Engineers' Club of Philadelphia* 8 (July 1891): 176–187.

Kneass, Strickland L. "High pressure steam tests of an injector." *Journal of the Franklin Institute* 162 (October 1906): 279–290.

Kneass, Strickland Landis. "Physical characteristics of certain bronzes for steam uses." *Journal of the Franklin Institute* 159 (January 1905): 65–76.

Kneass, Strickland Landis. *Practice and Theory of the Injector.* New York: Wiley, 1899.

Lanza, Gaetano. "The class room and the laboratory in their mutual adjustment to the end of the most efficient instruction." *SPEE Proceedings* 6 (1898): 37–48.

Lanza, Gaetano. "The organization and conduct of engineering laboratories." *SPEE Proceedings* 2 (1894): 149–161.

Long, Stephen Harriman. *Description of Col. Long's Bridges Together With a Series of Directions to Bridge Builders.* Concord, New Hampshire: John F. Brown, 1836.

Mahan, Dennis Hart. *An Elementary Course of Civil Engineering, for the Use of the Cadets of the United States Military Academy.* New York: Wiley and Putnam, 1837.

Malézieux, Emile. *Travaux publics des Etats-Unis d'Amérique en 1870.* Paris: Dunod, 1873.

Marx, C. W. "Amount and kind of shop work required in a mechanical engineering course." *SPEE Proceedings* 2 (1894): 207–216.

Millington, John. *Elements of Civil Engineering.* Philadelphia: J. Dobson, 1839.

Munroe, Henry S. "Courses in mining engineering." *Engineering News and American Railway Journal* 28 (November 3, 1892): 416.

"The New York anchorage of the East River Bridge." *Scientific American* 34 (January 8, 1876): 15–16.

Nissenson, George N. *Practical Treatise on Injectors.* New York: the author, 1890.

Nystrom, John W. "On the Giffard injector." *Journal of the Franklin Institute,* third series, 44, no. 6 (December 1862): 391–399.

Notice of Rensselaer Institute, 1835. Reproduced in "Engineering schools of the United States VI," *Engineering News and American Railway Journal* 27 (April 21 1892): 412.

Park, Roswell. *A Sketch of the History and Topography of West Point and the U.S. Military Academy.* Philadelphia: Henry Perkins, 1840.

Peale, Charles W. *An Essay on Building Wooden Bridges.* Philadelphia: the author, 1797. American Philosophical Society Pamphlet Collection, volume 4.

Pennsylvania General Assembly, Senate. Report on Roads, Bridges, and Canals, March 23, 1822. Philadelphia: C. Mowry, 1822.

Pope, Thomas. *A Treatise on Bridge Architecture.* New York: Alexander Niven, 1811.

Porter, Dwight. "The hydraulic laboratory of the Massachusetts Institute of Technology." *SPEE Proceedings* 1 (1893): 177–183.

"The Purdue University locomotive on grades." *Engineering News and American Railway Journal* 28 (September 22, 1892): 278.

Randolph, L. S. "The economic element in technical education." *SPEE Proceedings* 3 (1895): 181–188.

Raymond, William G. "The scope of an engineering college." *SPEE Proceedings* 3 (1895): 50–73.

Reed, Hugh T. *Cadet Life at West Point*. Chicago: the author, 1896.

"Reinforcing the Newburyport suspension bridge." *Engineering Record* 42 (October 6, 1900): 314–315.

"Report of the Committee on Science and the Arts on the steam injector and ejector of J. H. Irwin." *Journal of the Franklin Institute* 109 (February 1880): 106–113.

Rice, Arthur L. "Business methods in teaching engineering." *SPEE Proceedings* 8 (1900): 157–163.

Richard, Gustav. "Boiler feeding appliances: Injectors." *Journal of Railway Appliances* 13 (1893): 73–75, 89–90, 104–106, 120–122, 137–138, 152–155, 169, 177–178.

Röntgen, Robert. *Principles of Thermodynamics*. New York: J. Wiley, 1880.

Rousiers, Paul de. *La vie américaine: ranches, fermes, et usines*. Paris: Firmin-Didot, ca. 1896.

Schuerman, W. H. "Is not too much time given to merely manual work in the shops?" *SPEE Proceedings* 4 (1896): 340–356.

Sears, Robert. *Wonders of the World*. New York: E. Walker, 1847.

Sellers, Coleman. *American Supremacy in Applied Mechanics*. New York: Engineering Magazine Co., 1892.

Sellers, Coleman. Course of Lectures in the Department of Engineering Practice; Lecture Notes—Observations Made in Europe Bearing on Questions of Water-Wheels. Pamphlet in American Philosophical Society Library. Reprinted from *Stevens Indicator*, April 1892.

Sellers, Coleman. "Scientific method in mechanical engineering." *Journal of the Franklin Institute*, third series, 90 (December 1885): 420–438.

Sellers, Coleman. "Technical education. The mechanic arts abroad and at home." *Journal of the Franklin Institute*, third series, 89 (January 1885): 19–29.

Sellers, Coleman. "The Worcester County Free Institute of Industrial Science." *Journal of the Franklin Institute*, 65, no. 4 (April 1873): 217–221.

Sellers, George Escol. *Early Engineering Reminiscences*, ed. E. Ferguson. U.S. National Museum Bulletin, no. 238. Washington: Smithsonian Institution, 1965.

Sinclair, Angus. *Twentieth Century Locomotives*. New York: Railway and Locomotive Engineering, 1904.

Smith, Harry E. "Drawing, shop work and engineering laboratory practice in engineering colleges." *SPEE Proceedings* 7 (1899): 176–180.

Smith, Joseph. *Old Redstone; or, Historical Sketches of Western Presbyterianism*. Philadelphia: Lippincott, Grambo, & Co., 1854.

Spangler, H. W. "A course in mechanical engineering." *SPEE Proceedings* 3 (1895): 284–308.

"A statistical account of the Schuylkill permanent bridge." *Port Folio*, n.s. 5 (March 1808): 168–171, 182–187, 200–204, 222–224.

Steinman, David B. *A Practical Treatise on Suspension Bridges, Their Design, Construction, and Erection.* New York: J. Wiley, 1922.

Storrow, Charles. *A Treatise on Water-Works for Conveying and Distributing Supplies of Water.* Boston: Hilliard & Gray, 1835.

Swain, George F. "Comparison between American and European methods in engineering education." *SPEE Proceedings* 1 (1893): 75–117.

Talbot, Arthur N. "Requirements in mathematics for engineering education." *SPEE Proceedings* 1 (1893): 50–62.

Taylor, Frederick Winslow. *The Principles of Scientific Management.* New York: W. W. Norton, 1967.

"The thermal efficiency of an injector." *Sibley Journal of Engineering* 38, no. 1 (January 1924): 13–15.

Thurston, Robert H. "The modern mechanical laboratory especially as in process of evolution in America." *SPEE Proceedings* 8 (1900): 331–358.

Trautmann, George Henry. "A comparative test of steam injectors." *Bulletin of the University of Wisconsin*, Engineering Series 2, no. 2 (1897): 89–122.

Trautwine, John. *The Civil Engineer's Pocket Book.* Philadelphia: Claxton, Remsen, and Haffelfinger, 1872.

Vaughan, Pendred. *The Railway Locomotive—What It Is and Why It Is What It Is.* New York: Van Nostrand, 1908.

Veech, James. *The Monongahela of Old.* Pittsburgh: n.p., 1858–1892.

von Kármán, Theodore, et al. *The Failure of the Tacoma Narrows Bridge.* Pasadena: [n.p.], 1941.

Webb, J. Burkitt. "The mechanics of the injector." *American Society of Mechanical Engineers Transactions* 10 (1888–89): 339–348.

Weld, Theodore D. *First Annual Report of the Society for Promoting Manual Labor in Literary Institutions.* New York, 1833.

"William Sellers." *Journal of the Franklin Institute* 159 (May 1905): 365–381.

Whipple, Squire. *An Elementary and Practical Treatise on Bridge Building*, second edition. New York: Van Nostrand, 1873.

Wolff, Alfred R. "The value of the study of the mechanical theory of heat." *Journal of the Franklin Institute*, third series, 8 (January 1881): 34–42.

Wood, De Volson. "Presidential address." *SPEE Proceedings* 2 (1894): 21–38.

Yarmolinsky, Avrahm. *Picturesque United States of America, 1811, 1812, 1813.* New York: William Edwin Rudge, 1930.

Primary Sources: France

Anquetin, L. "Injecteur perfectionné de M. H. Giffard, construit par M. H. Flaud, à Paris." *Portefeuille économique des machines de l'outillage et du matériel* 17 (1872).

Armengaud, Jacques Eugène. "Notice biographique sur Eugène Pihet." *Le génie industriel* 32 (1866): 265–266.

Armengaud, Jacques Eugène. "Préface du tome premier de la deuxième édition de la Publication Industrielle." In *Traité pratique des moteurs hydrauliques et à vapeur.* Paris: the author, 1844.

Arrêté . . . portant réglement pour l'Ecole Royale des Mines. Paris: Imprimerie Royale, 1817. Manuscript 2196/C. 96. Ecole des Ponts et Chaussées.

"Article publié par l'administration dans le Moniteur du 29 Février 1828." *Journal du génie civil* 1 (1828): 444–452.

Bache, Alexander Dallas. *Report on Education in Europe to the Trustees of the Girard College for Orphans.* Philadelphia: Lydia R. Bailey, 1839.

Balzac, Honoré de. *The Village Rector.* Boston: Little, Brown, 1900.

Barnard, Henry. *Systems, Institutions, and Statistics of Scientific Instruction, Applied to National Industries in Different Countries.* New York: E. Steiger, 1872.

Barré de Saint-Venant. *Notice sur la vie et les ouvrages de Pierre-Louis-Georges, Comte du Buat.* Lille: L. Danel, 1866.

Becquey. "Discours." *Journal du génie civil* 1 (1828): 125–142.

Beillard, Alfred. *L'enseignement technique et les travailleurs.* Besançon: Imprimerie Franc-Comtoise, 1888.

Bélidor, Bernard Forest de. *Architecture hydraulique,* ed. C.-L.-M.-H. Navier. Paris: Firmin Didot, 1819.

Bertrand, J. *Eloge historique de Jean-Victor Poncelet.* Paris: Firmin Didot, 1872.

Bertrand, J. *Eloge de M. Charles Combes.* Paris: Firmin Didot, 1885.

Bougère, L. "Notice sur l'injecteur automoteur des chaudières à vapeur." *Annuaire de la Société des Anciens Elèves des Arts et Métiers* 12 (1859): 65–81.

Briot, Charles. *Théorie mécanique de la chaleur.* Paris: Gauthier-Villars, 1869.

Bruyère, L. *Etudes relatives à l'art des constructions* (2 volumes). Paris: Bance Aîné, 1823.

Carvallo, J. "Mémoire sur les lois mathématiques de l'écoulement et de la détente de la vapeur." *Comptes rendus de l'Académie des Sciences* 52 (January–June 1861): 684–688, 801–804.

[Choppard, Léon]. *Mémoire pour M. Paul Giffard contre l'administration des domaines représentant l'Etat.* Paris: Imprimerie de l'Etoile, 1883.

"Chute du pont suspendu de Broughton." *Annales des Ponts et Chaussées: mémoires,* 2ᵉ année, no. 1 (1832): 408–411.

Combes, E. "Note sur l'injecteur automoteur des chaudières à vapeur imaginé par M. Giffard et construit par M. H. Flaud." *Bulletin de la Société d'Encouragement pour l'Industrie Nationale* 58 (1859): 337–343.

Combes, E. "Rapport . . . sur les perfectionnements apportés à l'injecteur Giffard, par M. Turck." *Bulletin de la Société d'Encouragement pour l'Industrie Nationale,* second series, 11 (May 1864): 257–266.

"Considerations sur les conducteurs." *Journal du génie civil* 10 (1831): 25–30.

Corréard, A. "De la création immédiate d'écoles spéciales d'ingénieurs-mécaniciens. . . . " *Journal du génie civil* 14 (1847): 126–133.

Corréard, A. "De la nécessité de réorganiser le corps des ingénieurs des Ponts et Chaussées." *Journal du génie civil* 15 (1847): 92–111.

Corréard, Alexandre. "Observations sur le pont suspendu des Champs-Elysées." *Journal du génie civil* 8 (1830): 149–154.

Corréard, Alexandre. "Observations sur les ponts suspendus des Saintes-Pères et de la rue Belle-chasse." *Journal du génie civil* 10 (1831): 486–503.

Corréard, Alexandre. "Pont des Invalides." *Journal du génie civil* 1 (1828): 442–444.

Cotelle, T.-A. *Esquisse historique sur l'institution des Ponts et Chaussées.* Paris: Paul Dupont, 1849.

Couderc, J. *Essai sur l'administration . . . des Ponts et Chaussées.* Paris: Carilian-Gouery, 1829.

"Critique du projet de loi relatif à la réorganisation du Corps des Ponts et Chaussées." *Annuaire de la Société des Anciens Elèves des Ecoles d'Arts et Métiers* 2 (1849): 55–68.

"De l'organisation des conducteurs des Ponts et Chaussées." *Journal du génie civil* 15 (1847): 85–91.

"Décret réglant les conditions d'admission des conducteurs dans le corps des ingénieurs des Ponts et Chaussées." *La propagation industrielle* 3 (1868): 80–81.

Dejardin. "Simplification de quelques formules relatives à l'établissement des ponts suspendus." *Annales des Ponts et Chaussées: mémoires* 18 (1839): 253–287.

Delaistre, J. R. *Encyclopédie de l'ingénieur* (4 volumes). Paris: J. G. Dentu, 1812.

Delaistre, J. R. *La science de l'ingénieur* (2 volumes). Lyon: Brunet, 1825.

Deloy. "Expériences sur l'injecteur automoteur de M. Giffard." *Annales des mines*, fifth series, 17 (1860): 301–317.

"Description d'un pont suspendu, en fer, construit sur la rivière de Tweed, près de Berwick, en Angleterre." *Bulletin de la Société d'Encouragement pour l'Industrie Nationale* 22 (1823): 325–329.

Didion. *Notice sur la vie et les ouvrages de Général J.-V. Poncelet.* Paris: Gauthier-Villars, 1869.

Demoulin, Maurice. *Chemins de fer locomotive et matériel roulant.* Paris: Dunod, 1896.

Demoulin, Maurice. *Traité pratique de la locomotive.* Paris: Librairie Polytechnique, 1898.

Dudebout and Croneau. *Appareils accessoires des chaudières à vapeur.* Paris: Gauthier-Villars, 1895.

Duleau, A. *Essai théorique et expérimental sur la résistance du fer forgé.* Paris: Courçier, 1820.

Durand, J. N. L. *Précis des leçons d'architecture données à l'Ecole Polytechnique.* (two volumes). Paris: Ecole Polytechnique, 1802; second edition: Ecole Royale Polytechnique, 1819.

"Ecole Centrale des Arts et Manufactures: Prospectus." *Annales de l'industrie française et étrangère* 2 (1828): 380–477.

"Ecole Centrale Lyonnaise pour l'Industrie et le Commerce." *Annales du génie civil* 6 (1867): 592–596.

"Ecole Industrielle et Commerciale de Rouen." *Journal du génie civil* 14 (1847): 509–516.

Emmery. *Notice abrégée sur l'histoire, l'organisation et l'utilité sociale de l'institution des Ponts et Chaussées de France.* Paris: Imprimerie de Bourgogne, 1839.

Flachat, Eugène. "Discours—Société des Ingénieurs Civils, séance du 10 janvier 1862." *Annales du génie civil* 1, Part 2 (1862): 1–13.

Fourier. "Extrait de l'analyse des travaux de l'Académie Royale des Sciences . . . relative à l'Ecole Royale des Ponts et Chaussées de France." *Journal du génie civil* 5 (1829): 441–445.

Fourneyron, B. "Mémoire sur l'application en grand dans les mines et manufactures, des turbines hydrauliques ou roues à palettes courbes de Bélidor." *Bulletin de la Société d'Encouragement pour l'Industrie Nationale* 33 (1834): 3–17, 49–61, 87–94.

Freminville, Antoine J. *Cours pratique des machines à vapeur marines.* Paris: Arthur Bertrand, [n.d.].

Gauthey, Emiland-Marie. *Dissertation sur les dégradations survenues aux piliers du dôme du Panthéon français et sur les moyens d'y remédier.* Paris: H.-L. Perronneau, an VI.

Gauthey, Emiland-Marie. *Mémoire sur l'application des principes de la mécanique à la construciton des voûtes et des dômes.* Paris: C.-A. Jombert, 1771.

Gauthey, Emiland-Marie. *Traité de la construction des ponts* (two volumes), ed. C.-L.-M.-H. Navier. Paris: Firmin Didot, 1809–1813.

Giffard, Henri. *Application de la vapeur à la navigation aérienne.* Paris: Imprimerie de Pollet, 1851.

Giffard, Henri. "De la force dépensée pour obtenir un point d'appui dans l'air calme, au moyen de l'hélice." *Bulletin de la Société Aeronatique et Météorologiuqe de France* 1 (1852): 107–112.

Giffard, Henri. *Notice théorique et pratique sur l'injecteur automoteur.* Paris: H. Flaud, 1860.

Gouilly, Alexandre. "Analyse de l'oeuvre de Henri Giffard." *Mémoires et comptes rendus de la Société des Ingénieurs Civils* [n.v.] (September 1888): 365–504.

Guillemin, Jules. "Benôit Fourneyron—notice biographique." *Bulletin de la Société de l'Industrie Minérale* 12 (1867): 551–552.

Hirsch. *Notice nécrologique sur Gustave-Adolphe Hirn.* Paris: Société d'Encouragement pour l'Industrie Nationale, [n.d.].

"Les ingénieurs de l'état et les ingénieurs civils." *Chronique industrielle* 2 (16 February 1879): 41–43.

"Des ingénieurs en France." *Journal des chemins de fer* 4 (1845): 516–517.

"Injecteur Giffard—type de la marine construit par Flaud." *Revue industrielle* 5 (July 7, 1875): 246–247.

Janet, P. *Leçons d'électrotechnique générale,* fourth edition. Paris: Gauthier-Villars, 1920.

Karr, Eugène. "Sur les privilèges." *Annuaire de la Société des Anciens Elèves des Ecoles d'Arts et Métiers* 2 (1849): 49–54.

"Laboratoire des essais à l'Ecole des Mines." *Revue industrielle* 3 (1873): 379–383.

"Laboratoire d'essai de l'Ecole Impériale des Mines." *Cosmos* 4 (May 5, 1854): 521–523.

Lamé, Gabriel et al. *Vues politiques et pratiques sur les travaux publics.* Paris: Imprimerie d'Everat, 1832.

Leblanc. "Note sur quelques améliorations à apporter au système d'organisation du Corps Royal des Ponts et Chaussées." *Journal du génie civil* 4 (1829): 157–166.

Ledieu, A. *Les nouvelles machines marines.* Paris: Dunod, 1876.

Ledieu, A. "Réponse à quelques objections soulevées par nos récentes communications sur le rendement des injecteurs à vapeur." *Comptes rendus de l'Académie des Sciences* 81 (July–December 1875): 1023–1024.

Ledieu, A. "Sur le rendement des injecteurs à vapeur." *Comptes rendus de l'Académie des Sciences* 81 (July–December 1875): 711–715, 773–775. Also published in *Revue industrielle* 5 (November 1875): 437, 446, 487.

Le-Moyne, N. R. D. *Moyens faciles de parvenir à fixer les conditions de l'établissement des ponts suspendus.* Paris: Carilian-Gouery, 1825.

Léon. *Des changements projetés dans le mode de recrutement du Corps des Ponts et Chaussées.* Paris: Carilian-Goeury, 1849.

"Les travaux publics aux Etats-Unis d'Amérique: analyse d'un rapport de M. Malézieux." *Annales du génie civil,* second series, 4 (1875): 460.

"Lettre au directeur du Journal du génie civil." *Journal du génie civil* 4 (1829): 433–444.

"Lettre au directeur du Journal du génie civil." *Journal du génie civil* 4 (1829): 671–683.

"Lettre au directeur du Journal du génie civil adressée par un ingénieur des Ponts et Chaussées. . . . " *Journal du génie civil* 4 (1829): 143–157.

"Lettre au directeur du Journal du génie civil adressée par un ingénieur des Ponts et Chaussées. . . . " *Journal du génie civil* 4 (1829): 194–195.

Lévy, Théodore. *Mémoire sur l'administration des Ponts et Chaussées.* Paris: Dunod, 1872.

Martin, Emile. *Du fer dans les ponts suspendus.* Paris: Dondey-Dupré, 1832.

Martin, P. D. *Description du pont suspendu, construit sur la Garonne, à Langon.* Paris: Eberhart, 1832.

Mellet, François Nöel. "Des chemins de fer aux Etats-Unis d'Amérique." *Journal de l'industriel et du capitaliste* 2 (1836): 10–25.

Mellet and Henry. *L'arbitraire administratif des Ponts et Chaussées.* Paris: Imprimerie de Guiradet et Jouaust, 1835.

Minary and H. Résal. "Recherches expérimentales sur les propriétés physique du jet de l'injecteur automoteur." *Annales des mines,* sixth series, 1 (1862): 606–616.

Ministère de l'Agriculture, du Commerce, et des Travaux Publics. *Programmes de l'enseignement intérieur de l'Ecole des Ponts et Chaussées.* Paris: Thunot, 1867.

Ministère des Travaux Publics. *Laboratoire et atelier expérimental du nouveau dépôt de l'Ecole des Ponts et Chaussées.* Extrait des *Annales des ponts et chaussées* 1 (1871). Manuscript 137356, Bibliothèque Historique de la Ville de Paris.

Ministère des Travaux Publics. *Laboratoires de l'Ecole Nationale des Ponts et Chaussées: note sur leurs origines, leurs installations, les appareils et méthodes d'essai employés et leurs travaux.* Paris: Imprimerie Nationale, 1891.

Morin, Arthur. *Expériences sur . . . les turbines.* Metz: Thiel, 1838.

Morin, Arthur. *Hydraulique,* second edition. Paris: Hachette, 1858.

Navier, C. L. M. H. *De l'entreprise du pont des Invalides.* Paris: Firmin Didot, 1827.

Navier, C. L. M. H. "De l'exécution des travaux publics, et particulièrement des concessions." *Annales des Ponts et Chaussées: mémoires et documents* 1ᵉ trim. (1832): 1–31.

Navier, C. L. M. H. *Résumé des leçons données à l'École des Ponts et Chaussées sur l'application de la mécanique à l'établissement des constructions et des machines,* third edition. Paris: Dunod, 1864.

Navier, C. L. M. H. "De l'exécution des travaux publics et des concessions." *Annales des Ponts et Chaussées: mémoires* 1 (1832): 1–31.

Navier, C. L. M. H. "Détails historiques sur l'emploi du principe des forces vives dans la théorie des machines, et sur diverses roues hydrauliques." *Annales de chimie et de physique,* second series, 9 (1818): 146–159.

Navier, C. L. M. H. "Le pont de fer." *Le pandore* 1286 (25 November 1826): 2–3.

Navier, C. L. M. H. *Rapport à Monsieur Becquey et mémoire sur les ponts suspendus.* Paris: Imprimerie Royale, 1823.

Navier, C. L. M. H. *Rapport à Monsieur Becquey et mémoire sur les ponts suspendus,* second edition. Paris: Carilian-Goeury, 1830.

"Note relative à la disposition de chaînes du pont des Invalides." *Journal du génie civil* 2 (1829): 181–185.

"Note sur la nécessité de former en France une société portant la dénomination: Institut des Ingénieurs Civils de France." *Journal du génie civil* 13 (1846): 479–496.

"Notice sur les Sociétés d'Anciens Elèves des Ecoles d'Arts et Métiers." *Annuaire de la Société des Anciens Elèves des Ecoles d'Arts et Métiers* 1 (1848): 19–28.

Notice sur les travaux scientifiques de M. H. Résal. Paris: Gauthier-Villars, 1873.

Notice sur les travaux scientifiques de M. Reech. Paris: Gauthier-Villars, 1868.

"Nouveau mode d'admission pour le grade de conducteur des Ponts-et-Chaussées." *Journal des chemins de fer* 6 (1847): 621–623.

"Nouvelles réflexions relatives à l'administration des Ponts et Chaussées." *Journal du génie civil* 2 (1829): 350–359, 552–563

Observations . . . par la Société des Ingénieurs Civils sur le projet de décret relatif au mode de recrutement du Corps des Ponts et Chaussées. Paris: Imprimerie centrale de Napoléon Chaix, 1848.

Ordonnance du Roi portant création d'une Ecole de Mineurs à Saint-Etienne, département de la Loire. Paris: Imprimerie Royale, 1817. Manuscript 2198/C. 96. Ecole des Ponts et Chaussées.

Ordonnance du Roi relative à l'organisation . . . de l'Ecole des Mines. Paris: Imprimerie Royale, 1817. Manuscript 2194/C. 96. Ecole des Ponts et Chaussées.

Parallèle des édifices. Paris: M. Durand, 1801.

Plan du pont extraordinaire dit des Invalides. Paris, n.d. Manuscript 130394, Bibliothèque Historique de la Ville de Paris.

Perdonnet, Auguste. "De l'organisation du génie civil en France." *Journal des chemins de fer* 3 (1844): 29–30.

Poincaré, Henri. *Thermodyamique*, second edition. Paris: Gauthier-Villars, 1908.

Pont des Invalides: mémoire à consulter. [n.p.], 1826. Manuscript 134198, Bibliothèque Historique de la Ville de Paris.

"Pont suspendu des Champs-Elysées." *Journal du génie civil* 8 (1830): 144–149.

"Ponts en chaînes et en fils de fer." *Moniteur de l'industrie* 4 (1829): 100–101.

"Les Ponts et Chaussées et les entrepreneurs." *Journal des chemins de fer* 5 (1846): 736–737, 757–758, 805–807, 819–820; 6 (1847): 65–66.

Porée. *Notes sur l'organisation du corps des ingénieurs des Ponts et Chaussées.* Paris: Imprimerie Impériale, 1859. Manuscript 2154/C. 96. Ecole des Ponts et Chaussées.

Potiquet, Alfred. *Organisation des employés secondaires des Ponts et Chaussées,* second edition. Paris: Jousset, Clet, 1871.

Prony. "Notice biographique de Navier." *Annales des Ponts et Chaussées: mémoires et documents* 7, no. 1 (1837): 1–19.

Prony, G. Riche de. *Nouvelle architecture hydraulique* (2 volumes). Paris: Didot, 1790–1796.

Proposition de loi portant réorganisation du Corps des Ponts et Chaussées. Paris: Imprimerie du Journal Officiel, 1881. Manuscript 15842/C. 836. Ecole des Ponts et Chaussées.

"Quelques réflexions sur les mesures à prendre relativement à l'administration des Ponts et Chaussées." *Journal du génie civil* 2 (1829): 119–128.

Quenot, J. P. *Mémoire sur le pont suspendu en fil de fer construit sur la Charente à Jarnac.* Paris: Bachelier, 1828.

"Rapport fait au conseil de perfectionnement de l'Ecole Centrale des Arts et Manufactures, le 12 Juillet 1830." *Annales de l'industrie française et étrangère* 6 (1830): 190–198.

Reech. *Théorie de l'injecteur automoteur des chaudières à vapeur de M. H. Giffard.* Paris: Mallet-Bachelier, 1860.

"Réflexions sur l'adjudication des ponts suspendus de Bercy et de la rue des Saints-Pères." *Journal du génie civil* 10 (1831): 404–512.

"Remplacement des ponts d'Arcole et des Invalides, à Paris." *Annales des Ponts et Chaussées: mémoires et documents,* third series, 8 (1854): 246–247.

"Renseignements sur les écoles professionnelles en France: Ecole Impériale Centrale des Arts et Manufactures." *Annales du génie civil* 2, Part 2 (1863): 128–135.

"Renseignements sur les écoles professionnelles en France: Ecoles Impériales d'Arts et Métiers." *Annales du génie civil* 3 (1864): 582–586.

Résal, H. "Recherches théoriques sur les effets mécaniques de l'injecteur automoteur de M. Giffard." *Annales des mines,* sixth series, 1 (1862): 575–605. Extract in *Comptes rendus de l'Académie des Sciences* 53 (July–December 1861): 632–633.

Revue des systèmes d'injecteurs pour l'alimentation des chaudières et l'élévation de l'eau." *La propagation industrielle* 4 (1869): 21–22, 53–56, 113–116, 148–151, 179–181, 209–211, 360–363; 5 (1870): 24, 61–62, 93, 188–190, 236–238, 276–277, 312–315.

Seguin, Marc. *De l'influence des chemins de fer et de l'art de les tracer et de les construire.* Paris: Carilian-Goeury, 1839.

Seguin, Marc. *Description métrée et estimation d'un pont en fil de fer, construit sur la rivière de la Galore à Saint-Vallier, département de l'Isère.* Paris: Firmin Didot, 1824.

Seguin, Marc. *Des ponts en fil de fer,* second edition. Paris: Bachelier, 1826.

Seguin, Marc. *Mémoire sur les causes et sur les effets de la chaleur, de la lumière, et de l'électricité.* Paris: A. Tramblay, 1865.

Seguin, Marc et al. "Notice sur les ponts en fil de fer." *Annales de l'industrie nationale et étrangère* 12 (1823): 285–293.

Sganzin, Joseph Mathieu. *An Elementary Course of Civil Engineering,* third edition. Boston: Hilliard, Gray, Little, and Wilkins, 1827.

"Société des Ingénieurs Civils." *La propagation industrielle* 3 (1868): 28.

"Société des Ingénieurs Civils." *Revue industrielle* 8 (1878): 34.

"Statistique de l'industrie." *Le génie industriel* 18 (1859): 91–92.

Stewart, William. *Mémoire raisonné sur les ponts suspendus.* Paris: Paul Dupont, 1855.

Stourm. "Administration des Ponts et Chaussées." *Journal de l'industriel et du capitaliste* 3 (January 1837): 58.

Thirion, C. "Du role des ingénieurs civils dans l'industrie privée." *Le génie industriel* 27 (1864): 69. Also published in *La propagation industrielle* 1 (1865–66): 209–213, 247–251.

Thirion, C. "La succession de M. Perdonnet à la direction de l'Ecole Centrale des Arts et Manufactures." *La propagation industrielle* 2 (1867): 347–349.

Thomas, Emile. *Histoire des ateliers nationaux.* Paris: Michel Lévy, 1848.

Vallée, E. *Note sur quelques questions de travaux publics.* Paris: Librairie Lefrançois, 1877.

Vallée, L. L. *Améliorations à introduire dans les Ponts et Chaussées.* Paris: Carilian-Goeury, 1829.

Vignon, E.J.M. *Etudes historiques sur l'administration des voies publiques en France au XVIIᵉ et XVIIIᵉ siècle.* Paris: Dunod, 1862.

Villiers. "Résultats des expériences faites sur l'injecteur Giffard." *Annales des mines,* fifth series, 17 (1860): 357–366. Also in *Bulletin de la Société d'Encouragement pour l'Industrie Nationale* 60 (1861): 335–343.

Zuber, Ernest. "Rapport sur l'injecteur automateur des chaudières à vapeur de M. Giffard." *Bulletin de la Société Industrielle de Mulhouse* 29 (November 1859): 537–547.

Primary Sources: Britain and Elsewhere

An Account of Various Suspension Bridges, Particularly of Captain S. Brown's, Near Berwick-Upon-Tweed. (n.p., [ca. 1826]) Edwards manuscripts. National Museum of American History.

Barlow, Peter. *Essay on the Strength and Stress of Timber.* London: J. Taylor, 1817.

British Parliamentary Papers, House of Commons. Papers relating to building a Bridge over the Menai Strait near Bangor Ferry, Sess. 1819 (60), volume V.

British Parliamentary Papers, House of Commons. Third Report from the Select Committee on the Road from London to Holyhead; Menai Bridge, Sess. 1819 (256), volume V.

Brown, Samuel, ed. *Reports and Observations on the Patent Iron Cables Invented by Captain Samuel Brown.* London: T. Sotheran, 1815.

Brown, Samuel. "Description of the Trinity Pier of Suspension at Newhaven, near Edinburgh." *Edinburgh Philosophical Journal* 6 (October–April 1822): 22–28.

Buchanan, George. "Remarks on the strength of materials, with an account of several experiments on the transverse strength of wood and iron." *Edinburgh Philosophical Journal* 12 (1825): 154–163.

Buchanan, George. "Report on the present state of the wooden bridge at Montrose, and the practicability of erecting a suspended bridge of iron in its stead." *Edinburgh Philosophical Journal* 11 (April–October 1824): 140–156.

Chambers, William. *Designs of Chinese Buildings, Furniture, Dresses, Machines, and Utensils.* London: the author, 1757.

Chambers, William. *A Dissertation on Oriental Gardening.* London: W. Griffin, 1772.

Cordier, Joseph, trans. *Ponts et chaussées: essais sur la construction des routes, des ponts suspendus, des barrages, etc.; extraits de divers ouvrages anglais.* Lille: Reboux-Leroy, 1823.

Drewry, Charles Stewart. *A Memoir on Suspension Bridges.* London: Longman, Rees, et al., 1832.

Du Halde, Jean. *A Description of China.* London: T. Gardner, 1738.

Dupin, Pierre C.F. *The Commercial Power of Great Britain* (2 volumes). London: Charles Knight, 1825.

Fischer von Erlach, Johann Bernard. *Entwurff einer Historischen Architectur.* Leipzig: [n.p.], 1725.

Gilbert, Davies. "On some properties of the catenarian curve with reference to bridges of suspension." *Quarterly Journal of Science, Literature, and the Arts* (1821): 230–235.

Hutchinson, William. *The History and Antiquities of the County Palatine of Durham* (3 volumes). Carlisle, England: F. Jollie, 1785–1794.;

Hutton, Charles. *The Principles of Bridges.* Newcastle: T. Saint, 1772.

Juan, George, and Antonine de Ulloa. *Voyage Historique de l'Amérique Méridionale* (2 volumes). Amsterdam and Leipzig: Arkste'e and Merkus, 1752.

Provis, William Alexander. *An Historical and Descriptive Account of the Suspension Bridge over the Menai Strait in North Wales.* London: Ibotson & Palmer, 1828.

Rees, Abraham, ed. *The Cyclopaedia; or, Universal Dictionary of the Arts, Sciences, and Literature* (45 volumes). London: Longman, Hurst [1802–1819].

Rees, Abraham, ed. *The Cyclopaedia; or, Universal Dictionary of Arts, Sciences, and Miscellaneous Literature* (45 volumes; first American edition). Philadelphia: Samuel F. Bradford and Murray, Fairman, and Co., 1805–1822.

Schramm, Carl Christian. *Merkwürdigsten Brücken.* Leipzig: B. B. C. Breitkopf, 1735.

Sears, Robert. *Wonders of the World.* New York: E. Walker, 1847.

Stevenson, Robert. "Description of bridges of suspension." *Edinburgh Philosophical Journal* 5 (1821), April–October: 241–247.

Telford, Thomas. *Life of Thomas Telford.* London: James and Luke G. Hansard and Sons, 1838.

Telford, Thomas. Report, 1814, Respecting a Bridge over the River Mersey at Runcorn. Manuscript, Ironbridge Gorge Museum.

Turner, Samuel. *An Account of an Embassy to the Court of the Teshoo Lama in Tibet* (1800). New Delhi: Manjusri, 1971.

Ware, Samuel. *Tracts on Vaults and Bridges.* London: Thomas and William Boone, 1822.

Weale, John. *The Theory, Practice, and Architecture of Bridges of Stone, Iron, Timber, and Wire.* London: Architectural Library, 1843.

Woodcroft, B., ed. *Abridgements of Specifications Relating to Bridges, Viaducts, and Aqueducts, A.D. 1750–1866.* London: Office of the Commissioners of Patents for Inventions, 1868.

Young, Thomas. *A Course of Lectures on Natural Philosophy and the Mechanical Arts.* London: Joseph Johnson, 1807.

Secondary Sources: United States

Abraham, Evelyn. "Isaac Meason, the first ironmaster west of the Alleghenies." *Western Pennsylvania Historical Magazine* 20 (1937): 41–49.

Ahlstrom, Sydney. *A Religious History of the American People.* Yale University Press, 1972.

Ahlstrom, Sydney. "The Scottish philosophy and American theology." *Church History* 24 (1955), no. 3: 257–272.

Albright, Bear. "Palmer & Spofford: A heritage of building in the Merrimack River Valley." *Essex Institute Historical Collections* 122 (1986), no. 2: 142–164.

Allen, Richard Sanders. "Men who built bridges of wire and chain." *Consulting Engineer,* June 1962: 120–125.

Anderson, Edward Park. "The intellectual life of Pittsburgh, 1786–1836." Reprint from *Western Pennsylvania Historical Magazine,* 1931.

Baker, Ray Palmer. *A Chapter in American Education.* Scribner, 1924.

Baldwin, Leland D. *Whiskey Rebels: The Story of A Frontier Uprising.* University of Pittsburgh Press, 1939.

Ball, Rex Harrison. America in the French Liberal Mind, 1815–1871. Ph.D. dissertation, Harvard University, 1970.

Bartlett, Howard R. "The development of industrial research in the United States." In *Industrial Research: A National Resource.* U.S. Government Printing Office, 1941.

Bell, Whitfield J. "The scientific environment of Philadelphia, 1775–1790." *Proceedings of the American Philosophical Society* 92 (1948), March: 6–14.

Bining, Arthur Cecil. "Early ironmasters of Pennsylvania." *Pennsylvania History* 18 (1951), no. 2.

Bining, Arthur Cecil. "The iron plantations of early Pennsylvania." *Pennsylvania Magazine of History and Biography* 57 (1933), no. 2: 117–137.

Bining, Arthur Cecil. *Pennsylvania Iron Manufacture in the Eighteenth Century.* Harrisburg: Pennsylvania Historical Commission, 1973.

Bining, Arthur Cecil. "The rise of iron manufacturing in western Pennsylvania." *Western Pennsylvania Historical Magazine* 16 (1933): 235–256.

Birr, Kendall. *Pioneering in Industrial Research.* Public Affairs Press, 1957.

Birr, Kendall. "Science in American industry." In *Science and Society in the U.S.,* ed. D. Van Tassel and M. Hall. Homewood, Ill.: Dorsey, 1966.

Blumenthal, Henry. *American and French Culture, 1800–1900.* Louisiana State University Press, 1975.

Bode, Carl. *The American Lyceum: Town Meeting of the Mind.* Oxford University Press, 1956.

Boucher, John Newton. *History of Westmoreland County.* New York: Lewis, 1906.

Boyd, Julian P. ed. *The Papers of Thomas Jefferson.* Princeton University Press, 1956.

Boyer W. H., and Irving A. Jelly. "An early American suspension span." *Civil Engineering* 7 (1937), May: 338–340.

Bridenbaugh, Carl. *The Colonial Craftsman.* University of Chicago Press, 1950.

Bradford, James C., ed. *Captains of the Old Steam Navy: Makers of the American Naval Tradition, 1840–1880.* Naval Institute Press, 1986.

Bridenbaugh, Carl, and Jessica Bridenbaugh. *Rebels and Gentlemen.* New York: Reynal & Hitchcock, 1942.

Broderick, Francis L. "Pulpit, physics, and politics: The curriculum of the College of New Jersey, 1746–1794." *William and Mary Quarterly,* third series, 6 (1949): 42–68.

Buck, Solon J., and Elizabeth Hawthorn Buck. *The Planting of Civilization in Western Pennsylvania.* University of Pittsburgh Press, 1939.

Calhoun, Daniel H. *The American Civil Engineer, Origins and Conflict.* Harvard University Press, 1960.

Caller, Carmel. *Isaac Meason: The Man, Ironmaster and Businessman, His Mansion.* Connellsville, Pa.: Connellsville Historical Society, 1975.

Calvert, Monte. *The Mechanical Engineer in America, 1830–1910.* Johns Hopkins University Press, 1967.

Carlson, W. Bernard. "Academic entrepreneurship and engineering education: Dugald C. Jackson and the MIT-GE Cooperative Engineering Course, 1907–1932." *Technology and Culture* 29 (1988), no. 3: 536–567.

Carrell, William D. "American college professors: 1750–1800." *History of Education Quarterly* 8 (1968), fall: 289–305.

Carrell, William D. "Biographical list of American college professors to 1800." *History of Education Quarterly* 8 (1968), fall: 358–374.

Chalfant, Ella. *A Goodly Heritage.* University of Pittsburgh Press, 1955.

Chandler, Alfred D., Jr. *The Visible Hand: The Managerial Revolution in American Business.* Harvard University Press, 1977.

Condit, Carl. *American Building Art: The Nineteenth Century.* Oxford University Press, 1960.

Cooper, Carolyn. *Shaping Invention: Thomas Blanchard's Machinery and Patent Management in Nineteenth-Century America.* Columbia University Press, 1991.

Craighead, Rev. J. G. *Scotch and Irish Seeds in American Soil.* Philadelphia: Presbyterian Board of Publication, 1878.

Cremin, Lawrence A. *American Education: The Colonial Experience, 1607–1783.* Harper and Row, 1970.

Cronon, William. *Nature's Metropolis: Chicago and the Great West.* Norton, 1991.

Cronon, William. "Revisiting the vanishing frontier: The legacy of Frederick Jackson Turner." *Western Historical Quarterly* 18 (1987): 157–176.

Daniels, George. *American Science in the Age of Jackson.* Columbia University Press, 1968.

Danko, George Michael. The Evolution of the Simple Truss Bridge, 1790 to 1850: From Empiricism to Scientific Construction. Ph.D. dissertation, University of Pennsylvania, 1979.

Davis, Lance E., et al. *American Economic Growth: An Economist's History of the United States.* Harper and Row, 1972.

Day, Sherman. *Historical Collections of the State of Pennsylvania.* Philadelphia: George W. Gorton, 1843.

Dickson, R. J. *Ulster Emigration to Colonial America, 1718–1775.* Routledge, 1966.

Donehoo, George P., ed. *Pennsylvania: A History.* New York: Lewis Historical Publishing, 1926.

Duberman, David. American Civilization and French Travellers, 1865–1914. Ph.D. dissertation, University of Pennsylvania, 1963.

Dunaway, Wayland F. *The Scotch-Irish of Colonial Pennsylvania.* University of North Carolina Press, 1944.

Edwards, Llewellyn N. *A Record of History and Evolution of Early American Bridges.* University of Maine Press, 1959.

Edwards, Llewellyn N. "The evolution of early American bridges." *Newcomen Society Transactions* 11–13 (1930–1933): 95–116.

Eisenstadt, Abraham, ed. *Reconsidering Tocqueville's Democracy in America.* Rutgers University Press, 1988.

Elliott, Arlene Ann. The Development of the Mechanics' Institutes and Their Influence Upon the Field of Engineering: Pennsylvania, a Case Study. University of Southern California, 1972.

Ellis, Franklin, ed. *History of Fayette County, Pennsylvania.* Philadelphia: L. H. Everts, 1882.

Engle, Harry J. "Over 1000 Feet of continuity." *Engineering News-Record* 120 (1938), May 5: 651–653.

Everett, Edward. "John Smilie, forgotten champion of early western Pennsylvania." *Western Pennsylvania Historical Magazine* 33, no. 3–4 (1950): 77–89.

Ferguson, Eugene S. "The American-ness of American technology." *Technology and Culture* 20 (1979), no. 1: 3–24.

Ferguson, Russell J. *Early Western Pennsylvania Politics.* University of Pittsburgh Press, 1938.

Ferree, Barr. *Pennsylvania: A Primer.* New York: Leonard Scott, 1904.

Fisher, Marvin. *Workshops in the Wilderness: The European Response to American Industrialization, 1830–1860.* Oxford University Press, 1967.

Fleming, G. T. *History of Pittsburgh and Environs.* New York: American Historical Society, 1922.

Fletcher, Robert, and J. P. Snow. "A history of the development of wooden bridges." *Transactions of the American Society of Civil Engineers* 99 (1934): 325–332.

Ford, Henry Jones. *The Scotch-Irish in America*. Princeton University Press, 1966.

Ford, Peter A. "Charles S. Storrow, civil engineer: A case study of European training and technological transfer in the antebellum period." *Technology and Culture* 34 (1993), no. 2: 271–299.

Friedel, Robert, and Paul Israel. *Edison's Electric Light: Biography of an Invention*. Rutgers University Press, 1986.

Genovese, Eugene D. *The Political Economy of Slavery: Studies in the Economy and Society of the Slave South*. Vintage, 1965.

Goldstein, Jonathan. *Philadelphia and the China Trade, 1682–1846*. Pennsylvania State University Press, 1978.

Grayson, Lawrence P. "A brief history of engineering education in the United States." *Engineering Education* 68 (1977), no. 3: 246–263.

Greene, John C. *American Science in the Age of Jefferson*. Iowa State University Press, 1984.

Griess, Thomas E. Dennis Hart Mahan: West Point Professor and Advocate of Military Professionalism, 1830–1871. Ph.D. dissertation, Duke University, 1968.

Guthrie, Dwight Raymond. *John McMillan, The Apostle of Presbyterianism in the West, 1752–1833*. University of Pittsburgh Press, 1952.

Hadden, James. *A History of Uniontown, the County Seat of Fayette County, Pennsylvania*. Uniontown: James Hadden, 1913.

Hagner, Charles V. *Early History of the Falls of Schuylkill*. Philadelphia: Claxton, Remsen, and Haffelfinger, 1869.

Hahn, Edward. "Science in Pittsburgh, 1813–1848." *Western Pennsylvania Historical Magazine* 55 (1972): 66–75.

Hanna, Charles A. *The Wilderness Trail; Or, the Ventures and Adventures of the Pennsylvania Traders on the Allegheny Path*. Putnam, 1911.

Harding, Joseph W. *Hand Firing and Locomotive Attachments*. Scranton, Pa.: International Textbook, 1928.

Hart, Sidney. "'To encrease the comforts of life': Charles Willson Peale and the mechanical arts." *Pennsylvania Magazine of History and Biography* 110 (1986), no. 3: 323–357.

Hartz, Louis. *Economic Policy and Democratic Thought: Pennsylvania, 1776–1860*. Harvard University Press, 1948.

Hindle, Brooke, and Steven Lubar. *Engines of Change: The American Industrial Revolution, 1790–1860.* Smithsonian Institution Press, 1986.

Hindle, Brooke. *The Pursuit of Science in Revolutionary America, 1735–1789.* University of North Carolina Press, 1956.

Horn, W. F. *The Horn Papers: Early Westward Movement on the Monongahela and Upper Ohio, 1765–1795.* Hagstrom, 1945.

Hornberger, Theodore. *Scientific Thought in American Colleges, 1638–1800.* University of Texas Press, 1945.

Horowitz, Daniel. Insight into Industrialization: American Conceptions of Economic Development and Mechanization, 1865–1910. Ph.D. dissertation, Harvard University, 1967.

Howe, Malverd A. "An old chain suspension bridge." *Railroad Gazette* 32 (1900), January 5: 5.

Hubbard, Genevieve. French Travelers in America, 1775–1840: A Study of Their Observations. Ph.D. dissertation, American University, 1936.

Hughes, George W. "The pioneer iron industry in western Pennsylvania." *Western Pennsylvania Historical Magazine* 14 (1931): 207–224.

Hughes, Thomas P. *Networks of Power.* Johns Hopkins University Press, 1983.

Hughes, Thomas P. *American Genesis: A Century of Invention and Technological Enthusiasm.* Viking, 1989.

Hughes, Thomas P. *Elmer Sperry: Inventor and Engineer.* Johns Hopkins University Press, 1971.

Jakkula, Arne Arthur. "A history of suspension bridges in bibliographical form." *Texas Engineering Experiment Station Bulletin,* fourth series, 12 (1941), no. 7.

Jenkins, Howard M. *Pennsylvania Colonial and Federal: A History, 1608–1905.* Philadelphia: Pennsylvania Historical Publishing Association, 1905.

Jenkins, Reese V. *Images and Enterprise: Technology and the American Photographic Industry, 1839–1925.* Johns Hopkins University Press, 1975.

Jones, Russell Mosely. The French Image of America, 1830–2848. Ph.D. dissertation, University of Missouri, 1957.

Jordan, John W., ed. *Genealogical and Personal History of Fayette and Greene Counties.* New York: Lewis Historical Publishing, 1912.

Josephson, Matthew. *Edison.* McGraw-Hill, 1959.

Kasson, John F. *Civilizing the Machine: Technology and Republican Values in America, 1776–1900.* Penguin, 1976.

Kemp, Emory L. "Ellet's contribution to the development of suspension bridges." *Engineering Issues* 99 (1973), no. PP3: 331–351.

Kemp, Emory. "Thomas Paine and his 'pontifical matters.'" *Newcomen Society Transactions* 49 (1977–78): 21–40.

Kepner, Charles W. The Contributions Early Presbyterian Leaders Made in the Development of the Educational Institutions in Western Pennsylvania Prior to 1850. Ph.D. dissertation, University of Pittsburgh, 1942.

Klett, Guy S. *Presbyterians in Colonial Pennsylvania*. University of Pennsylvania Press, 1935.

"Kneass, Strickland Landis." *National Cyclopedia of Biography*.

Kouwenhoven, John A. *The Arts in Modern American Civilization*. Norton, 1967.

Kouwenhoven, John H. "The designing of the Eads Bridge." *Technology and Culture* 23 (1982): 535–568.

Latimer, Margaret, Brooke Hindle, and Melvin Kranzberg, eds. "Bridge to the future: A centennial celebration of the Brooklyn Bridge." *Annals of the New York Academy of Sciences* 424 (1984).

Layton, Edwin T. "American ideologies of science and engineering." *Technology and Culture* 17 (1976), no. 4: 688–701.

Layton, Edwin T. "European origins of the American engineering style of the nineteenth century." In *Scientific Colonialism*, ed. N. Reingold and M. Rothenberg. Smithsonian Institution, 1987.

Layton, Edwin T. "Millwrights and engineers, science, social roles, and the evolution of the turbine in America." In *The Dynamics of Science and Technology*, ed. W. Krohn et al. Reidel, 1978.

Layton, Edwin T. "Mirror image twins: The communities of science and technology in nineteenth-century America." *Technology and Culture* 12 (1971), October: 562–580.

Layton, Edwin T. *The Revolt of the Engineers: Social Responsibility and the American Engineering Profession*. Case Western Reserve University Press, 1971.

Layton, Edwin T. "Scientific technology, 1845–1900: The hydraulic turbine and the origins of American industrial research." *Technology and Culture* 20 (1979), January: 64–89.

Lewis, Gene D. *Charles Ellet, Jr.*. University of Illinois Press, 1968.

Leyburn, James G. *The Scotch-Irish: A Social History*. University of North Carolina Press, 1962.

Lowdermilk, Will H. *History of Cumberland (Maryland)*. Washington, D.C.: James Anglin, 1875.

Lundberg, David and Henry F. May. "The enlightened reader in America." *American Quarterly* 28 (1976), no. 2: 262–271.

MacReynolds, George. *Place Names in Bucks County, Pennsylvania,* second edition. Doylestown, Pa.: Bucks County Historical Society, 1976.

Martin, Asa Earl, and Hiram Herr Shenk. *Pennsylvania History Told by Contemporaries.* Macmillan, 1925.

May, Henry F. *The Enlightenment in America.* Oxford University Press, 1976.

McGaw, Judy. "Accounting for innovation: Technological change and business practice in the Berkshire County paper industry." *Technology and Culture* 26 (1985), no. 4: 703–725.

McGivern, James Gregory. *First Hundred Years of Engineering Education in the United States, 1807–1907.* Spokane: Gonzaga University Press, 1960.

McHugh, Jeanne. *Alexander Holley and the Makers of Steel.* Johns Hopkins University Press, 1980.

McMahon, A. *The Making of a Profession: A Century of Electrical Engineering in America.* IEEE Press, 1984.

Meier, Hugo A. The Technological Concept in American Social History, 1750–1860. Ph.D. dissertation, University of Wisconsin, 1956.

Meier, Hugo A. "Technology and democracy, 1800–1860." *Mississippi Valley Historical Review* 43 (1959), March: 618–640.

"Memoir of Thomas Gilpin (found among the papers of Thomas Gilpin, Jr.)," *Pennsylvania Magazine of History and Biography* 49 (1925), no. 4: 289–328.

Michael, Ronald L., and Ronald C. Carlisle. "A log settler's fort/home." *Pennsylvania Folklife* 25 (1976), no. 3: 39–46.

Miller, Howard. *The Revolutionary College: American Presbyterian Higher Education, 1707–1837.* New York University Press, 1976.

Miller, Howard S. *The Eads Bridge.* University of Missouri Press, 1979.

Mills, A. P. "Tests of wrought iron links after 100 years in service." *Cornell Civil Engineer* 19 (1911), April: 251–282.

Molloy, Peter Michael. Technical Education and the Young Republic: West Point as America's Ecole Polytechnique, 1802–1833. Ph.D. dissertation, Brown University, 1975.

Montgomery, Morton L. *History of Reading, Pennsylvania, 1748–1898.* Reading: Times Book Printer, 1898.

Moore, E. Earl. "An introduction to the Holker papers." *Western Pennsylvania Historical Magazine* 42 (1959): 225–239.

Morrison, James L., Jr. *"The Best School in the World": West Point, the Pre-Civil War Years, 1833–1866.* Kent State University Press, 1986.

Mulkearn, Lois, and Edwin V. Pugh. *A Traveler's Guide to Historic Western Pennsylvania.* University of Pittsburgh Press, 1954.

Multhauf, Robert P. *A Catalogue of Instruments and Models in the Possession of the American Philosophical Society.* American Philosophical Society, 1961.

Nelson, Lee H. *The Colossus of 1812: An American Engineering Superlative.* American Society of Civil Engineers, 1990.

Nettels, Curtis P. *The Emergence of a National Economy, 1775–1815.* Holt, Rinehart and Winston, 1962.

Nevin, Alfred, ed. *Encyclopaedia of the Presbyterian Church in the United States.* Philadelphia: Presbyterian Publishing, 1884.

Nichols, Roger L., ed. *American Frontier and Western Issues: A Historiographical Review.* Greenwood, 1986.

Noble, David F. *America by Design: Science, Technology, and the Rise of Corporate Capitalism.* Oxford University Press, 1977.

"A note on early American suspension bridges." *Engineering News* 53 (1905), no. 11: 269–270.

Nugent, Walter. *Structures of American Social History.* Indiana University Press, 1981.

Pease, George B. "Timothy Palmer: Bridgebuilder of the eighteenth century." *Essex Institute Historical Collections* 83 (1947): 97–111.

Petroski, Henry. *Engineers of Dreams: Great Bridge Builders and the Spanning of America.* Knopf, 1995.

Plane, Ann. Popular science, industry, and social class: Amos Eaton and the rise of technical education, 1820–1840. Unpublished paper, Yale University, 1985.

Powers, Fred Perry. "Historic bridges of Philadelphia." *Philadelphia Historical Society Papers* 11 (1917): 300–305.

Prescott, Samuel C. *When M.I.T. was "Boston Tech," 1861–1916.* Technology Press, 1954.

Rattner, Sidney, et al. *The Evolution of the American Economy: Growth, Welfare, and Decision Making.* Basic Books, 1979.

Reich, Leonard S. *The Making of American Industrial Research: Science and Business at GE and Bell, 1876–1926.* Cambridge University Press, 1985.

Reuss, Martin. "Andrew A. Humphreys and the development of hydraulic engineering: Politics and technology in the Army Corps of Engineers, 1850–1950." *Technology and Culture* 26 (1985): 1–33.

Reynolds, Terry S. "The education of engineers in America before the Morrill Act of 1862." *History of Education Quarterly* 32 (1992), no. 4: 465–465.

Roberts, Charles Rhoads. *History of Lehigh County, Pennsylvania.* Allentown: Lehigh Valley, 1914.

Rosenberg, Robert. "The origins of EE education: A matter of degree." *IEEE Spectrum* 21 (1984), no. 7: 60–68.

Ross, James Earle. *Democracy's College: The Land-Grant Movement in the Formative State.* Arno Press, 1969.

Rowe, James W. *Old Westmoreland in History.* Scottsdale, Pa., 1934.

Safford, Arthur T., and Edward P. Hamilton. "The American mixed-flow turbine and its setting." *Transactions of the American Society of Civil Engineers* 85 (1922): 1237–1356.

Sarrett, Agnes Lynch. *Through One Hundred and Fifty Years: The University of Pittsburgh.* University of Pittsburgh Press, 1937.

Seely, Bruce. "Research, engineering, and science in American engineering colleges." *Technology and Culture* 34, no. 2 (1993): 344–386.

Shallot, Todd. "Building waterways, 1802–1861: Science and the United States Army in early public works." *Technology and Culture* 31 (1990), no. 1: 18–50.

Sherman, Edward C. "The reconstruction of the old chain bridge at Newburyport, Massachusetts." *Engineering News* 70 (1913), September 25: 585–587.

Sinclair, Bruce. *A Centennial History of the American Society of Mechanical Engineers, 1880–1980.* University of Toronto Press and ASME, 1980.

Sinclair, Bruce. *Early Research at the Franklin Institute: The Investigation into the Causes of Steam Boiler Explosions, 1830–1837.* Franklin Institute, 1966.

Sinclair, Bruce. *Philadelphia's Philosopher Mechanics: A History of the Franklin Institute, 1824–1865.* Johns Hopkins University Press, 1974.

Shank, William H. *Historic Bridges of Pennsylvania,* second edition. American Canal and Transportation Center, 1974.

Sharp, Myron B. and William H. Thomas. "A guide to the old stone blast furnaces in western Pennsylvania: Part III: Fayette County." *Western Pennsylvania Historical Magazine* 48 (1965): 271–282.

Shyrock, Richard. "American indifference to basic science during the nineteenth century." *Archives internationales d'histoire des sciences* 28 (1948), no. 5: 50–65.

Sloan, Douglas. *The Scottish Enlightenment and the American College Ideal.* Columbia University Teachers' College Press, 1971.

Slocum, S. E. "A unique pioneer bridge." *Engineering News-Record* 110 (1933), May 25: 691.

Slotkin, Richard. *The Fatal Environment: The Myth of the Frontier in the Age of Industrialization, 1800–1890*. Atheneum, 1985.

Sokoloff, Kenneth L. "Inventive activity in early industrial America: evidence from patent records." *Journal of Economic History* 48, no. 3–4 (1988): 813–350.

Sokoloff, Kenneth L., and B. Zorina Khan. "The democratization of invention during early industrialization: Evidence from the United States, 1790–1846." *Journal of Economic History* 50, no. 1–2 (1990): 363–378.

Soltow, Lee. *Distribution of Wealth and Income in the United States in 1798*. University of Pittsburgh Press, 1989.

Struik, Dirk J. *Yankee Science in the Making*. Collier, 1962.

Taylor, George Rogers. *The Transportation Revolution, 1815–1860*. Holt, Rinehart and Winston, 1951.

Tinkcom, Harry Marlin. *The Republicans and Federalists in Pennsylvania, 1790–1801: A Study in National Stimulus and Local Response*. Pennsylvania Historical and Museum Commission, 1950.

Tocqueville, Alexis de. *Democracy in America*. Knopf, 1945.

Trachtenberg, Alan. *Brooklyn Bridge: Fact and Symbol*, second edition. University of Chicago Press, 1979.

Turner, Frederick Jackson. *Frontier and Section: Selected Essays of Frederick Jackson Turner*. Prentice-Hall, 1961.

Turner, Frederick Jackson. *The Frontier in American History* (1920). Reprinted by Holt, Rinehart and Winston, 1962.

Uberti, John Richard. Men, Manners, & Morals: The Young Man's Institute in Antebellum Philadelphia. Ph.D. dissertation, University of Pennsylvania, 1977.

Walkinshaw, Lewis Clark. *Annals of Southwestern Pennsylvania*. New York: Lewis Historical Publishing, 1939.

Webster, Richard. *A History of the Presbyterian Church in America*. Philadelphia: Joseph M. Wilson, 1857.

Wilson, Erasmus, ed. *Standard History of Pittsburgh, Pennsylvania*. Chicago: H. R. Cornell, 1898.

Woley, Richard T. *Monongahela, the River and Its Region*. Butler, Pa.: Ziegler, 1937.

Worner, William Frederick. "The Strasburg Scientific Society." *Lancaster County Historical Society Papers* 25 (1921), no. 9: 133–145.

Secondary Sources: France

Aguillon, Louis. *L'Ecole des Mines de Paris: Notice historique.* Paris: Dunod, 1889.

Aldrich, R. "Late-comer or early starter? New views on French economic history." *Journal of European Economic History* 16 (1987): 89–100.

Allen, Turner W. The Highway and Canal System in Eighteenth-Century France. Ph.D. dissertation, University of Kentucky, 1953.

Amouroux, Dominique and Bertrand Lemoine, "L'Age d'or des ponts suspendus en France." *Annales des Ponts et Chaussées,* new series, 19 (1981): 53–63.

Artz, Frederick B. *The Development of Technical Education in France, 1500–1850.* MIT Press, 1966.

Barrault, Emile. "The history of a great invention: The Giffard injector." *Scientific American* 47 (1882), September 9: 163–164.

Belhoste, Bruno, Amy Dahan Dalmedico, and Antoine Piconet al., eds. *La formation polytechnicienne, 1794–1994.* Paris: Dunod, 1994.

Belhoste, Bruno, Amy Dahan Dalmedico, Dominique Pestre, and Antoine Picon. *La France dex X: Deux siècles d'histoire.* Paris: Economica, 1995.

Beugnot, Auguste A. *Vie de Becquey.* Paris: Didot, 1852.

Blanchard, Anne. *Dictionnaire des ingénieurs militaires, 1691–1791.* Montpellier: Université Paul Valéry, 1981.

Blanchard, Anne. *Les ingénieurs du 'Roy' de Louis XIV à Louis XVI.* Collection du centre d'histoire militaire et d'études de défense nationale de Montpellier. Montpellier: Déhan, 1979.

Bourdieu, Pierre. *La noblesse d'état: grandes écoles et esprit de corps.* Paris: Editions du Minuit, 1989.

Bradley, Margaret. "Civil engineering and social change: The early history of the Paris *Ecole des Ponts et Chaussées.*" *History of Education* 14 (1985), no. 3: 171–183.

Brunot, A., and R. Coquand. *Le Corps des Ponts et Chaussées.* Paris: Editions du Centre National de la Recherche Scientifique, 1982.

Bucciarelli, Louis, and N. Dworsky. *Sophie Germain: An Essay in the History of the Theory of Elasticity.* Reidel, 1980.

Butricia, Andrew J. "The *Ecole Supérieure de Télégraphie* and the beginnings of French electrical engineering education." *IEEE Transactions on Education* E30 (1987), no. 3: 121–129.

Cahen, François. "L'évolution en France de l'enseignement de l'électrotechnique des courants forts." in *Soixanteaire de l'Ecole Supérieure d'Electricité.* Paris: Chaix, 1955.

Cameron, Rondo E. *France and the Economic Development of Europe, 1800–1914*. Princeton University Press, 1961.

Cameron, Rondo and Charles E. Freedeman. "French economic growth: A radical revision." *Social Science History* 7 (1983): 3–30.

Carter, Edward C., Robert Foster, and Joseph N. Moody, eds. *Enterprise and Entrepreneurs in Nineteenth and Twentieth Century France*. Johns Hopkins University Press, 1976.

Charmasson, Thérèse. *L'Enseignement technique de la révolution à nos jours*. Paris: Economica, 1987.

Chevalier, Louis. *Laboring Classes and Dangerous Classes in Paris During the First Half of the Nineteenth Century*. Princeton University Press, 1973.

Chicoteau, Y., A. Picon, and C. Rochant. "Gaspard Riche de Prony ou le génie 'appliqué." *Culture technique* 12 (1984), March: 176–177.

Clapham, J. H. *The Economic Development of France and Germany, 1815–1914*. Cambridge University Press, 1936.

Clough, S. B. "Retardative factors in French economic development in the 19th and 20th centuries." *Journal of Economic History* 6, Supp. (1946): 91–210.

Comberouse, Charles Julien Felix de. *Histoire de l'Ecole Centrale des Arts et Manufactures*. Paris: Gauthier-Villars, 1879.

Coste, Anne, et al., eds. *Un ingénieur des lumières: Emiland-Marie Gauthey*. Paris: Ecole des Ponts et Chaussées, 1993.

Cotte, Michel. "Seguin et Cie (1806–1824): Du négoce familial de drap à la construction du pont suspendu de Tournon-Tain." *History and Technology* 6 (1988): 95–144.

Cotte, Michel. "Le système technique des seguin en 1824–35." *History and Technology* 7 (1990), no. 2: 119–147.

Crafts, N. F. R. "Economic growth in France and Britain, 1830–1910: A review of the evidence." *Journal of Economic History* 44 (1984): 49–67.

Crosland, Maurice. *Science under Control: The French Academy of Sciences, 1795–1914*. Cambridge University Press, 1992.

Crosland, Maurice. *The Society of Arcueil: A View of French Science at the Time of Napoleon I*. Harvard University Press, 1967.

Crouzet, F. "England and France in the eighteenth century: A comparative analysis of two economic growths." In *The Causes of the Industrial Revolution*, ed. R. Hartwell. Methuen, 1967.

Daumard, Adeline. "Les élèves de l'Ecole Polytechnique de 1815 à 1848." *Revue d'histoire moderne et contemporaine* 5 (1958), July–September: 226–234.

Day, Charles R. "Des ouvriers aux ingéniers?" *Culture technique* 12 (1984), March: 281–291.

Day, Charles R. "The development of higher primary and intermediate technical education in France, 1800 to 1870." *Historical Reflections/Réflexions historiques* 3 (1976), no. 2: 49–67.

Day, Charles R. "The making of mechanical engineers in France: The Ecoles d'Arts et Métiers, 1803–1914." *French Historical Studies* 10 (1978), no. 3: 439–460.

Day, Charles R. *Education for the Industrial World: The Ecoles d'Arts et Métiers and the Rise of French Industrial Engineering.* MIT Press, 1987.

Delbecq, Jean-Michel. "Analyse de la stabilité des voûtes en maçonnerie de Charles Augustin Coulomb à nos jours." *Annales des Ponts et Chaussées* 1981, 3ᵉ trim.: 36–43.

Dunham, Arthur Louis. *The Industrial Revolution in France, 1815–1848.* New York: Exposition Press, 1955.

Edmonson, James M. From *Mécanicien* to *Ingénieur:* Technical Education and the Machine-Building Industry in Nineteenth-Century France. Ph.D. dissertation, University of Delaware, 1981.

Ernouf, Alfred Auguste, le Baron. *Histoire de quatre inventeurs français au 19ième siècle: Sauvage, Heilmann, Thimonnier, Giffard.* Paris: Hachette, 1884.

Fayet, Joseph. *La révolution française et la science, 1789–1795.* Paris: Marcel Rivière, 1960.

Feline-Romany. "Notice historique sur les ponts de Paris." *Annales des Ponts et Chaussées: mémoires et documents* 8 (1864): 170–177.

Fourcy, A. *Histoire de l'Ecole Polytechnique.* Ed. Jean Dhombres. Paris: Belin, 1987.

Fox, Robert, and Anna Guagnini, eds. *Education, Technology, and Industrial Performance in Europe, 1850–1939.* Cambridge University Press, 1993.

Fox, Robert, and George Weisz, eds. *The Organization of Science and Technology in France, 1808–1914.* Cambridge University Press, 1980.

Fox, Robert. "Science, industry, and the social order in Mulhouse, 1798–1871." *British Journal for the History of Science* 17 (1984): 127–169.

Geison, Gerald L., ed. *Professions and the French State, 1700–1900.* Philadelphia: University of Pennsylvania Press, 1984.

Gillispie, Charles Coulston. *Lazare Carnot, Savant.* Princeton University Press, 1971.

Gillispie, Charles Coulston. *The Montgolfier Brothers and the Invention of Aviation.* Princeton University Press, 1983.

Gillispie, Charles Coulston. *Science and Polity in France at the End of the Old Regime.* Princeton University Press, 1980.

Gillmor, C. Stewart. *Coulomb and the Evolution of Engineering in Eighteenth-Century France.* Princeton University Press, 1971.

Guillerme, André. *Corps à corps sur la route: Les routes, les chemins et l'organisation des services au XIXème siècle.* Paris: Ecole des Ponts et Chaussées, 1984.

Guillerme, André. "La formation des nouveaux édiles: Ingénieurs des Ponts et Chaussées et architectes, 1804–1815." *History and Technology* 5 (1988): 223–247.

Guinness, I. Grattan. "Modes and manners of applied mathematics: The case of mechanics," in *The History of Modern Mathematics*, volume 2, ed. D. Rowe and J. McCleary. Academic Press, 1989.

Guinness, I. Grattan. "The *ingénieur savant*, 1800–1830: A neglected figure in the history of French mathematics and science." *Science in Context* 6 (1993), no. 2: 405–433.

Guinness, I. Grattan. *Convolutions in French Mathematics, 1800–1840: From the Calculus and Mechanics to Mathematical Analysis and Mathematical Physics.* Basel: Birkhäuser Verlag, 1990.

Guinness, I. Grattan. *Joseph Fourier, 1768–1830.* MIT Press, 1972.

Hahn, Roger and René Taton, eds. *Ecoles techniques et militaires au XVIIIᵉ siècle.* Paris: Hermann, 1986.

Hahn, Roger. *The Anatomy of a Scientific Institution: The Paris Academy of Sciences, 1666–1803.* Berkeley: University of California Press, 1971.

Hamilton, Stanley B. "Charles Auguste de Coulomb: A bicentenary appreciation of a pioneer in the science of construction." *Newcomen Society Transactions* 17 (1936–37): 27–49.

Herivel, John. *Joseph Fourier: The Man and the Physicist.* Clarendon, 1975.

Herléa, Alexandre. "Préliminaires à la naissance des laboratoires publics de recherche industrielle en France." *Culture technique* 18 (1988), March: 220–231.

Horiuchi, Tatsuo. "La formation des ingénieurs civils en France, 1795–1848." *Historia Scientiarum: International Journal of the History of Science Society of Japan* 23 (1982), September: 45–61.

Jacomy, Bruno. "A la recherche de sa mission: La Société des ingénieurs civils." *Culture technique* 12 (1984), March: 209–219.

Kemp, Tom. *Economic Forces in French History.* Longman, 1972.

Keyder, Caglar, and P. K. O'Brien. *Economic Growth in Britain and France, 1780–1914: Two Paths to the 20th Century.* Allen and Unwin, 1978.

Kindleberger, Charles P. *The Economic Growth of France and Britain, 1851–1950*. Harvard University Press, 1964.

Kranakis, Eda. "The affair of the Invalides Bridge." *Jaarboek voor de Geschiedenis van Bedrijf en Techniek* 4 (1987): 106–130.

Kranakis, Eda. "Navier's theory of suspension bridges." In *From Ancient Omens to Statistical Mechanics: Essays on the Exact Sciences Presented to Asger Aaboe*, ed. J. Berggren and B. Goldstein (*Acta Historica Scientiarum Naturalium et Medicinalium* 39 (1987)).

Landes, David. *The Unbound Prometheus: Technological Change and Industrial Development in Western Europe from 1750 to the Present*. Cambridge University Press, 1969.

Langins, Janis. *La république avait besoin de savants: Les débuts de l'Ecole polytechnique: l'Ecole centrale des travaux publics et les cours révolutionnaires de l'an III*. Paris: Belin, 1987.

Lemoine, Bertrand. *La Tour de Monsieur Eiffel*. Gallimard, 1989.

Lévy-Leboyer, Maurice. "La croissance economique en France au XIXe siècle." *Annales: Economie, Société, Culture* 3 (1968), no. 4: 788–807.

Magraw, Roger. *France 1815–1914: The Bourgeois Century*. Fontana, 1987.

Mandelbaum, Jonathan R. The "Société Philomathique de Paris" from 1788 to 1835: A Study in the Institutional History and Collective Biography of a Parisian Scientific Society. Ph.D. dissertation, Ecole des Hautes Etudes en Sciences Sociales, 1980.

Marc Seguin. Annonay: Comité du Monument Marc Seguin, 1913.

Marcel, H. "Les débuts de Jean-Rodolphe Perronet ingénieur des Ponts et Chaussées de la généralité d'Alençon." *Annales des Ponts et Chaussées*, July–August 1948: 419–440.

Métivier, Michel, Pierre Costabel, and Pierre Dugac, eds. *Siméon-Denis Poisson et la science de son temps*. Palaiseau: Ecole Polytechnique, 1981.

Noblet, Jocelyn de, ed.. *Le génie civil. Culture technique* 26 (1992), December.

Nye, Mary Jo. *Science in the Provinces: Scientific Communities and Provincial Leadership in France, 1860–1930*. University of California Press, 1986.

Oliver, James Montgomery. The "*Corps des Ponts et Chaussées*," 1830–1848. Ph.D. dissertation, University of Missouri, 1967.

Paul, Harry W. "The issue of the decline of French science in the nineteenth century." *French Historical Review* 7 (1972): 416–450.

Payen, Jacques. "The role of the Conservatoire Nationale des Arts et Métiers in the development of technical education up to the middle of the 19th century." *History and Technology* 5 (1988): 95–138.

Pestre, Dominique. *Physique et physiciens en France, 1918–1940.* Paris: Editions des Archives Contemporaines, 1984.

Petot, Jean. *Histoire de l'administration des Ponts et Chaussées, 1599–1815.* Paris: Marcel Rivière, 1958.

Picard, Alfred. "Note historique sur l'administration des travaux publics." [n.p.], ca. 1885. Manuscript 15842/C. 844, Ecole des Ponts et Chaussées.

Picard, Emile. *Eloges et discours académiques.* Gauthier-Villars, 1931.

Picon, Antoine. "Les ingénieurs et l'idéal analytique à la fin du XVIIIᵉ siècle." *Sciences et techniques en perspective* 13 (1987–1988): 70–108.

Picon, Antoine. "Les ingénieurs et la mathématisation: L'exemple du génie civil et de la construction." *Revue d'histoire des sciences* 42, no. 1–2 (1989): 155–172.

Picon, Antoine. "Navier and the introduction of suspension bridges in France." *Construction History* 4 (1988): 21–34.

Picon, Antoine. *French Architects and Engineers in the Age of Enlightenment.* Cambridge University Press, 1992.

Picon, Antoine. *L'Invention de l'ingénieur moderne: L'Ecole des Ponts et Chaussées, 1747–1851.* Paris: Ecole des Ponts et Chaussées, 1992.

Picon, Antoine. "Les rapports entre sciences et techniques dans l'organisation du savoir." *Revue de synthèse* 4, no. 1–2 (1994): 103–120.

Plasseraud, Yves, and François Savignon. *Paris 1883: Genèse du droit unioniste des brevets.* Paris: Litec, 1983.

Pyenson, Susan Sheets. Low Scientific Culture in London and Paris, 1820–1875. Ph.D. dissertation, University of Pennsylvania, 1976.

Querrien, Anne. "Ecoles et corps: Le cas des Ponts et Chaussées, 1747–1848." *Les annales de la recherche urbaine* 5 (1979): 81–114.

Querrien, Anne. "Le travail quotidien de l'ingénieur des Ponts et Chaussées aux environs de 1830 dans le département des Côtes-du-Nord." *Annales des Ponts et Chaussées* 1981, 3ᵉ trim.: 26–31.

Redondi, Pietro. "Nation et entreprise: La Société d'encouragement pour l'industrie nationale, 1801–1815." *History and Technology* 5 (1988): 193–222.

Ribeill, Georges. "Les associations d'anciens élèves d'écoles d'ingénieurs des origines à 1914." *Revue française de sociologie* 27 (1986): 317–338.

Ribeill, Georges. *La révolution ferroviaire: la formation des compagnies de chemins de fer en France (1823–1870).* Paris: Belin, 1993.

Richards, Joan L. "Rigor and Clarity: Foundations of mathematics in France and England, 1800–1840." *Science in Context* 4 (1991), no. 2: 297–319.

Shinn, Terry. "Des corps de l'état au secteur industriel: Genèse de la profession d'ingénieur, 1750–1920." *Revue française de sociologie* 19 (1978): 39–71.

Shinn, Terry. "Des sciences industrielles aux sciences fondamentales: La mutation de l'Ecole Supérieure de Physique et de Chimie (1882–1970)." *Revue française de sociologie* 22 (1981): 167–182.

Shinn, Terry. "The French science faculty system, 1808–1914: Institutional change and research potential in mathematics and the physical sciences." *Historical Studies in the Physical Sciences* 10 (1979): 271–332.

Shinn, Terry. "The genesis of French industrial research, 1880–1940." *Social Science Information* 19 (1980), no. 3: 607–640.

Shinn, Terry. "Raisonnement déductif: Formation et déformation." *Bulletin de la Société française de physique* 42 (1981): 32–34.

Shinn, Terry. "Reactionary technologists: The struggle over the Ecole Polytechnique, 1880–1914." *Minerva* 22, no. 3–4 (1984): 329–345.

Shinn, Terry. *Savoir Scientific et Pouvoir Sociale: L'Ecole Polytechnique, 1794–1914.* Paris: Presses de la Fondation Nationale des Sciences Politiques, 1980.

Smith, Cecil O., Jr. "The longest run: Public engineers and planning in France." *American Historical Review* 95 (1990), no. 3: 657–692.

Société des Amis de l'Ecole Polytechnique. *L'Ecole Polytechnique.* Gauthier-Villars, 1932.

Soixantenaire de l'Ecole Supérieure d'Electricité, 1894–1954. Paris: Chaix, 1955.

Stein, Margo. *Working to Rule: The Social Origins of French Engine Drivers.* Ph.D. dissertation, Harvard University, 1977.

Taton, René, ed., *Enseignement et diffusion des sciences en France au XVIIIᵉ Siècle,* second edition. Paris: Hermann, 1986.

Taton, René. *L'oeuvre scientifique de Gaspard Monge.* Paris: Presses Universitaires de France, 1951.

Thépot, André. "Les ingénieurs du Corps des Mines." *Culture Technique* 12 (1984), March: 55–61.

Weiss, John H. "The lost baton: The politics of intraprofessional conflict in nineteenth-century french engineering." *Journal of Social History* 16 (1982), no. 1: 1–19.

Weiss, John H. *The Making of Technological Man: The Social Origins of French Engineering Education.* MIT Press, 1982.

Weisz, George. *The Emergence of Modern Universities in France, 1863–1914.* Princeton University Press, 1983.

Zeldin, Theodore. *France 1848–1945: Ambition and Love.* Oxford University Press, 1979.

Secondary Sources: General and Other

Alexander, W., and A. Street. *Metals in the Service of Man.* Penguin, 1954.

Asplund, Sven Olof. *On the Deflection Theory of Suspension Bridges.* Almqvist & Wiksell, 1943.

Beamish, Richard. *Memoir of the Life of Sir Marc Isambard Brunel.* Longman, Green, 1862.

Beckett, Derrick. *Great Buildings of the World: Bridges.* Hamlyn, 1969.

Bennett, Charles Alpheus. *History of Manual and Industrial Education up to 1870.* Peoria: Manual Arts Press, 1926.

Bijker, Wiebe E., and John Law, eds. *Shaping Technology/Building Society: Studies in Sociotechnical Change.* MIT Press, 1992.

Bijker, Wiebe E., Thomas P. Hughes, and Trevor J. Pinch, eds. *The Social Construction of Technological Systems: New Directions in the Sociology and History of Technology.* MIT Press, 1987.

Billington, David P. "History and Esthetics in Suspension Bridges." *Journal of the Structural Division: Proceedings of the American Society of Civil Engineers* 103 (1977), no. ST8: 13143–13197.

Billington, David P. *Robert Maillart's Bridges: The Art of Engineering.* Princeton University Press, 1979.

Billington, David P. *The Tower and the Bridge: The New Art of Structural Engineering.* Basic Books, 1983.

Bucciarelli, Louis L. *Designing Engineers.* MIT Press, 1994.

Bucciarelli, Louis L. "Engineering design process." In *Making Time,* ed. F. Dubinskas. Temple University Press, 1988.

Bucciarelli, Louis L. "An ethnographic perspective on engineering design." *Design Studies* 5 (1984), no. 3: 185–190.

Buonopane, Stephen G., and David P. Billington. "Theory and history of suspension bridge design from 1823 to 1940." *Journal of Structural Engineering* 119 (1993), no. 3: 954–977.

Callon, Michel. "Struggles and Negotiations to Define What is Problematic and What is Not: The Socio-logic of Translation." In *The Social Process of Scientific Investigation,* ed. K. Knorr et al. (*Sociology of the Sciences Yearbook* 4 (1980)).

Cant, R.G. "The Scottish universities and Scottish society in the eighteenth century." *Studies on Voltaire and the Eighteenth Century* 58 (1967): 1953–1966.

Cardwell, D.S.L. *From Watt to Clausius: The Rise of Thermodynamics in the Early Industrial Age.* Cornell University Press 1971.

Channell, David F. "The harmony of theory and practice: The engineering science of W. J. M. Rankine." *Technology and Culture* 23 (1982), no. 1: 39–52.

Charlton, T.M. *A History of Theory of Structures in the Nineteenth Century.* Cambridge University Press, 1982.

Chitnis, Anand C. *The Scottish Enlightenment: A Social History.* Croom Helm, 1976.

Cissel, James H. "Stiffness as a factor in long span suspension bridge design." *Roads and Streets,* April 1941: 64–72.

Collins, Harry M. "The TEA set: Tacit knowledge and scientific networks." *Science Studies* 4 (1974): 165–186.

Collinson, Robert. *Encyclopaedias: Their History Throughout the Ages.* New York: Hafner, 1964.

Constant, Edward. *The Origins of the Turbojet Revolution.* Johns Hopkins University Press, 1980.

Cooper, Carolyn, ed. "Patents and invention." *Technology and Culture* 32 (1991), no. 4.

Day, Thomas. "Samuel Brown: His influence on the design of suspension bridges." *History of Technology* 8 (1983): 61–90.

Day, Thomas. "Samuel Brown in north-east Scotland." *Industrial Archaeology Review* 7 (1985), no. 2: 154–170.

Deswarte, Sylvie and Bertrand Lemoine. *L'Architecture et les ingénieurs: Deux siècles de construction.* Paris: Moniteur, 1980.

Dorn, Harold. The Art of Building and the Science of Mechanics: A Study of the Union of Theory and Practice in the Early History of Structural Analysis. Ph.D. dissertation, Princeton University, 1970.

Dreicer, Gregory K. The Long Span: Intercultural Exchange in Building Technology: Development and Industrialization of the Framed Beam in Western Europe. Ph.D. dissertation, Cornell University, 1993.

Ferguson, Eugene. *Engineering and the Mind's Eye.* MIT Press, 1992.

Finch, James Kip. "Wind failures of suspension bridges; or evolution and decay of the stiffening truss." *Engineering News-Record* 126 (1941), March 13: 402–407.

Fugl-Meyer, H. *Chinese Bridges.* Shanghai: Kelly & Walsh, 1937.

Fujimura, Joan Hideko. Crafting Science, Transforming Biology: The Case of Oncogene Research. Unpublished book manuscript.

Geertz, Clifford. "Ideology as a cultural system." In *Ideology and Discontent,* ed. D. Apter. Free Press, 1964.

Gibb, Alexander. *The Story of Telford*. London: Alexander Maclehose, 1935.

Gibbons, Chester H. "History of testing machines for materials." *Transactions of the Newcomen Society* 15 (1934–35): 169–184.

Gies, Joseph. *Bridges and Men*. Doubleday, 1963.

Gingras, Yves. "Following scientists through society? Yes, but at arm's length!" *Cahiers d'Epistémologie* (Université de Québec à Montréal) 9203 (1992).

Gooding, David, Trevor Pinch, and Simon Schaffer, eds. *The Uses of Experiment: Studies in the Natural Sciences*. Cambridge University Press, 1989.

Goodwin, C. D. W., and I. B. Holley, Jr. *The Transfer of Ideas: Historical Essays*. Durham, N.C.: South Atlantic Quarterly, 1968.

Gordon, J. E. *Structures: or, Why Things Don't Fall Down*. Da Capo, 1978.

Harris, Steven. Jesuit Ideology and Jesuit Science: Religious Values and Scientific Activity in the Society of Jesus, 1540–1773. Ph.D. dissertation, University of Wisconsin, 1988.

Holmes, Frederic Lawrence. *Lavoisier and the Chemistry of Life: An Exploration of Scientific Creativity*. University of Wisconsin Press, 1985.

Hopkins, H. J. *A Span of Bridges*. Devon: David and Charles, 1970.

Iltis, Carolyn Merchant. The Controversy over Living Force: Leibniz to D'Alembert. Ph.D. dissertation, University of Wisconsin, 1967.

Jacobs, David, and Anthony E. Neville. *Bridges, Canals, and Tunnels*. American Heritage, 1968.

Johnson's New Illustrated Cyclopaedia. New York: Alvin A. Johnson, 1874.

Kemp, Emory. "Links in a chain: The development of suspension bridges, 1801–70." *Structural Engineer* 57A (1979), August: 255–263.

Kemp, Emory L. "Samuel Brown: Britain's pioneer suspension bridge builder." *History of Technology* 2 (1977): 1–37.

Kirby, Richard Shelton and Philip Laurson. *The Early Years of Modern Civil Engineering*. Yale University Press, 1932.

Kline, Morris. *Mathematical Thought from Ancient to Modern Times*. Oxford University Press, 1972.

Kline, Ronald. "Science and engineering theory in the invention and development of the induction motor, 1880–1900." *Technology and Culture* 28 (1987), no. 2: 283–313.

Kranakis, Eda. "The French connection: Giffard's injector and the nature of heat." *Technology and Culture* 23 (1982), no. 1: 3–38.

Kranakis, Eda. "Hybrid careers and the interaction of science and technology." In *Technological Development and Science in the Industrial Age,* ed. P. Kroes and M. Bakker. Kluwer, 1992.

Kranakis, Eda. "Social determinants of engineering practice: A comparative view of France and America in the nineteenth century." *Social Studies of Science* 19 (1989): 5–70.

Kranzberg, Melvin, ed. *Technological Education—Technological Style.* San Francisco Press, 1986.

Latour, Bruno. "Mixing humans and nonhumans together: The sociology of a door closer." *Social Problems* 35 (1988), no. 3: 298–310.

Latour, Bruno. *Science in Action: How to Follow Scientists and Engineers Through Society.* Open University Press, 1987.

Law, John, ed. *Power, Action and Belief: A New Sociology of Knowledge?* Routledge, 1985.

Law, John, and Michel Callon. "Engineering and sociology in a military aircraft project: A network analysis of technological change." *Social Problems* 35 (1988), no. 3: 284–297.

Layton, Edwin T. "Technology as knowledge." *Technology and Culture* 15 (1974), no. 1: 31–41.

Leinenkugel le Cocq. "Ponts suspendus." In *Encyclopédie Scientifique.* Paris: Octave Doin et fils, 1911.

Lenoir, Timothy. "Practice, reason, context: The dialogue between theory and experiment." *Science in Context* 2 (1988), no. 1: 3–22.

Lindqvist, Svante. *Technology on Trial: The Introduction of Steam Power Technology into Sweden, 1715–1736.* Almqvist & Wiskell, 1984.

Lundgreen, Peter. "Engineering education in Europe and the U.S.A., 1750–1930: The rise to dominance of school culture and the engineering professions." *Annals of Science* 47 (1990): 33–75.

MacKenzie, Donald, and Judy Wajcman, eds. *The Social Shaping of Technology: How the Refrigerator Got Its Hum.* Open University Press, 1985.

Needham, Joseph, et al. *Science and Civilization in China,* volume 4, part 3. Cambridge University Press, 1971.

Paxton, Roland A. "Menai Bridge (1818–1826) and its influence on suspension bridge development." *Transactions of the Newcomen Society* 49 (1977–78): 87–110.

Pease, George B. "Timothy Palmer: Bridgebuilder of the eighteenth century." *Essex Institute Historical Collections* 83 (1947): 97–111.

Penfold, Alastair, ed. *Thomas Telford: Engineer.* London: Thomas Telford, 1980.

Petroski, Henry. *Design Paradigms: Case Histories of Error and Judgment in Engineering.* Cambridge University Press, 1994.

Petroski, Henry. *To Engineer is Human: The Role of Failure in Successful Design.* Random House, 1992.

Pickering, Andy. "Knowledge, practice, and mere construction." *Social Studies of Science* 20 (1990), no. 4: 682–728.

Pickering, Andy. *Science as Practice and Culture.* University of Chicago Press, 1992.

Pugsley, Sir Alfred. *The Theory of Suspension Bridges,* second edition. Edward Arnold, 1968.

Ringer, Fritz K. *Education and Society in Modern Europe.* Indiana University Press, 1979.

Rolt, L.T.C. *Isambard Kingdom Brunel.* Penguin, 1972.

Rouse, Hunter, and Simon Ince. *History of Hydraulics.* Dover, 1963.

Salvadori, Mario. *Why Buildings Stand Up: The Strength of Architecture.* Norton, 1980.

Shils, Edward. "Ideology: The concept and function of ideology." In *International Encyclopedia of the Social Sciences.*

Smith, Denis. "The use of models in nineteenth century British suspension bridge design." *History of Technology* 2 (1977): 169–214.

Society for the Promotion of Engineering Education. *Report of the Investigation of Engineering Education, 1923–1929.* University of Pittsburgh, 1930.

Staudenmaier, John M. *Technology's Storytellers: Reweaving the Human Fabric.* MIT Press, 1985.

Steinman, David B., and Sara Ruth Watson. *Bridges and Their Builders,* second edition. Dover, 1957.

Steinman, David B. *The Builders of the Bridge.* Harcourt, Brace, 1950.

"Suspension bridges and wind resistance." *Engineering News-Record,* October 23, 1941: 97–100.

Timoshenko, Stephen P. *History of the Strength of Materials.* Dover, 1983.

Todhunter, Isaac, and Karl Pearson. *A History of the Theory of Elasticity and of the Strength of Materials from Galilei to Lord Kelvin.* Dover, 1960.

Truesdell, Clifford. *The Rational Mechanics of Flexible or Elastic Bodies, 1638–1788.* Zurich: Orell Fussli, 1960.

Tyrrell, Henry Grattan. *History of Bridge Engineering.* Chicago: G. B. Williams, 1911.

Vincenti, Walter. *What Engineers Know and How They Know It: Analytical Studies from Aeronautical History.* Johns Hopkins University Press, 1990.

Wengenroth, Ulrich. "The comparative study of the history of technology." Presented to Conference on Critical Problems and Research Frontiers in History of Technology and History of Science, Madison, Wisconsin, 1991.

Index